MODELLING AND HEDGING
EQUITY DERIVATIVES

MODELLING AND HEDGING EQUITY DERIVATIVES

Oliver Brockhaus
Andrew Ferraris
Christoph Gallus
Douglas Long
Reiner Martin
Marcus Overhaus

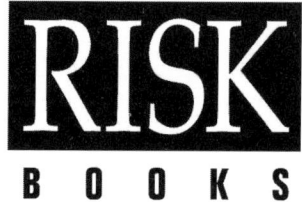

Published by Risk Books, a specialist division of Risk Publications.

Haymarket House
28-29 Haymarket
London SW1Y 4RX
Tel: +44 (0)171 484 9700
Fax: +44 (0)171 930 2238
E-mail: books@risk.co.uk
Home Page: http://www.riskpublications.com

© Financial Engineering Ltd 1999

ISBN 1 899332 34 0

British Library Cataloguing in Publication Data
A catalogue record for this book is available from the British Library

Risk Books Commissioning Editor: Laurie Donaldson
Copy-edited and typeset by The Geometric Press, Oxford

Printed and bound in Great Britain by Bookcraft
Covers printed by Bookcraft

Conditions of sale
All rights reserved. No part of this publication may be reproduced in any material form whether by photocopying or storing in any medium by electronic means whether or not transiently or incidentally to some other use for this publication without the prior written consent of the copyright owner except in accordance with the provisions of the Copyright, Designs and Patents Act 1988 or under the terms of a licence issued by the Copyright Licensing Agency Limited of 90, Tottenham Court Road, London W1P 0LP.

Warning: the doing of any unauthorised act in relation to this work may result in both civil and criminal liability.

Every effort has been made to ensure the accuracy of the text at the time of publication. However, no responsibility for loss occasioned to any person acting or refraining from acting as a result of the material contained in this publication will be accepted by Financial Engineering Ltd.

Many of the product names contained in this publication are registered trademarks, and Risk Books has made every effort to print them with the capitalisation and punctuation used by the trademark owner. For reasons of textual clarity, it is not our house style to use symbols such as ™, ®, etc. However, the absence of such symbols should not be taken to indicate absence of trademark protection; anyone wishing to use product names in the public domain should first clear such use with the product owner.

I am profoundly grateful to the most important person in my life for her inspiration and encouragement, without whom this work would not have become a reality.

Marcus

Contents

The Authors ix

Notation xi

Introduction xiii

1. Mathematical Fundamentals 1
   1.1 A review of probability theory and stochastic calculus 1
   1.2 The Black–Scholes equity model 19
   1.3 Extensions to Black–Scholes 26
   1.4 The Clark formula 32
   1.5 The hybrid model 35
   1.6 The multi-currency hybrid model 43

2. Closed-Form Solutions for Standard Products 49
   2.1 Basic products 49
   2.2 American options 52
   2.3 Digital options 58
   2.4 Barrier options 70
   2.5 Asian options 92

3. Closed-Form Solutions for Non-Standard Products 97
   3.1 Lookback options 97
   3.2 Fade-in options 104
   3.3 Fade-in barrier options 106
   3.4 Chooser options 109
   3.5 Prolongation options 111
   3.6 Improving options 112
   3.7 Power and powered options 113
   3.8 Compound options 115

4. Closed-Form Solutions for Multi-Asset Products 117
   4.1 Exchange options 117
   4.2 Relative digital options 118
   4.3 Relative outperformance options 119
   4.4 Outperformance options 120
   4.5 European digital option on best or worst of two assets 122
   4.6 Best or worst of several assets 124
   4.7 Basket options 128

	4.8	Hindsight options	132
	4.9	Outside barrier options	135
	4.10	Outside digital options	137
5.		Closed-Form Fixed Income and Hybrid Products	139
	5.1	Bond options and swaptions	139
	5.2	Caps and floors	141
	5.3	European options (Merton formula)	143
	5.4	Equity/bond outperformance options	143
6.		The Tree Approach	147
	6.1	Setting up the tree	148
	6.2	Option pricing using trees	154
	6.3	Barrier options	156
	6.4	Bermudan Asian options	161
	6.5	Convertible bonds	166
7.		Monte Carlo Methods	173
	7.1	The basic method	173
	7.2	Speeding up Monte Carlo	175
	7.3	Generic Monte Carlo pricing	179
	7.4	Hybrid Monte Carlo	180
	7.5	Monte Carlo for American options	181
8.		A Partial Differential Equation Solver	185
	8.1	Discretisation of the PDE	186
	8.2	Boundary conditions	189
	8.3	Moving barriers	192
	8.4	Range and fade-in options	194
	8.5	American options	196
	8.6	Discrete dividends	197
	8.7	Model calibration	200
9.		Further Modelling Issues	211
	9.1	Calibration of the extended Vasicek model	211
	9.2	Basket and Asian underlyings	213
	9.3	Volatility smile	216
10.		Hedging	221
	10.1	Hedging and risk management	221
	10.2	Pricing and hedging European options under transaction costs	233
	10.3	Hedging of specific products	236
11.		Implementation Issues	249
	11.1	The context of a model library	249
	11.2	Library interface design	251
	11.3	Internal design	259

Appendix: Useful Formulas 277

Bibliography 281

Index 285

The Authors

Marcus Overhaus is Global Head of Quantitative Research for Deutsche Bank's Global Equity Division based in London. Previously, he worked on interest rate and FX derivatives for the OTC Derivatives Group of Deutsche Bank in Frankfurt, and for the Product Development Group of Deutsche Bank Financial Products in New York. Marcus holds a PhD in mathematics from Bochum/Oxford, and a PhD in theoretical physics from ETH Zurich. He was a member of the Sonderforschungsbereich (SFB) Complex Manifolds.

Oliver Brockhaus joined the Quantitative Research Department of Deutsche Bank's Global Equity Division in 1996. His responsibilities within the team comprise skewed models, hybrid option calibration, credit risk and stochastic volatility. Before entering finance in 1996, Oliver was a consultant at Andersen Consulting. He was awarded a masters degree (DEA) in probability theory at the University P. et M. Curie in Paris in 1991, and holds a PhD in mathematics from the University of Bonn.

Andrew Ferraris works in the Global Equity Division's Quantitative Research Department for Deutsche Bank in London, with responsibilities that include the software architecture of the model library, and its integration into client applications. Prior to joining Deutsche Bank in 1995, he worked in the Equity Derivatives technology division of Banque Paribas, as a developer of the derivatives pricing library, and as an algorithm developer for BAeSema. Andrew holds a DPhil in experimental particle physics from the University of Oxford.

Christoph Gallus currently runs the exotic and structured derivatives book for Deutsche Bank in Germany. Prior to being a trader, he worked for two years in the Quantitative Research Department of Deutsche Bank's Global Equity Division. Christoph studied mathematics in Erlangen, Germany, and Cambridge. He specialised in probability theory and gained a PhD with a thesis on the robustness of models for pricing and hedging derivative contracts.

Douglas Long works in the Quantitative Research Department of Deutsche Bank's Global Equity Division, and has been in this role since August 1998. Previously he spent one and a half years working for Infinity and Renaissance Software as a quantitative analyst and financial engineer. Douglas's responsibilities included interest rate and FX exotic product development and risk management within Europe. He holds a PhD in theoretical physics from the University of Wales, Swansea.

Reiner Martin has been a member of the Quantitative Research Department of Deutsche Bank's Global Equity Division since 1995, and has worked in Frankfurt and London. Currently, he is based in New York. His modelling work includes hybrid products, stochastic volatility, Monte Carlo methods, generic pricing tools, credit risk and convertible bonds. Reiner holds a PhD in pure mathematics from the University of California, Los Angeles (UCLA).

Notation

A large part of this book is mathematical in nature. The understanding of this material is facilitated by a consistent notation. We here summarise some of the typical notation used.

$\mathbb{R}$	real numbers
$\mathbb{N}$	natural numbers $\{1, 2, 3, \ldots\}$
log	natural logarithm
$a \vee b$	$\max(a, b)$
$a \wedge b$	$\min(a, b)$
a^+	$\max(a, 0)$
Ω	probability space
$\mathcal{F}$	σ-algebra
$\mathcal{F}_t$	filtration
ω	elementary event
$\mathbb{P}$	standard risk-neutral probability measure
$\hat{\mathbb{P}}$	measure after Girsanov transformation
$\mathbb{E}$	expectation operator
X, Y, Z	standard normally distributed random variables
$\sigma(X)$	σ-algebra generated by a random variable X
$a(x), b(y)$	density of A, B with respect to Lebesgue measure at x, y
W_t	standard Brownian motion at time t
γ	drift of a Brownian motion
$N(x)$	univariate normal distribution
$N_2(x, y; \rho)$	bivariate normal distribution
$N_3(x, y, z; \rho_{12}, \rho_{13}, \rho_{23})$	trivariate normal distribution
S_t	asset at time t
$S_t^1, \ldots, S_t^n$	assets at time t for multi-asset situation
t, s	time indices

i, j	time sub-indices
m, n	spatial indices
T	maturity
$t = 0$	evaluation date
$P(s, t)$	zero bond value at s paying one unit at time t
$B(t)$	cash bond
r	constant interest rate (continuously compounded)
d	constant dividend yield (continuously compounded)
F_t	forward price maturing at time t
H	barrier
K	strike
X	payoff
U, L	upper, lower barrier
G	rebate
$w_1, w_2, \ldots$	weights
$C_{\cdots}^{\cdots}$	price of some type of call
$P_{\cdots}^{\cdots}$	price of some type of put
χ	1 for call and -1 for put
ρ	correlation

Introduction

The growth in equity derivatives markets has been explosive over the last decade, as investors have come increasingly to rely on equity-linked instruments to more finely tune their exposure to the equity markets. The equity derivatives market breaks down into two distinct categories: the listed market and the 'over the counter' (OTC) market, although today, of course, there are a myriad of different products which combine the features of both. These markets are now massive.

Listed derivatives are the standardised, tradable end of the market. Derivatives exchanges such as CBOE, DTB, Liffe and Eurex list derivative contracts on indices and single stocks, which can be bought and sold during trading hours. These instruments are also traded by market-makers such as investment banks.

The Bank for International Settlements (BIS) estimates that at year end 1997, the notional amount of equity futures contracts outstanding was US$16.4 trillion. This represented a 27% increase of 1996's total. On the single stock option side, there were US$13.0 trillion of notional contracts outstanding. Geographically, the US remains the world's biggest listed market, with 54% of the open interest. Europe has 27% of the market and Asia 15%. Overall, there were 115.9 million listed option contracts outstanding at the end of 1997.

The OTC market has evolved in order to give participants greater flexibility in structuring products tailored to meet specific requirements. A typical OTC transaction involves an investment bank structuring a product to give a client (typically a pension or insurance fund) a desired outcome (such as the performance of a certain index in a certain currency). These "securities" are not generally listed and are not tradable. Statistics on the OTC market are obviously very difficult to come by, but in recent years the size of the OTC market has eclipsed that of the listed business. We would estimate that the OTC market is now at least double the size of the listed market, implying that the total notional outstanding must now be in excess of US$60 trillion.

Given the size and sophistication of these markets, it is perhaps suprising that more literature on pricing and hedging is not already available. Our intention is to provide a self-contained, mostly stochastic, approach towards option pricing and hedging. We start with a review of probability theory and stochastic calculus, and based on this we cover closed-form and semi-closed-form solutions for equity, fixed income and hybrid products. For products which cannot be valued within this framework, we discuss alternative approaches, such as binomial and trinomial trees, Monte Carlo methods, and partial differential equation solvers. Finally, we present chapters on hedging of selected products and implementation issues.

We provide a consistent framework for hybrid products (ie products having equity and foreign exchange, as well as fixed income, exposure) and a variety of equity-related products. Our emphasis here is on new and generalised results in closed-form solutions, allowing time-dependent parameters, for products which are currently traded in the OTC market. In addition, we give a practical approach for hedging equity products beyond delta hedging, including a mathematical foundation of higher-order hedges. Readers familiar with stochastic calculus will see results and applications which are not usually found in finance literature, eg Clark's formula and Donsker's invariance principle. As another innovative feature, we discuss in detail practical software implementation issues.

We would like to thank the Global Equity Derivatives Team at Deutsche Bank for their strong support of this project, in particular quantitative research team members Michael Farkas and Christian Pietsch. Special thanks are due to Chris East and Jon Kinol. All names mentioned here are given in alphabetical order.

The Authors
London, June 1999

Chapter 1
Mathematical Fundamentals

1.1 A Review of Probability Theory and Stochastic Calculus

The basic mathematical tool for pricing derivatives is probability theory and in particular stochastic calculus. As a large part of this book makes use of these topics, the early sections develop and review basic concepts and facts from probability theory. At the same time we establish some notation. For additional details the reader is advised to consult any of the standard books on the subject, eg [67, 44, 59, 62, 63, 7].

1.1.1 Probability Spaces

Measurable spaces Before we can define probability spaces, we need to introduce the concept of measurability.

Given a set Ω, a *σ-algebra* $\mathcal{F}$ on Ω is a system of subsets of Ω with the following properties:

- It contains the empty set.

- If A is in $\mathcal{F}$, then its complement $\Omega - A$ is also in $\mathcal{F}$.

- If $A_1, A_2, \ldots$ is a sequence of sets in $\mathcal{F}$, then their union $A_1 \cup A_2 \cup \cdots$ is also in $\mathcal{F}$.

The space Ω, together with a σ-algebra $\mathcal{F}$ on Ω, is called a *measurable space* and is denoted by $(\Omega, \mathcal{F})$. The sets in $\mathcal{F}$ are called *$\mathcal{F}$-measurable* sets, or just *measurable* sets if it is clear to which σ-algebra they are referring.

Probabilities The basic concept of probability theory is the *probability space* $(\Omega, \mathcal{F}, \mathbb{P})$, which is a measurable space $(\Omega, \mathcal{F})$ together with a probability measure $\mathbb{P}$ on $\mathcal{F}$, ie a *countable additive* non-negative function on $\mathcal{F}$ with $\mathbb{P}(\Omega) = 1$. For $\mathbb{P}$ to be countable additive means that

$$\mathbb{P}\left(\bigcup_{n=1}^{\infty} A_n\right) = \sum_{n=1}^{\infty} \mathbb{P}(A_n)$$

holds for any sequence $A_1, A_2, \ldots$ of *disjoint* sets in $\mathcal{F}$. For simplicity, we will often just refer to the measure even if we mean the probability measure.

A probability space essentially represents a random experiment. The set Ω is the set of all *elementary events* or *outcomes*. For example, Ω could represent the set of all possible price paths of some asset between the times 0 and 1, and could therefore be the space $C[0, 1]$ of all continuous functions on the interval $[0, 1]$. In the probabilistic context, the elements of $\mathcal{F}$ are called *events*. The number $\mathbb{P}(A)$ is the *probability* of the event A.

Why do we introduce σ-algebras and not simply let $\mathcal{F}$ be the collection of all subsets of Ω? There is a technical and a non-technical reason for this. If we let $\mathcal{F}$ be the set of all subsets of the interval $[0, 1]$, say, then there does not exist any reasonable probability measure. However, we can choose $\mathcal{F}$ to be the smallest σ-algebra containing all open sets in $[0, 1]$. Then there does exist a probability measure on $\mathcal{F}$ with the property $\mathbb{P}([a, b]) = b - a$ for all $0 \leqslant a \leqslant b \leqslant 1$. Indeed, it requires some effort to construct a set that is *not* in $\mathcal{F}$. Note that we will always assume that any subset of $\mathbb{R}$ or $\mathbb{R}^n$ is equipped with the smallest σ-algebra containing all open sets, which is called the *Borel σ-algebra*. Any measure on a Borel σ-algebra is called a *Borel measure*. The non-technical reason for the introduction of σ-algebras is that they can, in the context of stochastic processes, be used to represent information available at a certain point in time (see Section 1.1.5).

1.1.2 Random Variables

Measurable maps Assume we have two measurable spaces $(\Omega_1, \mathcal{F}_1)$ and $(\Omega_2, \mathcal{F}_2)$. A map $f : \Omega_1 \to \Omega_2$ is called *measurable* if the pre-image $f^{-1}(A) = \{\omega \in \Omega_1 \mid f(\omega) \in A\}$ of any measurable set $A \in \mathcal{F}_2$ is again measurable, ie an element of $\mathcal{F}_1$.

Given $\mathcal{F}_2$ and f, there is always a smallest σ-algebra on Ω_1 such that f is measurable. This σ-algebra is denoted by $\sigma(f)$.

Definition of random variables The elements of Ω are complicated outcomes, eg whole price paths. However, often we are only interested in a single quantity or a finite number of quantities that depend on Ω. For this purpose we introduce *random variables*.

A real-valued function X on a probability space is called a random variable if it is measurable. Recall that tacitly we regard the real numbers to be equipped with the Borel σ-algebra. Occasionally, we extend this definition to mean functions taking values in $\mathbb{R}^n$.

If a measure $\mathbb{P}$ on $(\Omega, \mathcal{F})$ is given, two random variables X_1 and X_2 are said to be equal if they disagree with probability 0.

Distribution functions and densities For many purposes it is not necessary to know the random variable X as such. We consider instead its *distribution function* $F_X : \mathbb{R} \to [0, 1]$, defined by

$$F_X(x) = \mathbb{P}[X \leqslant x].$$

This is obviously a non-decreasing function taking values in $[0, 1]$. One can show that a Borel measure on $\mathbb{R}$ is completely determined by its values on all intervals. This implies that F_X determines the probability of any Borel set of $\mathbb{R}$.

In some cases we can write

$$F_X(x) = \int_{-\infty}^{x} \phi(u) \, du,$$

with some non-negative function $\phi : \mathbb{R} \to \mathbb{R}$. We then say that X has *density* ϕ (see Section 1.1.3). This readily generalises to higher dimensions.

Expectation and higher moments We will not review details of integration theory here. We just note that the integral of the random variable X is

$$\int_\Omega X(\omega) \mathbb{P}[d\omega] = \int_\mathbb{R} x \mathbb{P}[X \in dx],$$

and that the integral has the property that

$$\int_\Omega 1_A(\omega) \mathbb{P}[d\omega] = \mathbb{P}[A]$$

for each event A, where 1_A denotes the characteristic function given by

$$1_A(\omega) = \begin{cases} 1 & \text{if } \omega \in A, \\ 0 & \text{otherwise.} \end{cases}$$

Given a measure $\mathbb{P}$ and a random variable X on Ω, we define the *expectation* of X with respect to $\mathbb{P}$ by

$$\mathbb{E}_\mathbb{P}[X] = \mathbb{E}[X] = \int_\Omega X(\omega) \mathbb{P}[d\omega].$$

Note that the integral may not exist or may be infinite. The *n-th moment* of X is defined by $\mathbb{E}[X^n]$, while the *n-th central moment* is given by $\mathbb{E}[(X - \mathbb{E}[X])^n]$. So the first moment is just the expectation, while the second central moment is the variance of the random variable X.

Jensen's inequality One of the simplest and most frequently used inequalities involving the expectation is *Jensen's inequality*. Assume we have a given convex function $f : \mathbb{R} \to \mathbb{R}$. This means that $f\left(\frac{1}{2}(x+y)\right) \leqslant \frac{1}{2}[f(x) + f(y)]$ for all x and y. Then we have

$$f(\mathbb{E}[X]) \leqslant \mathbb{E}[f(X)]$$

for any random variable X. As a simple special case, Jensen's inequality implies

$$\mathbb{E}[X] \leqslant \left(\mathbb{E}[X^2]\right)^{1/2}.$$

1.1.3 Independence and Conditional Expectations

Independence Sets $A_1, \ldots, A_n \in \mathcal{F}$ are called *independent* if

$$\mathbb{P}[A_1 \cap \cdots \cap A_n] = \mathbb{P}[A_1] \cdots \mathbb{P}[A_n].$$

Note that pairwise independence does not imply independence. This concept readily generalises to σ-algebras and random variables: a sequence of σ-algebras $\mathcal{F}_1, \ldots, \mathcal{F}_n \subset \mathcal{F}$ is independent if $A_1 \in \mathcal{F}_1, \ldots, A_n \in \mathcal{F}_n$ implies the independence of $A_1, \ldots, A_n$. Finally, random variables $X_1, \ldots, X_n$ on the same probability space $(\Omega, \mathcal{F})$ are called independent if

$$\mathbb{P}[X_1 \in A_1, \ldots, X_n \in A_n] = \mathbb{P}[X_1 \in A_1] \cdots \mathbb{P}[X_n \in A_n]$$

for all sets $A_1, \ldots, A_n$ in $\mathcal{F}$. This is equivalent to the independence of

$\sigma(X_1), \ldots, \sigma(X_n)$. If $X_1, \ldots, X_n$ are independent, then

$$\mathbb{E}[X_1 \cdots X_n] = \mathbb{E}[X_1] \cdots \mathbb{E}[X_n],$$

if all these expectations exist.

The Radon–Nikodym theorem Let X denote a random variable on a probability space $(\Omega, \mathcal{F}, \mathbb{P})$ with $X \geq 0$ and $\mathbb{E}[X] = 1$. Then a new measure $\hat{\mathbb{P}}$ is defined by

$$\hat{\mathbb{P}}[A] = \hat{\mathbb{E}}[X 1_A], \quad A \in \mathcal{F}.$$

The random variable X is called the *density* of $\mathbb{P}$ with respect to $\hat{\mathbb{P}}$. Another notation for the density X is

$$\frac{d\hat{\mathbb{P}}}{d\mathbb{P}} = X. \tag{1.1}$$

We remark that $\mathbb{P}[A] = 0$ implies $\hat{\mathbb{P}}[A] = 0$.

The *Radon–Nikodym theorem* is the converse of this statement. Let two measures $\mathbb{P}$ and $\hat{\mathbb{P}}$ on the same measurable space $(\Omega, \mathcal{F})$ be given with the property that $\mathbb{P}[A] = 0$ implies $\hat{\mathbb{P}}[A] = 0$. Then there exists a unique random variable X such that (1.1) holds.

Conditional expectations Fix a random variable X on a probability space $(\Omega, \mathcal{F}, \mathbb{P})$. The $\mathcal{F}$-measurability of X can be stated as follows. If, for all events $A \in \mathcal{F}$ and $\omega \in \Omega$, we know whether $\omega \in A$ or $\omega \notin A$, then we know X.

Now, if our information is limited to a σ-algebra $\hat{\mathcal{F}} \subset \mathcal{F}$, what do we know about X? In other words, what is the best estimator $\hat{X}$ of X given $\hat{\mathcal{F}}$? An estimator in that sense should satisfy:

- $\hat{X}$ is $\hat{\mathcal{F}}$-measurable;
- $\mathbb{E}[\hat{X} 1_A] = \mathbb{E}[X 1_A]$ for all $A \in \hat{\mathcal{F}}$.

A random variable $\hat{X}$ satisfying these two conditions is called a *conditional expectation* of X, given $\hat{\mathcal{F}}$, and denoted by $\mathbb{E}[X \mid \hat{\mathcal{F}}]$. Clearly, if we have no information at all, ie $\hat{\mathcal{F}} = \{\Omega, \varnothing\}$, the conditional expectation is simply the expectation $\hat{X} = \mathbb{E}[X]$.

Now let an arbitrary $\hat{\mathcal{F}} \subset \mathcal{F}$ be given and let $X \geq 0$ and $\mathbb{E}[X] = 1$. Then a measure $\hat{\mathbb{P}}$ on $\mathcal{F}$ can be defined via (1.1). Obviously, both $\mathbb{P}$ and $\hat{\mathbb{P}}$ can be regarded as measures on $\hat{\mathcal{F}}$. Therefore the existence of $\hat{X}$ is a consequence of the Radon–Nikodym theorem.

A random variable X is *independent* of $\hat{\mathcal{F}}$ if the knowledge of $\hat{\mathcal{F}}$ does not give rise to an estimator different from the trivial one, ie if

$$\mathbb{E}[X \mid \hat{\mathcal{F}}] = \mathbb{E}[X].$$

1.1.4 The Normal Distribution

The distribution function that is the most useful to us is the normal distribution. We will denote the one-dimensional cumulative standard normal distribution function by N and its density by φ. So we have

$$\varphi(x) = \frac{1}{\sqrt{2\pi}} \exp(-\tfrac{1}{2} x^2), \quad x \in \mathbb{R},$$

Mathematical Fundamentals

and

$$N(x) = \int_{-\infty}^{x} \varphi(u)\, du, \quad x \in \mathbb{R}.$$

For implementations of pricing methods it is often necessary to evaluate $N(\cdot)$. There do exist fast and accurate numerical approximations for $N(\cdot)$ [61].

We denote by N_2 the standard bivariate normal distribution

$$N_2(x, y; \rho) = \int_{-\infty}^{y} \int_{-\infty}^{x} \varphi_2(u, v, \rho)\, du\, dv, \quad x, y \in \mathbb{R},$$

with covariance ρ and with density

$$\varphi_2(x, y; \rho) = \frac{1}{2\pi\sqrt{1-\rho^2}} \exp\left(-\frac{x^2 - 2\rho xy + y^2}{2(1-\rho^2)}\right), \quad x, y \in \mathbb{R}.$$

For multi-asset options we often need the general n-dimensional normal distribution. For this, let R be a positive definite symmetric $n \times n$ matrix. Then the n-dimensional distribution with density

$$\varphi(x, R) = \frac{(\det R)^{1/2}}{(2\pi)^{n/2}} \exp(-\tfrac{1}{2} x^{\mathsf{T}} R x), \quad x \in \mathbb{R}^n,$$

is called the n-dimensional normal distribution N_n with zero mean and covariance matrix R. Its cumulative distribution function is denoted by $N_n(x; R)$.

1.1.5 Stochastic Processes

In Section 1.1.2, experiments with uncertain outcome were formalised. One may think of the result of tossing a coin, throwing a dice, the lottery number as of next Wednesday or the price of a share of company XYZ as quoted in tomorrow's newspapers. The last two examples both contain a time element. Nevertheless they are quite different: the knowledge of lottery numbers of the last three months will not help in predicting the next outcome, whereas tomorrow's share price is believed to depend somehow on the history of the share.

A family $X = (X_t, t \in \mathcal{T})$ of random variables on a probability space $(\Omega, \mathcal{F}, \mathbb{P})$ is called a *stochastic process*. Here the indexing set $\mathcal{T}$ represents time. For our purposes, we assume $\mathcal{T}$ to be an interval $[0, T]$ bounded by some terminal time T or a finite sequence of numbers $(0, 1, \ldots, N)$.

For fixed $t \in \mathcal{T}$, we can apply the concepts related to random variables as developed in Section 1.1.2. If we fix an elementary event $\omega \in \Omega$, then $X(\omega) = (X_t(\omega), t \in \mathcal{T})$ is a function from $\mathcal{T}$ into $\mathbb{R}$, called the *path* of the stochastic process X. In the case of $\mathcal{T} = [0, T]$, a process is said to be *continuous* if *almost all* paths of X are continuous, meaning that

$$\mathbb{P}[X \text{ is continuous}] = 1.$$

In this book all processes are assumed to be continuous. In a similar way, a process is *bounded* (or starts at 0 or crosses a barrier) if almost all paths have this property.

The concept of time as introduced via the index $t \in \mathcal{T}$ would not be complete without a mathematical notion formalising knowledge of the past and uncertainty about the future. If today is $t \in \mathcal{T}$, we "know" the process $(X_s, s \leqslant t)$ up to t and we may have more information about the future $(X_s, s \geqslant t)$ of the process than we had at time 0.

Let $(\mathcal{F}_t, t \in \mathcal{T})$ denote an increasing family of σ-algebras satisfying $\mathcal{F}_T \subset \mathcal{F}$. Then $(\mathcal{F}_t, t \in \mathcal{T})$ is called *filtration*. The process X is said to be *adapted* to that filtration if X_t is $\mathcal{F}_t$-measurable for all $t \in \mathcal{T}$. The σ-algebra $\mathcal{F}_t$ thus represents the knowledge at time t which includes the process $(X_s, s \leqslant t)$ up to time t. We remark that there is always a smallest filtration such that X is adapted to that filtration, namely $(\mathcal{F}_t, t \in \mathcal{T})$ with $\mathcal{F}_t = \sigma(X_s, s \leqslant t)$.

1.1.6 Martingales

A particularly important class of stochastic processes are martingales. A stochastic process $M = (M_t, t \in \mathcal{T})$ is called a *martingale* with respect to a filtration $(\mathcal{F}_t, t \in \mathcal{T})$ if M is adapted to that filtration (M_t is $\mathcal{F}_t$-measurable for all $t \in \mathcal{T}$), if M is integrable, ie

$$\mathbb{E}[|M_t|] < \infty, \quad t \in \mathcal{T},$$

and if M has the martingale property

$$\mathbb{E}[M_t \mid \mathcal{F}_s] = M_s, \quad s \leqslant t.$$

Some examples in discrete time are:

- Random walk

Let $(X_n, n \in \mathbb{N})$ be a sequence of independent random variables with $\mathbb{E}[|X_n|] < \infty$. Then $S = (S_n, n \in \mathbb{N})$, with

$$S_n = \sum_{i=0}^{n} (X_i - \mathbb{E}[X_i]),$$

is a martingale.

- Classic random walk

In the special case of

$$\mathbb{P}[X_i = 1] = \mathbb{P}[X_i = -1] = \tfrac{1}{2},$$

the martingale S can be viewed as the path obtained by successively throwing a fair coin and moving up in the case of head and down in the case of tail.

- Gambling system

Let $V = (V_n, n \in \mathbb{N})$ be an adapted process, and let $X = (X_n, n \in \mathbb{N})$ denote a martingale. Assume all V_n to be bounded. Then VX, defined by

$$VX_n = X_0 + \sum_{i=1}^{n} V_{i-1}(X_i - X_{i-1}),$$

is a martingale.

- Martingale system

A special case is given by assuming X to be a classic random walk and V the gambling strategy defined by

$$V_n = \begin{cases} 2^n & \text{if } X_0 = -1, X_1 = -1, \ldots, X_n = -1, \\ 0 & \text{otherwise.} \end{cases}$$

This amounts to doubling the bet on tail until head occurs. This strategy pays 1 for *almost all* paths.

Examples of continuous martingales will be given in Sections 1.1.9 and 1.1.12. A *semi-martingale* X is a continuous process that has a decomposition $X = M + V$, with M denoting a continuous martingale and V being a *process of bounded variation*, ie a process that is the difference of two continuous non-decreasing processes.

1.1.7 Quadratic Variation

Assume $\mathcal{T} = [0, T]$ and a continuous martingale M with $\mathbb{E}[M_T^2] < \infty$ to be given. Let τ be a finite subset of $\mathcal{T}$ containing 0. We introduce a continuous process V^τ via

$$V_t^\tau = \sum_{s \in \tau, s \leqslant t} (M_{s' \wedge t} - M_s)^2.$$

Here s' denotes the successor of s in τ. We remark that $M^2 - V^\tau$ is a continuous martingale since

$$M_t^2 - V_t^\tau = M_0^2 + 2 \sum_{s \in \tau, s \leqslant t} M_s(M_{s' \wedge t} - M_s).$$

Moreover, V^τ restricted to τ is increasing. It can be shown, using a limit procedure with an increasing sequence $(\tau_n, n \in \mathbb{N})$ of subsets of $\mathcal{T}$, that there is a unique continuous increasing process $\langle M \rangle$ such that $M^2 - \langle M \rangle$ is a martingale. The uniqueness part in the proof uses the fact that a martingale of bounded variation is constant. The process $\langle M \rangle$ is called *quadratic variation* of M. If a second continuous martingale N with $\mathbb{E}[N_T^2] < \infty$ is given, then a sequence of processes of the form

$$W_t^\tau = \sum_{s \in \tau, s \leqslant t} (M_{s' \wedge t} - M_s)(N_{s' \wedge t} - N_s),$$

where τ runs through an increasing sequence $(\tau_n, n \in \mathbb{N})$, gives rise to a process of bounded variation. This process, called the *quadratic covariation* $\langle M, N \rangle$ of M and N, is the unique process of bounded variation making $MN - \langle M, N \rangle$ a martingale. Clearly, we have

$$\langle M, M \rangle = \langle M \rangle.$$

We can extend the definition of quadratic variation to continuous semi-martingales. It is just the quadratic variation of the martingale summand in its decomposition.

1.1.8 Stochastic Integration

Let f and g be two functions on $\mathcal{T} = [0, T]$. Then the integral

$$\int_0^t f \, dg$$

is understood to be the limit $\lim_{n \to \infty} I_n$ (if it exists) of sums of the form

$$I_n = \sum_{s \in \tau_n, s \leqslant t} f_s(g_{s' \wedge t} - g_s),$$

where $(\tau_n, n \in \mathbb{N})$ is a sequence of finite subsets of $\mathcal{T}$ satisfying
$$\lim_{n \to \infty} |\tau_n| = 0,$$
with $|\tau|$ denoting the mesh width
$$|\tau| = \sup\{s' - s, s \in \tau\} \tag{1.2}$$
of τ (s' again being the successor of s in τ). We now replace g by a continuous martingale M satisfying $\mathbb{E}[M_T^2] < \infty$ and f by a locally constant process H of the form
$$H_t(\omega) = \sum_{s \in \tau} h_s(\omega) I_{(s,s']}(t)$$
for some finite $\tau \subset \mathcal{T}$, where the h_s are assumed to be $\mathcal{F}_s$-measurable for all $s \in \tau$. By analogy, the stochastic integral of H with respect to M is defined to be the process $\int H \, dM$ given by
$$\int_0^t H_s \, dM_s = \sum_{s \in \tau, s \leq t} h_s(M_{s' \wedge t} - M_s)$$
for all $t \in \mathcal{T}$. As already mentioned in the gambling system example in Section 1.1.6, assuming all h_s ($s \in \tau$) to be bounded, this process is a martingale.

Moreover, we have
$$\left\langle \int H \, dM \right\rangle_t = \int_0^t H_s^2 \, d\langle M \rangle_s.$$

Using this identity, the stochastic integral with respect to M can be extended to more general processes H, in particular continuous adapted processes. We remark finally that, if N is another continuous martingale with $\mathbb{E}[N_T^2] < \infty$, we have
$$\left\langle \int H \, dM, N \right\rangle_t = \int_0^t H_s \, d\langle M, N \rangle_s.$$

This property can be used to show the uniqueness of the stochastic integral.

1.1.9 Brownian Motion

Definition One of the central concepts in stochastic calculus is Brownian motion. It is also the basic building block for asset price processes. A real-valued random process $W = (W_t, t \in [0, \infty))$ is called *Brownian motion* if it has the following properties:

- Almost all paths of W are continuous.

- The random variable W_t is normally distributed, with mean 0 and variance t, for each $t \geq 0$.

- W has stationary increments.

- W has independent increments.

To say W has *stationary increments* means that, for $0 \leq s < t$, the distribution of $W_t - W_s$ depends only on $t - s$. Also, W_t has *independent increments* if, for $0 \leq t_0 < t_1 < \cdots < t_k$, the variables $W_{t_i} - W_{t_{i-1}}$ ($i = 1, \ldots, k$) are independent.

Mathematical Fundamentals

This definition generalises directly to the n-dimensional (independent) Brownian motion.

Time inversion and scaling Brownian motion is preserved under scaling and time inversion. This means that, given a standard Brownian motion W, the process $\tilde{W}$ is also a standard Brownian motion, where $\tilde{W}$ is defined via one of the following:

- $\tilde{W}_t = tW_{1/t}$ for $t > 0$ and $\tilde{W}_0 = 0$;
- $\tilde{W}_t = W_{ct}/\sqrt{c}$ for any fixed $c > 0$.

Sample path properties To get a feeling for the fine structure of Brownian motion, we now list some properties that hold for almost all sample paths $t \mapsto W_t(\omega)$:

- The zero set $\{t \geqslant 0 \mid W_t(\omega) = 0\}$ has measure zero, is unbounded, and has no isolated points.

- The path up to t has quadratic variation t, ie $\langle W \rangle_t(\omega) = t$, or equivalently

$$\lim_{n \to \infty} \sum_{s \in \tau_n, s \leqslant t} [W_{s' \wedge t}(\omega) - W_s(\omega)]^2 = t \tag{1.3}$$

if $\lim_{n \to \infty} |\tau_n| = 0$. Here τ_n denotes a sequence in $[0, T]$ with $T \geqslant t$, s' denotes the successor of s in τ_n, and $|\tau_n|$ is the mesh width of τ_n as defined in equation (1.2). In particular, $W(\omega)$ is of unbounded variation and is therefore neither monotone on any interval nor differentiable.

- The set of local maxima is countable and dense in $[0, \infty)$.

- We have

$$\limsup_{t \to 0+} \frac{W_t(\omega)}{\sqrt{2t \log \log(1/t)}} = 1 \quad \text{and} \quad \limsup_{t \to \infty} \frac{W_t(\omega)}{\sqrt{2t \log \log t}} = 1.$$

This is called the *law of the iterated logarithm*.

1.1.10 Itô's Lemma

Let g be a twice continuously differentiable function g, and let f be a function on $[0, \infty)$. The chain rule states that

$$g(f_T) = g(f_0) + \int_0^T \frac{\partial g}{\partial x}(f_t) \, df_t$$

and the integral with respect to df is well defined if f has bounded variation. If f is replaced by a typical Brownian path W, this assumption no longer holds. However, given a sequence $(\tau_n, n \in \mathbb{N})$ as in equation (1.3), we can define an integral of g with respect to W via

$$\int_0^T \frac{\partial g}{\partial x}(W_u) \, dW_u = \lim_{n \to \infty} \sum_{t \in \tau_n, t \leqslant T} \frac{\partial g}{\partial x}(W_t)(W_{t' \wedge T} - W_t),$$

and we obtain the formula

$$g(W_T) = g(W_0) + \int_0^T \frac{\partial g}{\partial x}(W_t)\,dW_t + \tfrac{1}{2}\int_0^T \frac{\partial^2 g}{\partial x^2}(W_t)\,dt$$

using the Taylor expansion

$$g(W_{t'}) = g(W_t) + \frac{\partial g}{\partial x}(W_t)(W_{t'} - W_t) + \tfrac{1}{2}\frac{\partial^2 g}{\partial x^2}(W_t)(W_{t'} - W_t)^2$$

$$+ \text{terms of order 3 and higher.}$$

In the case of semi-martingales X, as defined in Section 1.1.6, one can show that

$$dg(X_t) = \frac{\partial g}{\partial x}(X_t)\,dX_t + \tfrac{1}{2}\frac{\partial^2 g}{\partial x^2}(X_t)\,d\langle X \rangle_t,$$

with $\int \cdot dX_t$ denoting the stochastic integral defined in Section 1.1.8. In particular, if X is an n-dimensional Itô process, ie

$$dX_t^i = \mu^i(t, X_t)\,dt + \sum_{j=1}^n \sigma^{ij}(t, X_t)\,dW_t^j, \quad i = 1, 2, \ldots, n,$$

with n independent Brownian motions ($W^i, i = 1, 2, \ldots, n$), and g is a twice differentiable function on $\mathbb{R}^{n+1}$, then, omitting the arguments of μ and σ, we have

$$dg(t, X_t) = \left(\frac{\partial g}{\partial t}(t, X_t) + \sum_{i=1}^n \frac{\partial g}{\partial x_i}(t, X_t)\mu^i + \tfrac{1}{2}\sum_{i,j,k=1}^n \sigma^{ik}\sigma^{jk}\frac{\partial^2 g}{\partial x_i \partial x_j}(t, X_t)\right)dt$$

$$+ \sum_{i,j=1}^n \frac{\partial g}{\partial x_i}(t, X_t)\sigma^{ij}\,dW_t^j. \quad (1.4)$$

This formula, known as *Itô's lemma*, plays a similar role in stochastic calculus as the chain rule does in ordinary calculus.

1.1.11 Stochastic Differential Equations

In Section 1.1.10, we defined Itô processes using the stochastic differential equation

$$dX_t^i = \mu^i(t, X_t)\,dt + \sum_{j=1}^n \sigma^{ij}(t, X_t)\,dW_t^j, \quad i = 1, 2, \ldots, n,$$

with n independent Brownian motions ($W^i, i = 1, 2, \ldots, n$). It can be shown [44] that, if μ and σ satisfy some technical conditions, there exists a unique process X adapted to the filtration of W satisfying the above equation. The solution X may be viewed as a function $F(W)$ of the n-dimensional Brownian motion W.

Example: Lognormal process Let W denote a Brownian motion. Then Itô's lemma shows that X with

$$X_t = X_0 \exp\left(\int_0^t \sigma_s\,dW_s + \int_0^t (\mu_s - \tfrac{1}{2}\sigma_s^2)\,ds\right),$$

where μ_s and σ_s are deterministic, is a solution to the equation

$$\frac{dX_t}{X_t} = \mu_t\, dt + \sigma_t\, dW_t.$$

1.1.12 Girsanov's Theorem

Let $(\Omega, \mathcal{F})$ be a probability space with a filtration $(\mathcal{F}_t, t \in [0, T])$ satisfying $\mathcal{F}_T = \mathcal{F}$. Assume a measure $\mathbb{P}$ as well as a non-negative continuous martingale $D = (D_t, t \in [0, T])$ with $\mathbb{E}[D_T] = 1$ to be given. Then

$$\frac{d\hat{\mathbb{P}}}{d\mathbb{P}} = D_T$$

defines a new measure $\hat{\mathbb{P}}$ on $(\Omega, \mathcal{F})$ (see Section 1.1.3) and the densities of $\hat{\mathbb{P}}$ with respect to $\mathbb{P}$ on $\mathcal{F}_t$ are given by D_t ($t \in [0, T]$). The definition is consistent, as we have $\mathbb{P}[D_t A] = \mathbb{P}[D_s A]$ for $t < s$ and $A \in \mathcal{F}_t \subset \mathcal{F}_s$.

Now let a P-martingale $M = (M_t, t \in [0, T])$ be given. What can be said about M with respect to $\hat{\mathbb{P}}$? The answer to this question is the content of *Girsanov's theorem*, which states that the new process $\tilde{M}$, defined by

$$d\tilde{M} = dM - \frac{1}{D} d\langle M, D \rangle,$$

is a martingale with respect to $\hat{\mathbb{P}}$. In other words, M with respect to $\hat{\mathbb{P}}$ is the sum of a martingale and a drift term, ie a semi-martingale with $\int d\langle M, D \rangle / D$ as a finite variation process. The proof is based on Itô's lemma. The equations

$$\begin{aligned} d(\tilde{M}D) &= d(MD) - d\langle M, D \rangle \\ &= M\, dD + D\, dM + d\langle M, D \rangle - d\langle M, D \rangle \\ &= M\, dD + D\, dM \end{aligned}$$

show that $\tilde{M}D$ is a martingale with respect to $\mathbb{P}$. This in turn is equivalent to $\tilde{M}$ being a $\hat{\mathbb{P}}$-martingale.

Application: Brownian motion with drift Let $W = (W_t, t \in [0, T])$ be a Brownian motion under a measure $\mathbb{P}$, and let γ_t be a suitable integrable deterministic function. Now, $D = (D_t, t \in [0, T])$, with

$$D_t = \exp\left(-\int_0^t \gamma_s\, dW_s - \tfrac{1}{2} \int_0^t \gamma_s^2\, ds\right),$$

is a positive continuous martingale with $\mathbb{E}[D_T] = 1$. Therefore

$$\frac{d\hat{\mathbb{P}}}{d\mathbb{P}} = D_T$$

defines a measure $\hat{\mathbb{P}}$ on $\mathcal{F} = \mathcal{F}_T$. Moreover, $dD = -\gamma D\, dW$ (cf Section 1.1.11) and hence

$$dW - \frac{1}{D} \langle dW, -\gamma D\, dW \rangle = dW + \gamma\, dt$$

defines a martingale with respect to $\hat{\mathbb{P}}$. In other words, $(W_t + \int_0^t \gamma_s\, ds, t \in [0, T])$ is a Brownian motion with respect to $\hat{\mathbb{P}}$. Hence, given a Brownian motion with drift, it is possible to change the probability measure in such a way that it becomes a Brownian motion *without* drift with respect to the new measure.

1.1.13 Donsker's Invariance Principle

The central limit theorem states that, given a sequence $(X_i, i \in \mathbb{N})$ of independent identically distributed random variables X_i with $\mathbb{E}[X_i] = 0$ and $\text{Var}[X_i] = 1$, the distribution of the normalised sums

$$\frac{1}{\sqrt{n}} \sum_{i=1}^{n} X_i$$

converges, as n tends to infinity, to a standard normal distribution. The importance of Brownian motion in probability theory is due to the fact that, on the process level, a similar result shows that every random walk converges to a Brownian motion. This result is known as *Donsker's invariance principle*.

We now give an outline of the result. Given the sequence $(X_i, i \in \mathbb{N})$, it is possible to construct a random walk $Y = (Y_t, t \geq 0)$ by setting $Y_m = \sum_{i=1}^{m} X_i$ and interpolating linearly between Y_m and Y_{m+1} for all $m \in \mathbb{N}$. Viewing the process from increasing distance, ie considering a sequence of processes $(Y^n, n \in \mathbb{N})$ defined by

$$Y_t^n = \frac{1}{\sqrt{n}} Y_{nt}, \quad t \geq 0,$$

the central limit theorem shows that the distribution of Y_t^n converges to a normally distributed random variable with variance t as n tends to infinity. Hence, for fixed t, the limit distribution agrees with the distribution of W_t, with W denoting a Brownian motion. In order to prove the convergence on the process level, we first have to show the convergence of the distribution of vectors of the form $(Y_{t_0}^n, Y_{t_1}^n - Y_{t_0}^n, \ldots, Y_{t_m}^n - Y_{t_{m-1}}^n)$ for arbitrary sequences $(t_i, i = 1, 2, \ldots, m)$ and $m \in \mathbb{N}$ to the corresponding Gaussian vector $(W_{t_0}, W_{t_1} - W_{t_0}, \ldots, W_{t_m} - W_{t_{m-1}})$ The continuity of the process under the limit distribution follows from the tightness of the sequence of random walk distributions. For details, we refer the reader to [44, 60].

As an application of Donsker's theorem, we obtain the existence of Brownian motion and the convergence of the trees constructed in Chapter 6.

1.1.14 The Reflection Principle

The *reflection principle* states that, given a Brownian motion W starting in 0 at time $t = 0$ and two real numbers x and y with $y \geq 0$ and $y \geq x$, the events

- W hits y at some instant $t \in [0, 1]$ and $W_1 < x$,
- W_1 is greater than $2y - x$,

have equal probability. The basic principle is illustrated in Figure 1.1.

We exploit this result to derive the distribution of the pair (X, X_*) where

$$X = \sigma W_1 + \nu,$$
$$X_* = \min\{\sigma W_u + \nu u : 0 \leq u \leq 1\},$$

with W denoting a Brownian motion with respect to $\mathbb{P}$.

Mathematical Fundamentals

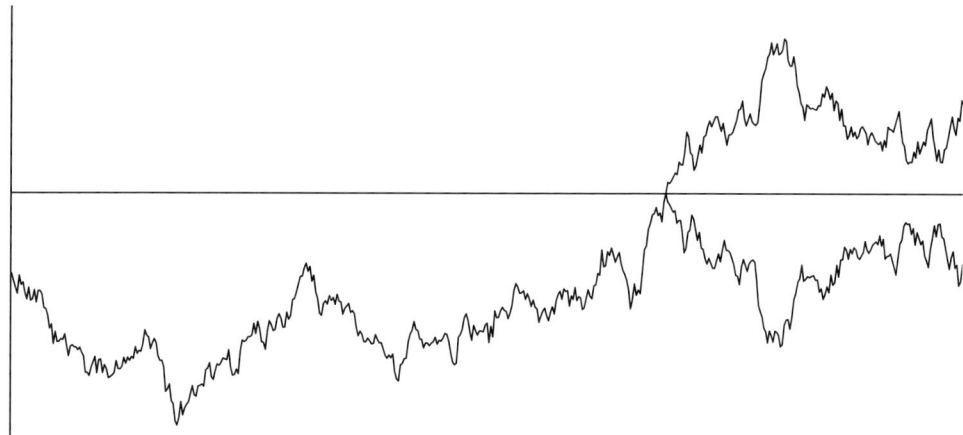

1.1 The reflection principle.

The case $v = 0$ In the following argument we use the strong Markov property of Brownian motion, which can be stated as

$$\mathbb{E}[f(W_1) \mid \mathcal{F}_\tau] = \mathbb{E}_{W_\tau}[f(W_{1-\tau})].$$

Here τ denotes a stopping time bounded by 1, and $\mathbb{P}_x$ is the distribution of $W + x$ with respect to $\mathbb{P}$. In our situation we use the *hitting time* τ_y defined by

$$\tau_y = \inf\{t \geqslant 0 \mid W_t = y\}. \tag{1.5}$$

We assume $y \leqslant 0$ as well as $y \leqslant x$ and hence $2y - x \leqslant y$. Then, using the normalisation $W_* = X_*/\sigma$, we have

$$\begin{aligned}
\mathbb{P}[W_1 > x, W_* < y] &= \mathbb{P}[W_1 > x, \tau_y < 1] \\
&= \mathbb{P}\big[\mathbb{P}[W_1 > x \mid \mathcal{F}_{\tau_y}], \tau_y < 1\big] \\
&= \mathbb{P}\big[\mathbb{P}[W_{1-\tau_y} + y > x], \tau_y < 1\big] \\
&= \mathbb{P}\big[\mathbb{P}[W_{1-\tau_y} + y < 2y - x], \tau_y < 1\big] \\
&= \mathbb{P}\big[\mathbb{P}[W_1 < 2y - x \mid \mathcal{F}_{\tau_y}], \tau_y < 1\big] \\
&= \mathbb{P}[W_1 < 2y - x].
\end{aligned}$$

Deriving the standard normal distribution function at $2y - x$ with respect to x and y and then incorporating the factor σ again gives the density function

$$\mathbb{P}[X \in dx, X_* \in dy] = \frac{2(2y - x)}{\sqrt{2\pi\sigma^2}} \exp\left(-\frac{(2y - x)^2}{2\sigma^2}\right) dx\, dy.$$

The general case Given the density function above, Girsanov's theorem (see Section 1.1.12) yields, for $y < 0$, $y < x$, and all v,

$$\mathbb{P}[X \in dx, X_* \in dy] = \frac{2(2y - x)}{\sqrt{2\pi\sigma^2}} \exp\left(-\frac{(2y - x)^2}{2\sigma^2} + \frac{v}{\sigma}x - \frac{v^2}{2\sigma^2}\right) dx\, dy.$$

Some corollaries Integration of the last equation with respect to x now gives

$$\mathbb{P}[X \geqslant x, X_* \in dy] = 2\exp\left(\frac{2\nu y}{\sigma^2}\right)\left[\varphi\left(\frac{2y-x+\nu}{\sigma}\right) + \frac{\nu}{\sigma}N\left(\frac{2y-x+\nu}{\sigma}\right)\right]dy,$$

and another integration with respect to y yields

$$\mathbb{P}[X \geqslant x, X_* \geqslant y] = N\left(\frac{-x+\nu}{\sigma}\right) - \exp\left(\frac{2\nu y}{\sigma^2}\right) N\left(\frac{-x+2y+\nu}{\sigma}\right)$$

if $y \leqslant 0$ and $y \leqslant x$. Girsanov's theorem (see Section 1.1.12) allows us to generalise this to

$$\mathbb{E}[e^{\alpha X} 1_{\{X \geqslant x, X_* \geqslant y\}}]$$
$$= \exp(\tfrac{1}{2}\alpha^2\sigma^2)\left\{N\left(\frac{-x+\nu+\alpha\sigma^2}{\sigma}\right) \exp\left(\frac{2\nu y}{\sigma^2} + 2\alpha y\right) N\left(\frac{-x+2y+\nu+\alpha\sigma^2}{\sigma}\right)\right\}.$$
(1.6)

Setting $x = y$ yields the density function of the maximum of drifted Brownian motion

$$\mathbb{E}[e^{\alpha X} 1_{\{X_* \geqslant y\}}]$$
$$= \exp(\tfrac{1}{2}\alpha^2\sigma^2)\left\{N\left(\frac{-y+\nu+\alpha\sigma^2}{\sigma}\right) - \exp\left(\frac{2\nu y}{\sigma^2} + 2\alpha y\right) N\left(\frac{y+\nu+\alpha\sigma^2}{\sigma}\right)\right\}.$$
(1.7)

1.1.15 *Hitting Times*

One barrier In equation (1.5) we introduced the hitting time τ_y of Brownian motion. More generally, given a constant drift ν, let τ_y^ν be defined by

$$\tau_y^\nu = \inf\{\, t \geqslant 0 \mid W_t + \nu t = y \,\}.$$

Clearly $\tau_0^\nu = 0$ holds. According to (1.7) we have

$$\mathbb{P}[\tau_y^\nu \leqslant T] = \mathbb{P}[\min\{W_u + \nu u : 0 \leqslant u \leqslant T\} \leqslant y]$$
$$= N\left(\frac{y-\nu T}{\sqrt{T}}\right) + e^{2\nu y} N\left(\frac{y+\nu T}{\sqrt{T}}\right),$$

if $y < 0$. Therefore the density of τ_y^ν is given by

$$\mathbb{P}[\tau_y^\nu \in dt] = \frac{|y|}{\sqrt{2\pi t^3}} \exp\left(-\frac{(y-\nu t)^2}{2t}\right) dt,$$

if $y \neq 0$.

Two barriers We now consider the situation when two barriers to be hit are involved. Let $L < 0 < U$. One can show that

$$\mathbb{P}[T_L \in dt, T_L \leqslant T_U] = \frac{1}{\sqrt{2\pi t^3}} \sum_{n=-\infty}^{\infty} a_n \exp\left(-\frac{a_n^2}{2t}\right) dt,$$

with $a_n = 2n(U - L) - L$ (see [44]). Therefore

$$\mathbb{P}[\,T_L \leqslant T, T_L \leqslant T_U\,] = \sum_{n=-\infty}^{\infty} \int_0^T \frac{a_n}{\sqrt{2\pi t^3}} \exp\left(-\frac{a_n^2}{2t}\right) dt$$

$$= 2\sum_{n=0}^{\infty} N\left(\frac{-a_n}{\sqrt{T}}\right) - 2\sum_{n=-\infty}^{-1} N\left(\frac{a_n}{\sqrt{T}}\right).$$

1.1.16 Hitting Times for Brownian Bridges

It is often useful to know the probability of a stock price reaching a certain level during the life of an option, conditional on the stock price at maturity. This can, for example, be used when performing Monte Carlo simulations of barrier options. Consider the Brownian motion with drift X_t ($t \geqslant 0$) where

$$X_t = \sigma W_t + vt,$$

with $\sigma > 0$ and $v \in \mathbb{R}$. We assume that either $x, y < U$ or $x, y > U$. Then

$$\mathbb{P}\big[\,X_u = U \text{ for some } u \in [0, 1] \,\big|\, X_0 = x, X_1 = y\,\big]$$

$$= \mathbb{P}\left[\,W_u = \frac{U}{\sigma} \text{ for some } u \in [0, 1] \,\bigg|\, W_0 = \frac{x}{\sigma}, W_1 = \frac{y}{\sigma}\,\right]$$

Here we used the identity of distributions

$$\big(W_u - z, u \in [0, 1] \,\big|\, W_0 = x + z, W_1 = y + z\big)$$

$$\stackrel{(d)}{=} \big(W_u + (1 - u)x + uy, u \in [0, 1] \,\big|\, W_0 = 0, W_1 = 0\big).$$

Now we apply the formula

$$\mathbb{P}\big[\,W_u = U \text{ for some } u \in [0, 1] \,\big|\, W_0, W_1\,\big] = \exp[-2(U - W_0)(U - W_1)]$$

given in [44] and generalise to an arbitrary interval $[s, t]$ to obtain

$$\mathbb{P}\big[\,X_u = U \text{ for some } u \in [s, t] \,\big|\, X_s, X_t\,\big] = \exp\left(-\frac{2}{\sigma^2(t-s)}[(U - X_s)(U - S_t)]\right), \tag{1.8}$$

if either $X_s, X_t \leqslant U$ or $X_s, X_t \geqslant U$.

1.1.17 Extremum and Terminal Value

In Section 1.1.15 we calculated some probabilities related to

$$X = \sigma W_1 + v,$$

$$X_* = \min\{\sigma W_u + vu : 0 \leqslant u \leqslant 1\},$$

with W denoting a Brownian motion with respect to $\mathbb{P}$. In this section we derive some further expressions, which will be used in Section 3.1 to obtain closed-form formulas for lookback options. We leave it to the reader to establish the results below with

$$X^* = \max\{\sigma W_u + vu : 0 \leqslant u \leqslant 1\}$$

instead of X_* and with terminal time t rather than 1.

We fix two real numbers γ and λ. Using the formulas established in Section 1.1.14 and assuming $\gamma \leqslant 0 \leqslant \lambda$, one has

$$\mathbb{E}[\,1_{\{X_* \leqslant \gamma,\, X \geqslant X_* + \lambda\}}\, e^{\eta X_*}\,] = \int_{-\infty}^{\gamma} e^{\eta y}\, \mathbb{P}[\, X \geqslant y + \lambda,\, X_* \in dy\,]\, dy.$$

With the aid of the formulas provided in Appendix A.4, one obtains

$$\mathbb{E}[\,1_{\{X_* \leqslant \gamma,\, X \geqslant X_* + \lambda\}}\, e^{\eta X_*}\,] = \left(1 + \frac{\eta \sigma^2}{2\tilde{\nu}}\right) \exp\left(\eta \tilde{\nu} + \frac{2\tilde{\nu}\lambda}{\sigma^2}\right) N\left(\frac{\gamma - \lambda + \nu - 2\tilde{\nu}}{\sigma}\right)$$
$$+ \left(1 - \frac{\eta \sigma^2}{2\tilde{\nu}}\right) \exp\left(\frac{2\tilde{\nu}\gamma}{\sigma^2}\right) N\left(\frac{\gamma - \lambda + \nu}{\sigma}\right),$$

where $\tilde{\nu} := \nu + \tfrac{1}{2}\sigma^2 \eta$ is assumed to be different from zero. In the case of $\tilde{\nu} = 0$, a careful limit procedure yields

$$\mathbb{E}[\,1_{\{X_* \leqslant \gamma,\, X \geqslant X_* + \lambda\}}\, e^{\eta X_*}\,]$$
$$= 2N\left(\frac{\gamma - \lambda + \nu}{\sigma}\right)$$
$$+ \lim_{\tilde{\nu} \to 0} \frac{\eta \sigma^2}{2\tilde{\nu}} \left\{ \exp\left(\frac{2\tilde{\nu}\gamma}{\sigma^2}\right) \left[\exp\left(\eta \tilde{\nu} + \frac{2\tilde{\nu}(\lambda - \gamma)}{\sigma^2}\right) \right.\right.$$
$$\left.\left. \cdot N\left(\frac{\gamma - \lambda + \nu - 2\tilde{\nu}}{\sigma}\right) - N\left(\frac{\gamma - \lambda + \nu}{\sigma}\right) \right] \right\}$$
$$= \left(2 + \frac{\eta^2 \sigma^2}{2} + \eta(\lambda - \gamma)\right) N\left(\frac{\gamma - \lambda + \nu}{\sigma}\right) - \eta \sigma \varphi\left(\frac{\gamma - \lambda + \nu}{\sigma}\right).$$

1.1.18 Partial Extremum and Terminal Value

In this section we calculate probabilities related to

$$X_1 = \sigma_1 W_1 + \nu_1,$$
$$X_* = \min\{\sigma_1 W_u + \nu_1 u : 0 \leqslant u \leqslant 1\},$$
$$X_2 = \sigma_2 (W_2 - W_1) + \nu_2,$$

with $\sigma_1 \neq 0 \neq \sigma_2$ and W denoting a Brownian motion with respect to $\mathbb{P}$. The expressions derived here and in the previous section will be used as building blocks for partial lookback option prices in Section 3.1. Again we leave it to the reader to:

- establish the results with

$$X^* = \max\{\sigma_1 W_u + \nu_1 u : 0 \leqslant u \leqslant 1\}$$

  rather than X_*;

- generalise the formulas to arbitrary s and t with $s < t$ rather than 1 and 2;

- swap the time intervals, ie consider $(X_2, \tilde{X}_*, X_1)$, where

$$X_2 = \sigma_2(W_2 - W_1) + v_2,$$
$$\tilde{X}_* = \min\{\sigma_2(W_u - W_1) + v_2(u-1) : 1 \leq u \leq 2\},$$
$$X_1 = \sigma_1 W_1 + v_1,$$

rather than (X_1, X_*, X_2).

In the following, we use the notation $v = v_1 + v_2$ and $\sigma^2 = \sigma_1^2 + \sigma_2^2$. Recall that

$$\mathbb{P}[X_2 \geq z] = N\left(\frac{-z+v_2}{\sigma_2}\right) \quad \text{and} \quad \mathbb{P}[X_2 \in dz] = \frac{1}{\sigma_2}\varphi\left(\frac{-z+v_2}{\sigma_2}\right).$$

Hence, assuming $\gamma \leq 0 \leq \lambda$ and $\rho \leq \lambda$, we have

$$\mathbb{E}[e^{\eta X_*} 1_{\{X_2 \leq \rho, X_* \leq \gamma, X_1 + X_2 \geq X_* + \lambda\}}]$$

$$= \int_{-\infty}^{\rho} \mathbb{E}[e^{\eta X_*} 1_{\{X_* \leq \gamma, X_1 \geq \lambda + X_* - z\}}] \mathbb{P}[X_2 \in dz]$$

$$= \left(1 + \frac{\eta \sigma_1^2}{2\tilde{v}_1}\right) \exp\left[\frac{2\tilde{v}_1}{\sigma_1^2}\left(\lambda - v_2 + v_1\frac{\sigma_2^2}{\sigma_1^2} + \frac{1}{2}\eta\sigma^2\right)\right]$$

$$\cdot N_2\left(\frac{\rho - v_2 + 2\tilde{v}_1\frac{\sigma_2^2}{\sigma_1^2}}{\sigma_2}, \frac{\gamma - \lambda + v - 2\tilde{v}_1\frac{\sigma^2}{\sigma_1^2}}{\sigma}; -\frac{\sigma_2}{\sigma}\right)$$

$$+ \left(1 - \frac{\eta\sigma_1^2}{2\tilde{v}_1}\right)\exp\left(\frac{2\tilde{v}_1\gamma}{\sigma_1^2}\right) N_2\left(\frac{\rho - v_2}{\sigma_2}, \frac{\gamma - \lambda + v}{\sigma}; -\frac{\sigma_2}{\sigma}\right)$$

if $\tilde{v}_1 = v_1 + \frac{1}{2}\eta\sigma_1^2 \neq 0$. Here we used results from Appendix A.4. In the case of $2v_1 = -\eta\sigma_1^2$, we obtain

$$\mathbb{E}[e^{\eta X_*} 1_{\{X_2 \leq \rho, X_* \leq \gamma, X_1 \geq X_* + \lambda - X_2\}}]$$

$$= \int_{-\infty}^{\rho} \mathbb{E}[e^{\eta X_*} 1_{\{X_* \leq \gamma, X_1 \geq \lambda + X_* - z\}}] \mathbb{P}[X_2 \in dz]$$

$$= -\eta\sigma\varphi\left(\frac{\gamma - \lambda + v}{\sigma}\right) N\left(\frac{\sigma^2(\rho - v_2) + (\gamma - \lambda + v)\sigma_2^2}{\sigma\sigma_1\sigma_2}\right)$$

$$+ \left(2 + \frac{\eta^2\sigma_1^2}{2} + \eta(\lambda - \gamma - v_2)\right) N_2\left(\frac{\rho - v_2}{\sigma_2}, \frac{\gamma - \lambda + v}{\sigma}; -\frac{\sigma_2}{\sigma}\right)$$

$$+ \eta\sigma_2\varphi\left(\frac{\rho - v_2}{\sigma_2}\right) N\left(\frac{\rho + \gamma - \lambda + v_1}{\sigma_1}\right).$$

We continue with the results:

$$\mathbb{P}[X_2 \leq \rho, X_* > \gamma, X_1 + X_2 > \gamma + \lambda] = N_2\left(\frac{\rho - v_2}{\sigma_2}, \frac{-\lambda - \gamma + v}{\sigma}, -\frac{\sigma_2}{\sigma}\right)$$

$$- \exp\left(\frac{2v_1\gamma}{\sigma_1^2}\right) N_2\left(\frac{\rho - v_2}{\sigma_2}, \frac{-\lambda + \gamma + v}{\sigma}, -\frac{\sigma_2}{\sigma}\right),$$

again using Appendix A.4. Then,

$$\mathbb{E}[e^{\eta X_* + \xi X_2} 1_{\{X_2 \geq \rho, X_* + X_2 \leq \gamma\}}]$$

$$= \int_\rho^\infty e^{\xi z} \mathbb{E}[e^{\eta X_*} 1_{\{X_* \leq \gamma - z\}}] \mathbb{P}[X_2 \in dz]$$

$$= \left(1 + \frac{\eta \sigma_1^2}{2\tilde{v}_1}\right) \exp\left(\eta \tilde{v}_1 + \xi v_2 + \frac{1}{2}\xi^2 \sigma_2^2\right)$$

$$\cdot N_2\left(\frac{-\rho + v_2 + \xi \sigma_2^2}{\sigma_2}, \frac{\gamma - v - \eta \sigma_1^2 - \xi \sigma_2^2}{\sigma}; -\frac{\sigma_2}{\sigma}\right)$$

$$+ \left(1 - \frac{\eta \sigma_1^2}{2\tilde{v}_1}\right) \exp\left[\frac{2\tilde{v}_1 \gamma}{\sigma_1^2} + v_2 \xi - \frac{2\tilde{v}_1 v_2}{\sigma_1^2} + \sigma_2^2\left(\frac{2\tilde{v}_1^2}{\sigma_1^4} - \frac{2\tilde{v}_1 \xi}{\sigma_1^2} + \frac{\xi^2}{2}\right)\right]$$

$$\cdot N_2\left(\frac{-\rho + (v_2 - 2v_1 \frac{\sigma_2^2}{\sigma_1^2}) - (\eta - \xi)\sigma_2^2}{\sigma_2}, \frac{\gamma - (v_2 - 2v_1 \frac{\sigma_2^2}{\sigma_1^2}) + v_1 + (\eta - \xi)\sigma_2^2}{\sigma}; -\frac{\sigma_2}{\sigma}\right)$$

if $\tilde{v}_1 = v_1 + \frac{1}{2}\eta \sigma_1^2 \neq 0$, where we used Appendix A.4. In the case of $\tilde{v}_1 = 0$, we have

$$\mathbb{E}[e^{\eta X_* + \xi X_2} 1_{\{X_2 \geq \rho, X_* + X_2 \leq \gamma\}}]$$

$$= \int_\rho^\infty e^{\xi z} \mathbb{E}[e^{\eta X_*} 1_{\{X_* \leq \gamma - z\}}] \mathbb{P}[X_2 \in dz]$$

$$= e^{\xi v_2} e^{\xi^2 \sigma_2^2/2} \left\{ (2 + \tfrac{1}{2}\eta^2 \sigma_1^2 + \eta v_2 - \eta \gamma + \eta \xi \sigma_2^2) \right.$$

$$\cdot N_2\left(\frac{-\rho + v_2 + \xi \sigma_2^2}{\sigma_2}, \frac{\gamma - v_2 + v_1 - \xi \sigma_2^2}{\sigma}; -\frac{\sigma_2}{\sigma}\right)$$

$$- \eta \sigma \varphi\left(\frac{\gamma - v_2 + v_1 - \xi \sigma_2^2}{\sigma}\right) N\left(\frac{\sigma^2(v_2 - \rho) + \sigma_1^2 \sigma_2^2 \xi + \sigma_2^2(\gamma - v_2 + v_1)}{\sigma \sigma_1 \sigma_2}\right)$$

$$\left. + \eta \sigma_2 \varphi\left(\frac{-\rho + v_2 + \xi \sigma_2^2}{\sigma_2}\right) N\left(\frac{-\rho + \gamma + v_1}{\sigma_1}\right) \right\}.$$

1.1.19 Correlated Extremum and Terminal Value

In this section we consider two correlated Brownian motions with drifts X and Y. For given x and y, we want to compute the probability of Y reaching level y at time T and X never falling below level x between time 0 and T. More formally, let

$$X_* = \min\{\sigma_1 W_u^1 + v_1 u : 0 \leq u \leq T\},$$

$$Y_T = \sigma_2 W_T^2 + v_2 T,$$

with $\sigma_1 \neq 0 \neq \sigma_2$ and W^1, W^2 denoting two Brownian motions with correlation ρ. The expressions derived in this section will be used as building blocks for outside barrier option prices in Section 4.9.

We replace (W^1, W^2) by $(W, \rho W + \sqrt{1 - \rho^2}\, \tilde{W})$, with $W, \tilde{W}$ denoting two independent Brownian motions. In the following, we assume $\sigma_1 = \sigma_2 = T = 1$

and $v_1 = v$, $v_2 = 0$. We remark that

$$\mathbb{P}[X_* \geq x, X_1 \geq z] = N(-(x \vee z) + v) - e^{2vx} N(-(x \vee z) + 2x + v)$$

if $x \leq 0$ (see Section 1.1.14). Differentiating with respect to z yields

$$\mathbb{P}[X_* \geq x \mid X_1 \in dz] = [\varphi(-z + v) - e^{2vx} \varphi(-z + 2x + v)] 1_{\{z \geq x\}} dz.$$

On the other hand, we have

$$\mathbb{P}[Y_1 \geq y \mid X_1 = z] = \mathbb{P}\left[\tilde{W}_1 \geq \frac{y - \rho(z - v)}{\sqrt{1 - \rho^2}}\right]$$

$$= N\left(\frac{-y + \rho(z - v)}{\sqrt{1 - \rho^2}}\right).$$

Since X^* and Y_1 are independent, given X_1, we have

$$\mathbb{P}[X_* \geq x, Y_1 \geq y] = \int_x^\infty \mathbb{P}[Y_1 \geq y \mid X_1 = z] \mathbb{P}[X_* \geq x, X_1 \in dz]$$

$$= N_2(-x + v, -y; \rho) - e^{2vx} N_2(x + v, -y + 2\rho x; \rho),$$

where we have used formulas provided in Appendix A.2. Using Girsanov's theorem (Section 1.1.12), this can be generalised to

$$\mathbb{E}[e^{\alpha Y_1} 1_{\{X_* \geq x, Y_1 \geq y\}}] = e^{\alpha^2/2} N_2(-x + (v + \rho\alpha), -y + \alpha; \rho)$$

$$- \exp\left(\tfrac{1}{2}\alpha^2 + 2(v + \rho\alpha)x\right) N_2(x + (v + \rho\alpha), -y + \alpha + 2\rho x; \rho).$$

The formula with arbitrary $\sigma_i > 0$, v_i, and $T > 0$ reads

$$\mathbb{E}[e^{\alpha Y_T} 1_{\{X_* \geq x, Y_T \geq y\}}]$$

$$= \exp(\tfrac{1}{2}\alpha^2 \sigma_2^2 T + \alpha v_2 T)$$

$$\cdot N_2\left(\frac{1}{\sigma_1 \sqrt{T}}[-x + (v_1 + \rho\alpha\sigma_1\sigma_2)T], \frac{1}{\sigma_2 \sqrt{T}}[-y + (v_2 + \alpha\sigma_2^2)T]; \rho\right)$$

$$- \exp\left[\tfrac{1}{2}\alpha^2 \sigma_2^2 T + \alpha v_2 T + 2(v_1 + \rho\alpha\sigma_1\sigma_2)x/\sigma_1^2\right]$$

$$\cdot N_2\left\{\frac{1}{\sigma_1 \sqrt{T}}[x + (v_1 + \rho\alpha\sigma_1\sigma_2)T], \frac{1}{\sigma_2 \sqrt{T}}\left[-y + \left(v_2 + \alpha\sigma_2^2 + 2\rho\frac{\sigma_2}{\sigma_1}x\right)T\right]; \rho\right\}.$$

(1.9)

1.2 The Black–Scholes Equity Model

We begin our look at the Black–Scholes model by first considering the behaviour of a continuous market using arbitrage pricing theory and providing a general framework for valuing derivative securities. An application of these results will result in the Black–Scholes model.

Assume that our market contains N traded assets, $Z_1(t), \ldots, Z_N(t)$, each satisfying the stochastic differential equation

$$dZ_i(t) = \mu_i \, dt + \sigma_i \, dW_i, \qquad (1.10)$$

on the probability space $(\Omega, \mathcal{F}, \mathbb{P})$. These assets follow a Brownian motion process with a drift μ_i and a volatility σ_i. Note that μ_i does not have to be a function of $Z_i(t)$ and t. We will generally consider one of the traded assets, say $\mathcal{Z} = Z_1$, as a numéraire (positive valued and non-dividend paying) and denominate all prices in this asset. These relative price processes are denoted by

$$Z'_i(t) = \frac{Z_i(t)}{\mathcal{Z}(t)}.$$

To keep things simple, we assume for now that the assets do not pay any income, such as dividends or securities lending income to the holder. Furthermore, we assume no transaction costs.

A general *trading (portfolio) strategy* involves holding certain positions in the traded assets and is represented by the N-dimensional stochastic process

$$\varphi(t) = [\varphi_1(t), \ldots, \varphi_N(t)],$$

where $\varphi_i(t)$ is the holding in the asset i at time t. The process $\varphi_i(t)$ will be assumed to be predictable, because all the trader's decisions at time t have to be based on information acquired up to this time only. For technical reasons, it will also be assumed to be suitably integrable, so that $\int \varphi \, dZ$ is well defined.

The value $V(\varphi, t)$ of the portfolio is then given by

$$V(\varphi, t) = \sum_{i=1}^{N} \varphi_i(t) Z_i(t).$$

A strategy whose value changes depend only on the changes in the assets' values is called a *self-financing trading strategy*. In such a strategy, funds are neither removed from nor added to the portfolio, and the strategy has the property

$$V(\varphi, s) - V(\varphi, t) = \sum_{i=1}^{N} \int_{t}^{s} \varphi_i(u) \, dZ_i(u)$$

for $s > t$. It is easy to show that this portfolio remains self-financing after a change of numéraire. That is, if we look at the value V' in terms of the numéraire $\mathcal{Z}$ defined as

$$V'(\varphi, t) = \frac{V(\varphi, t)}{\mathcal{Z}(t)} = \sum_{i=1}^{N} \varphi_i(t) Z'_i(t),$$

then the strategy is self-financing if and only if

$$V'(\varphi, s) - V'(\varphi, t) = \sum_{i=1}^{N} \int_{t}^{s} \varphi_i(u) \, dZ'_i(u).$$

Note that, in this formulation, the value of $\varphi_1(u)$ is irrelevant for the right-hand side of the above equation, as $dZ'_1 = 0$. In particular, for a self-financing

Mathematical Fundamentals 21

strategy φ, the value of $\varphi_1(t)$ (usually interpreted as the balance of the cash account Z at time t) can be computed from $(\varphi_2, \ldots, \varphi_N)$ according to

$$\varphi_1(t) = \varphi_1(0) + \sum_{i=2}^{N} \int_0^t \varphi_i(u) \, dZ'_i(u) - \sum_{i=2}^{N} \varphi_i(t) Z'_i(t)$$

for $t > 0$.

The stochastic integral $\int \varphi_i \, dZ_i$ can be interpreted as the cumulative gains and losses incurred from trading asset i according to strategy φ_i. This is clear if φ_i is $1_{(t,s]}$, ie just a buy-and-hold strategy, as $\int \varphi_i \, dZ_i = Z_s - Z_t$; in general, one can approximate a complex trading strategy by a sequence of linear combinations of buy-and-hold strategies, which justifies regarding $\int \varphi_i \, dZ_i$ as the cumulative profit and loss, even for complex strategies.

We can now introduce the concept of *arbitrage*: an arbitrage opportunity is a self-financing trading strategy φ' that generates wealth from nothing. That is, if $V(\varphi', 0) = 0$, then φ' is an arbitrage opportunity if

$$P[V(\varphi', T) \geqslant 0] = 1 \quad \text{and} \quad P[V(\varphi', T) > 0] > 0.$$

If we assume equation (1.10) for the assets Z, then we will show below that no such arbitrage opportunities can exist. However, the questions become very complex if we don't start with the explicit equation (1.10) for Z but just assume that Z is a semi-martingale under $\mathbb{P}$. We will just attempt to give a brief overview of the theory in this general setting. A measure $\tilde{\mathbb{P}}$ on $(\Omega, \mathcal{F})$ will be called an *equivalent martingale measure* if the following two conditions hold:

- $\tilde{\mathbb{P}}$ is equivalent to $\mathbb{P}$, ie both measures have the same null sets.

- The relative price processes Z'_i are martingales under $\tilde{\mathbb{P}}$ for all i,

$$\mathbb{E}_{\tilde{\mathbb{P}}}[Z'_i(s) \mid \mathcal{F}_t] = Z'_i(t), \quad s \geqslant t.$$

If there is an equivalent martingale measure $\tilde{\mathbb{P}}$, it is obvious that no arbitrage opportunities φ' can exist. Usually one is interested in the converse: if the model is arbitrage-free, does there always exist an equivalent martingale measure? Unfortunately the answer is no. We can, however, establish the existence of an equivalent martingale measure $\tilde{\mathbb{P}}$ for an adapted continuous process Z' if we assume that Z' allows *no free lunch with vanishing risk* (NFLVR). The condition of NFLVR is somewhat stronger than the no-arbitrage condition: whereas the no-arbitrage condition just requires that a certain set of strategies shall be empty, NFLVR requires that the closure of this set with respect to a certain topology shall be empty. For a discussion of this problem, see the work of Delbaen and Schachermayer [16].

From now on, we assume that there is no free lunch with vanishing risk and hence at least one equivalent martingale measure $\tilde{\mathbb{P}}$ exists. We now consider a European option with a payoff X at time T. We say that X is *attainable* if there is a self-financing strategy φ such that $V(\varphi, t)$ ($t > 0$) is a $\tilde{\mathbb{P}}$ martingale and

$$V(\varphi, T) = X.$$

If X is attainable, then the trading strategy φ will generate a portfolio whose value equals the payoff of the option X for almost every random evolution of the market. Thus, φ can be regarded as a perfect hedge of X. Since φ is self-financing, the cost of performing this hedge is just given by the number $V(\varphi, 0)$. Obviously $V(\varphi, 0)$ must be the fair price for the option X. If X could be sold for more than $V(\varphi, 0)$, we would sell X and hedge by replicating the payoff through φ, thus creating a risk-free profit. If X could be bought for less than $V(\varphi, 0)$, we would buy X and hedge by trading $-\varphi$, again making a risk-free profit.

Thus, for an attainable option X, we just have to compute $V(\varphi, 0)$ to obtain the fair price of X at time $t = 0$, as we have

$$\frac{X}{\mathcal{Z}(T)} = \frac{V(\varphi, T)}{\mathcal{Z}(T)} = V'(\varphi, T)$$

$$= \sum_{i=0}^{N} \int_0^T \varphi_i(u)\, dZ'_i(u) + V'(\varphi, 0).$$

It follows from the martingale property that

$$V(\varphi, t) = \mathcal{Z}(t) V'(\varphi, t) = \mathcal{Z}(t) \mathbb{E}_{\tilde{\mathbb{P}}}\left[\left. \frac{X}{\mathcal{Z}(T)} \right| \mathcal{F}_t \right] \qquad (1.11)$$

for all $t < T$. This is the mathematical framework for computing the fair price of the option X. Much of the book will be devoted to explicitly computing this expected value for various option payoffs X.

Of course, equation (1.11) only makes sense for attainable options. Fortunately, in the Brownian model (1.10) it can be shown that every integrable option X is indeed attainable and hence the model is called *complete*. This follows from the martingale representation property of Brownian motion (see Section 1.4).

Furthermore, it can be shown that the model is complete if and only if the equivalent martingale measure $\tilde{\mathbb{P}}$ is unique. This result is derived from the fact that that martingale representation property of $\tilde{\mathbb{P}}$ is related to $\tilde{\mathbb{P}}$ being extremal in the set of all measures that make Z' a martingale. See [31, 32] for the original work.

In the Brownian world with one source of uncertainty (ie Brownian motion) per tradable asset, as in (1.10), we therefore have an arbitrage-free and complete market. The unique equivalent martingale measure can be explicitly computed via a Girsanov transformation, which we now show for the Black–Scholes world.

One-dimensional Black–Scholes model The Black–Scholes world [5, 52] consists of two traded assets, a riskless money market account B_t and a risky equity asset S_t. These assets are assumed to satisfy the stochastic differential equations

$$\frac{dB_t}{B_t} = r_t\, dt, \qquad B_0 = 1, \qquad (1.12)$$

$$\frac{dS_t}{S_t} = \mu_t\, dt + \sigma_t\, dW_t, \qquad (1.13)$$

where r is assumed to be deterministic and μ is a possibly random drift term. We assume for now that S does not pay dividends or other yields like lending income.

Mathematical Fundamentals

The money market account is assumed to earn the risk-free instantaneous interest rate r_t (also called the *short rate*) and has a value 1 at time $t = 0$ and

$$B_t = \exp\left(\int_0^t r_u \, du\right).$$

at time t. Here B_t plays the role of the numéraire $\mathcal{Z}(t)$.

The risky equity asset is assumed to follow a geometric Brownian motion with drift μ_t and local volatility σ_t. The process W_t is a standard Brownian motion under the probability measure $\mathbb{P}$. This measure represents the true market distribution of the asset.

If we consider the money market account as the numéraire, then the relative price process $Z'_t = S_t/B_t$ is the discounted asset process. Using Itô's lemma (1.4), we obtain

$$\frac{dZ'_t}{Z'_t} = (\mu_t - r_t)\, dt + \sigma_t \, dW_t.$$

For the absence of arbitrage opportunities in our Black–Scholes market, we require an equivalent martingale measure. Using Girsanov's theorem (see Section 1.1.12), we find that the new process $\tilde{W}_t$, and corresponding probability measure $\tilde{\mathbb{P}}$ is achieved by the transformation $\gamma_t = (\mu_t - r_t)/\sigma_t$. The process $\tilde{W}_t$ is a Brownian motion under $\tilde{\mathbb{P}}$.

Under this new probability measure $\tilde{\mathbb{P}}$, the discounted asset process is a martingale as required. The asset process is then given by

$$\frac{dS_t}{S_t} = r_t \, dt + \sigma_t \, d\tilde{W}_t. \tag{1.14}$$

It can be shown that, for every $\tilde{\mathbb{P}}$ square integrable European contingent claim X, the discounted process is a martingale. That is, the arbitrage-free price of the claim at time t is given by

$$V(S_t, t) = B_t \, \mathbb{E}_{\tilde{\mathbb{P}}}[\, B_T^{-1} X \mid \mathcal{F}_t\,]. \tag{1.15}$$

Moreover, as shown in Section 1.4, in our set-up every sufficiently integrable functional of S and B is an attainable claim.

Unless otherwise stated, it is assumed that all calculations are done with respect to the risk-neutral probability measure.

Black–Scholes partial differential equation We shall now show that this price satisfies a partial differential equation – the Black–Scholes equation.

Using Itô's lemma (1.4), the discounted contingent claim process $X'(S_t, t) = B_t^{-1} V(S_t, t)$ can be written as

$$d[X'(S_t, t)] = B_t^{-1}\left[-r_t V \, dt + \frac{\partial V}{\partial t}\, dt + \frac{\partial V}{\partial S}\, dS_t + \tfrac{1}{2}\sigma_t^2 S^2 \frac{\partial^2 V}{\partial S^2}\, dt\right],$$

and because we are working in the risk-neutral probability measure we use equation (1.14) to obtain

$$d[X'(S_t, t)] = B_t^{-1}\left[-r_t V + \frac{\partial V}{\partial t} + r_t S \frac{\partial V}{\partial S} + \tfrac{1}{2}\sigma_t^2 S^2 \frac{\partial^2 V}{\partial S^2}\right] dt + B_t^{-1} \sigma_t S \frac{\partial V}{\partial S}\, d\tilde{W}_t.$$

However, under $\tilde{\mathbb{P}}$, this is a martingale and, as such, the drift term is zero. So

$$\frac{\partial V}{\partial t} + r_t S \frac{\partial V}{\partial S} + \tfrac{1}{2}\sigma_t^2 S^2 \frac{\partial^2 V}{\partial S^2} = r_t V.$$

This gives us our Black–Scholes partial differential equation for the price of the contingent claim. The extension to include the presence of an instantaneous dividend yield d_t is straightforward and, in the risk-neutral measure,

$$\frac{dS_t}{S_t} = (r_t - d_t)\, dt + \sigma_t\, d\tilde{W}_t, \qquad (1.16)$$

where we assume the parameters r_t, d_t and σ_t to be deterministic. The price of European contingent claims then satisfy the Black–Scholes equation

$$\frac{\partial V}{\partial t} + (r_t - d_t) S \frac{\partial V}{\partial S} + \tfrac{1}{2}\sigma_t^2 S^2 \frac{\partial^2 V}{\partial S^2} - r_t V = 0 \qquad (1.17)$$

under appropriate boundary conditions.

Illustration In order to illustrate some of the principles used in the above work, we now provide a simple example of option pricing in a discrete one-step binomial arbitrage-free pricing framework.

Consider an asset S_0 at time t_0. At time t_1, the asset can jump up to uS_0 or fall to dS_0. Our option has a payoff $V_1(S_1)$, and we want to know its value V_0 today.

We can calculate this by constructing a trading strategy: sell the option for V_0, buy Δ_0 shares and invest (or borrow) the remaining $V_0 - \Delta_0 S_0$ in the money market account. Our wealth at time t_1 should then cover any costs incurred by the option irrespective of whether the asset goes up or down, V_1^u and V_1^d respectively. This is our no-arbitrage condition. We therefore have two equations

$$V_1^u = \Delta_0 u S_0 + (1+r)(V_0 - \Delta_0 S_0),$$
$$V_1^d = \Delta_0 d S_0 + (1+r)(V_0 - \Delta_0 S_0),$$

which we can solve for Δ_0 and V_0:

$$\Delta_0 = \frac{V_1^u - V_1^d}{uS_0 - dS_0},$$

$$V_0 = \frac{1}{1+r}[\tilde{p} V_1^u + \tilde{q} V_1^d],$$

where

$$\tilde{p} = \frac{1+r-d}{u-d}, \qquad \tilde{q} = 1 - \tilde{p}.$$

Hence we have the value and hedge for our option. In fact you can see that this value is nothing more than the discounted expected value of the option payoff under the probabilities $\tilde{p}$ and $\tilde{q}$. These are the risk-neutral probabilities, and have nothing to do with the actual probabilities that the asset will move up or down. These ideas extend naturally to an N-step binomial model, which in the limit of large N converges to the continuous model.

The same approach is often used to derive the Black–Scholes equation in continuous time. A risk-free portfolio consisting of a short position in the contingent claim and a long position in the underlying stock is constructed and it is then assumed that the strategy is self-financing [5, 36]. Although this approach is rigorous in a discrete setting, it is incorrect when one considers the continuous Black–Scholes model. This is because the assumption of self-financing fails when the claim satisfies the Black–Scholes equation.

1.2.1 *The Basic Model*

We can now assume that the equity price process S_t with $t \geq 0$ follows the stochastic differential equation (1.16). We denote non-local quantities using double subscripts. So, for $0 < s < t$, we have

$$r_{st} = \frac{1}{t-s} \int_s^t r_u \, du,$$

$$d_{st} = \frac{1}{t-s} \int_s^t d_u \, du,$$

$$\sigma_{st}^2 = \frac{1}{t-s} \int_s^t \sigma_u^2 \, du.$$

Using this notation, we can write the asset price process explicitly as

$$S_t = S_0 \exp\left((r_{0t} - d_{0t} - \tfrac{1}{2}\sigma_{0t}^2)t + \int_0^t \sigma_u \, dW_u \right).$$

This follows from Itô's lemma, which implies that

$$d\left[\log \frac{S_t}{S_0} \right] = (r_t - d_t - \tfrac{1}{2}\sigma_t^2) \, dt + \sigma_t \, dW_t.$$

Note that the random variable $\int_s^t \sigma_u \, dW_u$ is normally distributed with mean 0 and variance $\sigma_{st}^2(t-s)$.

We now introduce further notation, which will be used throughout this book:

$$R_{st} = \int_s^t r_u \, du = r_{st}(t-s),$$

$$D_{st} = \int_s^t d_u \, du = d_{st}(t-s),$$

$$\mathcal{M}_{st} = R_{st} - D_{st} = \int_s^t (r_u - d_u) \, du,$$

$$\Sigma_{st}^2 = \int_s^t \sigma_u^2 \, du = \sigma_{st}^2(t-s),$$

$$\mathcal{N}_{st} = \mathcal{M}_{st} - \tfrac{1}{2}\Sigma_{st}^2,$$

$$\tilde{\mathcal{N}}_{st} = \mathcal{M}_{st} + \tfrac{1}{2}\Sigma_{st}^2.$$

In the case that the first subindex of any of these six quantities is zero (which corresponds to the evaluation date), we simply drop it. For example, we have

$$\Sigma_t^2 = \int_0^t \sigma_u^2 \, du.$$

Thus we can rewrite the asset price process more succinctly as

$$S_t = S_0 \exp\left(\mathcal{M}_t - \tfrac{1}{2}\Sigma_t^2 + \int_0^t \sigma_u \, dW_u\right)$$
$$= S_0 \exp\left(\mathcal{N}_t + \int_0^t \sigma_u \, dW_u\right).$$

For some path-dependent options, we cannot allow the parameters r_t, d_t and σ_t to vary with time. In this case we will denote them simply by r, d and σ.

1.3 Extensions to Black–Scholes

The standard Black–Scholes model has a limited range of applicability, and therefore we introduce generalisations to this framework to allow us to price a much wider range of instruments. In this section we also introduce stochastic exchange rate processes into our stochastic equity models.

1.3.1 *Multiple Assets*

Often we have to deal with the situation of several correlated assets. Denoting the asset price processes by $S_t^1, \ldots, S_t^n$, we assume they satisfy

$$\frac{dS_t^i}{S_t^i} = (r_t - d_t^i)\, dt + \sigma_t^i \, dW_t^i, \quad i = 1, \ldots, n,$$

where $W_t^1, \ldots, W_t^n$ are standard Brownian motions instantaneously correlated via

$$d\langle W_t^i, W_t^j \rangle = \rho_t^{ij} \, dt.$$

In other words, W_t^i and W_t^j are instantaneously correlated with correlation ρ_t^{ij}. So we have a time-dependent instantaneous correlation matrix $\Gamma_t = [\rho_t^{ij}]$.

We adapt the notation introduced in Section 1.2.1 by adding subscripts giving the index of the asset that the quantity is referring to. For example, we have

$$(\Sigma_{st}^i)^2 = \int_s^t (\sigma_u^i)^2 \, du = (\sigma_{st}^i)^2 (t - s).$$

We also define

$$\Psi_{st}^{ij} = \int_s^t \rho_u^{ij} \sigma_u^i \sigma_u^j \, du.$$

So $\Psi_t^{ij} = \Psi_{0t}^{ij}$ is the covariance of $\log S_t^i$ and $\log S_t^j$. In particular, the correlation between these two random variables is given by

$$\frac{\Psi_t^{ij}}{\Sigma_t^i \cdot \Sigma_t^j}.$$

Note that, when σ_u^i, σ_u^j and $\rho_u^{ij} = \rho^{ij}$ are all constant in time, this fraction simplifies to ρ^{ij}. However, even if the instantaneous correlation is equal to some constant value over time, the correlation of the logarithms of two assets will in general not be equal to this value.

The covariance also has the following properties:

$$\Psi_{st}^{ii} = (\Sigma_{st}^i)^2 \quad \text{and} \quad \Psi_{st}^{ij} = \Psi_{st}^{ji}.$$

Thus, in cases where there is no ambiguity, we will drop the superscript and write the covariance as Ψ_{st}.

1.3.2 Discrete Dividends

For our basic stock price model, we have so far assumed a continuous dividend yield. In reality, dividends are, of course, paid discretely. Our model can easily be adjusted to take such dividends into account if it is assumed that the dividend amount is proportional to the stock price at the dividend payment date.

This is, however, not a very realistic assumption. It is more common to assume that the dividend amount paid is known in advance (usually reported by an equity research department). We will present three different ways of modelling such discrete dividends. These methods have very distinct features, and can yield quite different prices. It is therefore important for the hedging of a derivative position that a consistent model be used throughout the life of the position.

In what follows, let $0 < t_1 < t_2 < \cdots$ be the dividend payment days and let $\delta_1, \delta_2, \ldots$ be the corresponding amounts paid.

Method I This method depends on the maturity of the instrument in question. This is certainly unsatisfactory. Nevertheless, the method is popular in some markets.

We model the asset price minus the present value of all dividends yet to be paid until maturity T using a standard Itô process. So

$$S_t = \tilde{S}_t + D_t,$$

where $\tilde{S}_t$ follows some Itô process, and where

$$D_t = \sum_{t < t_i \leqslant T} e^{-R_{tt_i}} \delta_i.$$

For a European plain vanilla option, this method amounts to replacing the spot S_0 by $S_0 - D_0$. We also note that it is well suited for non-path-dependent closed-form solutions, for trees and for Monte Carlo pricing.

Method II Here we model the asset price minus all dividends already paid as an Itô process. So

$$S_t = \hat{S}_t - \hat{D}_t,$$

where $\hat{S}_t$ is an Itô process and

$$\hat{D}_t = \sum_{0 < t_i \leqslant t} e^{R_{t_i t}} \delta_i.$$

Note that $\hat{D}_0 = 0$.

For a European plain vanilla option, this amounts to replacing the strike K by $K + \hat{D}_T$. Again, this method is well suited for non-path-dependent closed-form solutions, for trees and for Monte Carlo pricing.

Method III Here we subtract the dividends as they come and model lognormally between them. More formally, S_t is a lognormal process on each $[t_{i-1}, t_i)$ and $S_{t_i} - \lim_{t \to t_i-} S_t = D_i$.

Put differently, we model

$$S_t = S'_t \cdot \prod_{0 < t_i \leq t} \left(1 - \frac{\delta_{t_i}}{S_{t_i}}\right),$$

with S'_t being a geometric Brownian motion.

This method is particularly suitable for Monte Carlo methods, but unsuitable for closed-form solutions.

1.3.3 Forward Start Options

In practice, options are often *forward started*. The precise meaning of this expression depends on the particular product. If the product has a strike, in the forward-started variant it is determined at some future date $t > 0$ as the then current underlying price multiplied by some previously agreed factor.

For example, a forward-started plain vanilla call has the payoff

$$(S_T - \alpha S_t)^+,$$

where T is the maturity date and α a fixed factor.

The pricing of a forward-start option can in many cases be reduced to that of the corresponding non-forward-start product with forward market parameters. To indicate the approach, consider a plain vanilla call and note that

$$\mathbb{E}\left[(S_T - \alpha S_t)^+\right] = \mathbb{E}[S_t] \cdot \mathbb{E}\left[\left(\frac{S_T}{S_t} - \alpha\right)^+\right]$$

by independence. The term $\mathbb{E}[S_t]$ is just the forward at time t. We also note that $\log(S_T/S_t)$ is normally distributed with mean $\mathcal{N}_{tT}$ and variance Σ^2_{tT}.

1.3.4 Quanto and Composite Options

Equity option contracts containing a foreign exchange rate component are commonly termed "quanto options". This section explains the relationship with their non-quanto counterparts. We denote by S the price of an underlying equity expressed in its own domestic currency, eg in EUR for the DAX. Furthermore, the exchange rate from domestic currency into some foreign currency is given by X, which is of dimension "foreign currency/domestic currency (eg USD/EUR)".

An investor whose accounting currency is the domestic currency of the underlying equity is called a *domestic investor*; an investor who uses the foreign currency as numéraire is called a *foreign investor*.

What is commonly called a "quanto option" appears in two variants:

- as a *composite* option, where the payoff depends on the underlying equity price translated into the foreign currency at the current exchange rate – mathematically this means that the payoff of a composite option is a functional of the stochastic process SX;

- as a *quanto* option, where the payoff is given as the payoff of the corresponding non-quanto option converted into the foreign currency at a previously stipulated constant exchange rate $\bar{x}$ with dimension "foreign currency/domestic currency".

Option pricing Valuation of option contracts is done on the basis of the usual no-arbitrage argument, which shows that the option should be priced as the expected value under the risk-neutral probability measure of its discounted payoff in accounting currency. The risk-neutral measure is the measure under which the price of the traded asset discounted by the cost-of-carry is a martingale, the cost-of-carry being the difference between the interest paid to finance the asset and the income earned on the asset.

We assume that the payoff of the non-quanto option is given by a path functional f of the underlying price process S and that the payment is due at a stopping time τ, which is bounded by a maturity date T. Of course, $f(S)$ has to be $\mathcal{F}_\tau$-measurable since the amount due has to be known at the time of payoff. Typically, τ will be the first time when S intersects a given deterministic function of time.

We assume that we already know how to calculate the price

$$\pi_f(S_0, r^d, d, \sigma^S) = \mathbb{E}_{Q^d}\left[e^{-R_\tau^d} f(S)\right]$$

of the non-quanto option at $t = 0$. Here Q^d denotes the risk-neutral measure of the domestic investor.

We will now show how to express the price of the quanto and composite versions of this option in terms of the function π_f. The composite version of the option pays the amount $f(SX)$ in foreign currency. The quanto version of the option pays $f(S)\bar{x}$ in foreign currency. We will give evaluations for the composite and quanto options as they are seen by a domestic and by a foreign investor. The results do, of course, agree, but the foreign market method is more natural, as a change of numéraire leads to simpler calculations.

Mathematical model We assume a Black–Scholes model for both the asset S and the exchange rate process X, ie

$$\frac{dS_t}{S_t} = \mu_t^S dt + \sigma_t^S dW_t^S,$$

$$\frac{dX_t}{X_t} = \mu_t^X dt + \sigma_t^X dW_t^X.$$

Here μ^S and μ^X denote drift processes of finite variation, and σ^S and σ^X denote the volatilities of the equity price and the exchange rate process. Furthermore, (W^S, W^X) is a two-dimensional correlated Brownian motion with correlation ρ on some probability space.

The continuously compounded interest rate of the domestic and of the foreign currency will be denoted by r^d and r^f, and the continuously compounded dividend yield is given by d.

Domestic approach For a domestic investor, S and $Y = 1/X$ are tradable assets, whereas X cannot be traded directly.

By applying the no-arbitrage argument developed above, the drift μ^S of the asset price S is seen to be

$$\mu^S = r^d - d$$

and W^S is a Brownian motion with respect to Q^d. Similarly, our mathematical model and the no-arbitrage argument imply that the exchange rate process $Y = 1/X$ satisfies

$$\frac{dY_t}{Y_t} = (r_t^d - r_t^f)\,dt + \sigma_t^Y\,dW_t^Y.$$

Here we have $\sigma^X = \sigma^Y$ and the Q^d-Brownian motion W^Y is linked to W^S by a correlation factor of $-\rho$. Applying Itô's lemma (1.4), we find that

$$\begin{aligned}\frac{d(S_t/Y_t)}{S_t/Y_t} &= -\frac{dY_t}{Y_t} + \frac{d\langle Y\rangle_t}{Y_t^2} + \frac{dS_t}{S_t} + \frac{Y_t}{S_t}d\left\langle S, \frac{1}{Y}\right\rangle_t \\ &= (r_t^f - d_t - \rho_t\sigma_t^S\sigma_t^Y + \sigma_t^Y\sigma_t^Y)\,dt + \sigma_t^S\,dW_t^S - \sigma_t^Y\,dW_t^Y \\ &= (r_t^f - d_t)\,dt + \sigma_t^S(dW_t^S - \rho_t\sigma_t^Y\,dt) - \sigma_t^Y(dW_t^Y - \sigma_t^Y\,dt).\end{aligned}$$

We remark that $\sigma_t^S\,dW_t^S - \sigma_t^Y\,dW_t^Y = \tilde{\sigma}_t^S\,dW_t$, with $\tilde{\sigma}^S$ defined by

$$\tilde{\sigma}_t^S = \sqrt{(\sigma_t^S)^2 - 2\rho_t\sigma_t^S\sigma_t^Y + (\sigma_t^Y)^2},$$

defines a Q^d-Brownian motion W.

The composite option gives the payoff $f(S/Y)$ at time τ in foreign currency. To find its price Π^C (in domestic currency), a domestic investor has to convert that amount into domestic currency and discount it to $t = 0$. That means we have to compute

$$\Pi^C = \mathbb{E}_{Q^d}\left[e^{-R_\tau^d}f(S/Y)Y_\tau\right].$$

To evaluate this expectation, we introduce a new probability measure Q given by $dQ = M_T\,dQ^d$ on $\mathcal{F}_T$, with

$$M_T = e^{-R_T^d + R_T^f}\frac{Y_T}{Y_0}.$$

We find that, with respect to Q, $dW^1 = dW_t^Y - \sigma^Y\,dt$ and $dW^2 = dW_t^S - \rho_t\sigma_t^Y\,dt$ defines a pair (W^1, W^2) of Brownian motions with correlation $-\rho$. Thus S/Y with respect to Q is a geometric Brownian motion with drift $r^f - d$, and hence

$$\begin{aligned}\Pi^C &= \mathbb{E}_{Q^d}\left[e^{-R_\tau^d}f(S/Y)Y_\tau\right] \\ &= Y_0\mathbb{E}_{Q^d}\left[e^{-R_\tau^d}f(S/Y)e^{R_\tau^d - R_\tau^f}M_\tau\right] \\ &= Y_0\mathbb{E}_Q\left[e^{-R_\tau^f}f(S/Y)\right] = Y_0\pi_f(S_0/Y_0, r^f, d, \tilde{\sigma}^S).\end{aligned}$$

The quanto option has the price

$$\Pi^Q = \mathbb{E}_{Q^d}\left[e^{-R_\tau^d}\frac{f(S)}{\bar{x}}Y_\tau\right]$$

Mathematical Fundamentals

in domestic currency. Using the Girsanov transformation and Q above, we obtain

$$\Pi^Q = \mathbb{E}_{Q^d}\left[e^{-R_\tau^d}\frac{f(S)}{\bar{x}}Y_\tau\right]$$
$$= \frac{X_0}{\bar{x}}\mathbb{E}_Q\left[e^{-R_\tau^f}f(S)\right]$$
$$= \frac{X_0}{\bar{x}}\pi_f(S_0, r^f, r^f - r^d + d - \rho\sigma^S\sigma^X, \sigma^S)$$

because S with respect to Q has the drift $r^d - d + \rho\sigma^S\sigma^X$.

Foreign approach We consider the contract from the standpoint of a foreign investor. For a foreign investor, SX and X are tradable assets, whereas S cannot be traded directly. The dynamics of X with respect to Q^f, the risk-neutral measure of the foreign investor, is given by

$$\frac{dX_t}{X_t} = (r_t^f - r_t^d)\,dt + \sigma_t^X\,dW_t^X,$$

with W^X denoting a Q^f-Brownian motion. We have

$$\frac{d(S_t X_t)}{S_t X_t} = \frac{dX_t}{X_t} + \frac{dS_t}{S_t} + \frac{d\langle S, X\rangle_t}{X_t S_t}$$
$$= \sigma_t^S\,dW_t^S + \sigma_t^X\,dW_t^X + (r_t^f - d_t + \rho_t\sigma_t^S\sigma_t^X)\,dt.$$

The composite option has the payoff $f(SX)$ in foreign currency. This gives the price Π^C of the composite option in foreign currency as

$$\Pi^C = \mathbb{E}_{Q^f}\left[e^{-R_\tau^f}f(SX)\right]$$
$$= \pi_f(S_0 X_0, r^f, d, \tilde{\sigma}^S),$$

since SX has drift $r^f - d$ and volatility $\tilde{\sigma}^S$ under Q^f.

The quanto option has the payoff $f(S)\bar{x}$ at time T. A foreign investor therefore prices the quanto option as the risk-neutral expectation Π^Q of the discounted payoff in his accounting currency. This gives a price of the quanto option in foreign currency as

$$\Pi^Q = \bar{x}\mathbb{E}_{Q^f}\left[e^{-R_\tau^f}f(S)\right]$$
$$= \bar{x}\pi_f(S_0, r^f, r^f - r^d + d - \rho\sigma^S\sigma^X, \sigma^S),$$

since S has drift $r^d - d + \rho\sigma^S\sigma^X$ under Q^f.

Remark Obviously, the results derived using the domestic approach agree with the results of the foreign approach. After finishing the foreign approach, it is easy to see that the measure Q defined by the Girsanov transform in the domestic approach is just the foreign risk-neutral measure Q^f.

American quanto options are also covered by the approach described above. The foreign approach provides a direct justification for the intuitive way to deal with American quanto options by constructing the modelling tree for the underlying process SX or S under the measure Q^f.

Examples We consider $S = \text{DAX}$, EUR as domestic currency and USD as foreign currency. We fix

$$K = 3000\,\text{EUR}, \qquad \bar{K} = 2000\,\text{USD}, \qquad \bar{x} = 1.5\,\text{EUR/USD},$$

and define as before

$$\tilde{\sigma}^S = \sqrt{(\sigma^S)^2 - 2\rho\sigma^S\sigma^X + (\sigma^X)^2}.$$

European call

- The composite option pays $(S_T/Y_T - \bar{K})^+$ at maturity time $\tau = T$. We obtain the price in USD by using the Black–Scholes call option formula with spot S_0/Y_0, strike $\bar{K}$, interest rate r^f, dividend yield d and volatility $\tilde{\sigma}^S$.

- The quanto option pays $(S_T - K)^+/\bar{x}$ at maturity T. We obtain the option price in USD if we use the Black–Scholes call option formula with spot S_0, strike K, interest rate r^f, dividend yield $r^f - r^d + d - \rho\sigma^S\sigma^Y$ and volatility σ^S, and then convert the resulting price into USD by dividing by $\bar{x}$.

American digital call with payment at hit

- The composite option pays 1 USD at time τ if $\tau < T$, with

$$\tau = \inf\{t \mid S_t/Y_t > \bar{K}\};$$

otherwise it pays nothing. We obtain the price in USD by using the American digital call option formula with spot S_0/Y_0, strike $\bar{K}$, interest rate r^f, dividend yield d and volatility $\tilde{\sigma}^S$.

- The quanto option pays $1/\bar{x}$ EUR at time τ if $\tau < T$, with

$$\tau = \inf\{t \mid S_t > K\};$$

otherwise it pays nothing. We obtain the price in USD if we use the American digital call option formula with spot S_0, strike k, interest rate r^f, dividend yield $r^f - r^d + d - \rho\sigma^S\sigma^Y$ and volatility σ^S, and then convert the resulting price into USD by dividing by $\bar{x}$.

1.4 The Clark Formula

Option pricing is based on the assumption of an absence of arbitrage opportunities ("no free lunch", NFLVR). Given a self-financing portfolio strategy ξ^F which replicates a claim F, the fair price for that claim is equal to the amount needed to set up the portfolio. In Section 1.2 it was shown that this value can be obtained by calculating the expectation of the claim with respect to the risk-neutral measure. In this section we discuss the existence and construction of a strategy replicating a given claim.

1.4.1 Existence of a Hedging Strategy

We assume the one-dimensional Black–Scholes world introduced in Section 1.2 with constant deterministic drift μ and volatility σ, ie

$$\frac{dS_t}{S_t} = \mu\, dt + \sigma\, dW_t,$$

$$\frac{dB_t}{B_t} = r\, dt, \quad t \in [0, T],$$

with respect to some measure. As shown above, using the cash process B as numéraire, removing the drift μ and applying Girsanov's theorem from Section 1.1.12, Itô's lemma 1.1.10 shows that $Z = S/B$ satisfies the equation

$$\frac{dZ}{Z} = \sigma\, dW$$

with respect to a measure $\mathbb{P}$ on the same probability space. Since the solution to this system is given by a lognormal asset process and a deterministic bond process, we may view a claim F as a deterministic function of $W = (W_t, t \in [0, T])$, and the fair price is given as $\mathbb{E}[F/B_T]$, where T is the maturity of the claim. In particular, F is $\mathcal{F}$-measurable, where $\mathcal{F}$ denotes the σ-algebra generated by W. We have the following result (see [44] or [62]).

Representation property of Brownian motion Given a bounded $\mathcal{F}$-measurable function F, there exists an adapted process ξ such that

$$F = \mathbb{E}[F] + \int_0^T \xi_t\, dW_t$$

One can apply this result to our situation to obtain

$$F = \mathbb{E}_F + \int_0^T \frac{\xi_t}{\sigma Z_t}\, dZ_t.$$

This shows the existence of a hedging strategy.

1.4.2 Malliavin Calculus

Calculus on the Wiener space has been introduced by Malliavin [50]. In this section we state some basic results, omitting proofs and technicalities. Given a claim F, we will construct the hedging strategy ξ^F. From a mathematical point of view, we aim at presenting Clark's formula.

For a rigorous account of the subject, we refer the reader to the books by Ikeda and Watanabe [39] and Nualart [56], as well as the papers by Ocone [57] on integral representation, Ocone and Karatzas [58] on the Clark formula, and Colwell, Elliott and Kopp [13] on applications to hedging. The first two examples below are taken from a lecture given by Föllmer [23].

Assume that F is differentiable at $\omega \in \Omega = C([0, T], \mathbb{R})$ in the sense that there exists a linear continuous function $DF(\omega)$ on Ω such that

$$F(\omega + h) = F(\omega) + DF(\omega)(h) + o(|h|).$$

It is known that linear continuous functions on Ω may be identified with signed finite measures on $[0, T]$ in the sense that

$$DF(\omega)(h) = \int_0^T h_s DF(\omega)(ds).$$

If $h \in \Omega$ can be written as an integral

$$h_t = \int_0^t u_s \, ds, \quad t \in [0, T],$$

then, by applying Fubini's theorem, we have

$$DF(\omega)(h) = \int_0^T h_s DF(\omega)(ds)$$
$$= \int_0^T u_t DF(\omega)((t, T]) \, dt.$$

We introduce the set $\mathcal{H}$ of functions $h \in \Omega$ satisfying the additional condition that $\int_0^T u_t^2 \, dt < \infty$. The following result is due to Bismut [4].

A partial integration theorem Let U be a bounded continuous adapted process with paths in $\mathcal{H}$ and let F denote a bounded $\mathcal{F}$-measurable function. Then, with u denoting the derivative of U, we have

$$\mathbb{E}\left[\int_0^T u_t DF(dt)\right] = \mathbb{E}\left[F \int_0^T u_t \, dW_t\right]$$

An application of this formula is the Clark formula.

The Clark formula Let F be an $\mathcal{F}$-measurable function with derivative DF. Then

$$F = \mathbb{E}[F] + \int_0^T \mathbb{E}[DF((t, T]) \mid \mathcal{F}_t] \, dW_t.$$

To illustrate the result, we give some examples. We denote the coordinate mapping W_t from Ω to $\mathbb{R}$ by $W_t(\omega) = \omega(t)$. With respect to $\mathbb{P}$, W is a Brownian motion starting at 0.

- For $F = \int_0^T W_t \, dt$, we have $DF = dt$ and therefore

$$\int_0^T W_t \, dt = \int_0^T (T - t) \, dW_t.$$

- *Terminal payoff*: If F is given by a function f on W_T with derivative ∂f, we obtain

$$DF = \partial f(W_T) \delta_T(dt).$$

In particular, if f is given by

$$f(x) = (x - K)^+,$$

we have

$$DF((0, T]) = 1_{\{W_T \geq K\}}$$

for all $t \in [0, T)$, and therefore

$$\xi_t = \mathbb{P}[1_{\{W_T \geq K\}} \mid \mathcal{F}_t] = N\left(\frac{W_t - K}{\sqrt{T-t}}\right).$$

Hence, a "call option" on a Brownian motion can be hedged by holding a portfolio of ξ_t assets at time t. Obviously, the fair price is given by

$$C = \varphi(K) - KN(-K).$$

- *Black and Scholes*: Reformulating the problem in terms of the Black–Scholes model and starting with a function $f(S_T)$, one obtains immediately

$$DF = \partial f(S_T)\sigma S_T \delta_T(dt).$$

In particular, our European call payoff $f(S_T) = (S_T - K)^+$ yields

$$DF = 1_{\{\sigma W_T > \log(K/S_0) + (d - r + \frac{1}{2}\sigma^2)T\}}\sigma S_T \delta_T(dt),$$

and hence (viewed from maturity T)

$$(S_T - K)^+ = \mathbb{E}\big[(S_T - K)^+\big]$$
$$+ \int_0^T \mathbb{E}\left[1_{\{\sigma(W_T - W_t) > \log(K/S_t) + (d - r + \frac{1}{2}\sigma^2)(T-t)\}}\frac{Z_T}{Z_t}\bigg|\mathcal{F}_t\right]B_T\,dZ_t$$
$$= B_T C + \int_0^T \tilde{\xi}_t \frac{B_T}{B_t}\,dS_t + \int_0^T \tilde{\eta}_t \frac{B_T}{B_t}\,dB_t,$$

where C is the Black–Scholes price for the call paid at $t = 0$, and with $\tilde{\xi}_t$ and $\tilde{\eta}_t$ given by

$$\tilde{\xi}_t = N\left(\frac{\log(S_t/K) + (r - d + \frac{1}{2}\sigma^2)(T - t)}{\sigma\sqrt{T-t}}\right).$$

and $\tilde{\eta}_t = -\tilde{\xi}_t S_t/B_t$. Therefore the self-financing strategy is given, for all $t \in [0, T)$, by holding $\tilde{\xi}_t$ units of asset S_t and $\tilde{\eta}_t$ units of cash.

1.5 The Hybrid Model

So far we have kept interest rates deterministic. This assumption is completely sufficient for many equity derivative applications. However, if the product in question also has a substantial fixed income component, it can be necessary to model interest rates using a stochastic model. Such products are called *hybrid* products. Typical examples are convertible bonds, bond/equity outperformance options and equity-triggered swaps.

One reasonable choice for the interest rate model is the extended Vasicek model (also known as the Hull and White model). It models the short interest rate (the instantaneous rate) using a Gaussian process. Today's zero rates are matched perfectly by this approach. To keep the exposition simple, we restrict ourselves exclusively to this model.

1.5.1 *The Extended Vasicek Model*

We first present the interest rate modelling framework without any reference to the equity process, which we will fix later. The purpose is to present the basic properties of the model and derive some useful formulas.

The extended Vasicek model for the short rate process The model assumption is that the short rate process r_t satisfies the stochastic differential equation

$$dr_t = (\theta_t - \lambda_t r_t)\, dt + \sigma_t\, dW_t \tag{1.18}$$

under the measure $\mathbb{P}$, with λ_t, σ_t and θ_t deterministic, and such that discounted zero bonds are martingales.

This short rate process is a Gaussian process. This implies that the short rate can become negative with a non-zero probability. However, in most realistic circumstances the probability of negative rates will be quite small, and so this effect does not constitute a major drawback of the model.

The distribution of the short rate Since the process for the short rate is a Gaussian process, the short rate at a given time t is normally distributed. We shall now compute the mean and variance of this distribution.

For the computation, consider the deterministic process Λ_t defined by

$$\Lambda_t = \exp\left(\int_0^t \lambda_s\, ds\right). \tag{1.19}$$

Applying Itô's lemma to $\Lambda_t r_t$, we get the stochastic differential equation

$$d(\Lambda_t r_t) = \Lambda_t(\theta_t\, dt + \sigma_t\, dW_t).$$

Thus

$$\Lambda_{t_2} r_{t_2} - \Lambda_{t_1} r_{t_1} = \int_{t_1}^{t_2} \Lambda_s \theta_s\, ds + \int_{t_1}^{t_2} \Lambda_s \sigma_s\, dW_s$$

is normally distributed with mean and variance given by

$$\int_{t_1}^{t_2} \Lambda_s \theta_s\, ds \quad \text{and} \quad \int_{t_1}^{t_2} \Lambda_s^2 \sigma_s^2\, ds.$$

In particular, we have

$$\mathbb{E}[r_t] = \Lambda_t^{-1}\left(r_0 + \int_0^t \Lambda_s \theta_s\, ds\right)$$

and

$$\text{Var}[r_t] = \Lambda_t^{-2} \int_0^t \Lambda_s^2 \sigma_s^2\, ds.$$

The cash bond process The process for the *cash bond* B_t is given by $B_0 = 1$ and

$$dB_t = r_t B_t\, dt.$$

Explicitly, one can write

$$B_t = \exp\left(\int_0^t r_s\, ds\right).$$

In practice, the cash bond can be regarded as a money market account in which money accumulates according to the one-day interest rates.

Contingent claim pricing It follows from the usual replication argument that the price at time t of a contingent claim paying the amount X at some later time T (X is $\mathcal{F}_T$-measurable, ie it is determined by the information available at time T) is given by

$$B_t \mathbb{E}[\, B_T^{-1} X \mid \mathcal{F}_t \,].$$

In particular, today's price of the claim is given by

$$\mathbb{E}[\, B_T^{-1} X \,].$$

So, effectively, the deterministic discounting in the Black–Scholes framework has been replaced by stochastic discounting.

Zero bonds The simplest possible contingent claim is a zero bond paying the amount 1 at time T. The price of this zero bond at time $t \leqslant T$, which we will denote by $P(t, T)$, is given by

$$B_t \mathbb{E}[\, B_T^{-1} \mid \mathcal{F}_t \,] = \mathbb{E}\left[\, \exp\!\left(-\int_t^T r_s\, ds\right) \,\bigg|\, \mathcal{F}_t \,\right].$$

Our model assumption was that these processes are martingales.

Using the stochastic version of Fubini's theorem, which allows us to interchange the order of integration, we get

$$\log(B_{t_2}/B_{t_1}) = \int_{t_1}^{t_2} r_s\, ds$$

$$= r_{t_1} B_1(t_1, t_2) + \int_{t_1}^{t_2} \theta_s B_1(s, t_2)\, ds + \int_{t_1}^{t_2} \sigma_s B_1(s, t_2)\, dW_s,$$

where we set

$$B_1(t_1, t_2) = \Lambda_{t_1} \int_{t_1}^{t_2} \Lambda_s^{-1}\, ds. \tag{1.20}$$

Conditional on $\mathcal{F}_{t_1}$, the random variable $\log(B_{t_2}/B_{t_1})$ is normally distributed with mean

$$r_{t_1} B_1(t_1, t_2) + \int_{t_1}^{t_2} \theta_s B_1(s, t_2)\, ds$$

and variance

$$B_2(t_1, t_2) = \int_{t_1}^{t_2} \sigma_s^2 B_1^2(s, t_2)\, ds. \tag{1.21}$$

Using the fact that $\mathbb{E}[\exp(\mu + \sigma X)] = \exp(\mu + \tfrac{1}{2}\sigma^2)$ for a standard normally distributed variable X, we get

$$P(t, T) = \exp\!\left(-r_t B_1(t, T) - \int_t^T \theta_s B_1(s, T)\, ds + \tfrac{1}{2} B_2(t, T)\right). \tag{1.22}$$

Note that the expectation of $\log B_T$ can be expressed as

$$\mathbb{E}[\log B_T] = -\log P(0, T) - \tfrac{1}{2} B_2(0, T).$$

Forward rates The forward rate $f(t, T)$ at time T seen at time t is defined by

$$f(t, T) = -\frac{\partial}{\partial T} \log P(t, T), \tag{1.23}$$

or equivalently by

$$P(t, T) = \exp\left(-\int_t^T f(t, s)\, ds\right).$$

Substituting (1.22) into (1.23) gives, after a short computation,

$$f(t, T) = \Lambda_T^{-1}\left(r_t \Lambda_t + \int_t^T \theta_s \Lambda_s\, ds - \int_t^T \sigma_s^2 \Lambda_s B_1(s, T)\, ds\right).$$

From this, it follows that

$$\mathbb{E}[r_T \mid \mathcal{F}_t] = f(t, T) + \Lambda_T^{-1} \int_t^T \sigma_s^2 \Lambda_s B_1(s, T)\, ds.$$

Using forward rates, one can also write the zero bond as

$$P(t, T) = \frac{P(0, T)}{P(0, t)} \exp\{[f(0, t) - r_t] B_1(t, T) - C\}, \tag{1.24}$$

where

$$C = \tfrac{1}{2}[B_2(0, T) - B_2(0, t) - B_2(t, T)] - \Lambda_t^{-1} B_1(t, T) \int_0^t \sigma_s^2 \Lambda_s B_1(s, t)\, ds. \tag{1.25}$$

The dynamics of zero bond prices We can also compute a stochastic differential equation for the zero bond process. Applying Itô's lemma to (1.24) and (1.18) gives

$$\frac{dP(t, T)}{P(t, T)} = r_t\, dt - \sigma_t B_1(t, T)\, dW_t.$$

Thus

$$P(t, T) = P(0, T) \exp\left(-\int_0^t \sigma_s B_1(s, T)\, dW_s + \int_0^t [r_s - \tfrac{1}{2}\sigma_s^2 B_1^2(s, T)]\, ds\right),$$

and therefore

$$P(t, T) =$$

$$\frac{P(0, T)}{P(0, t)} \exp\left(-\int_0^t \sigma_s [B_1(s, T) - B_1(s, t)]\, dW_s - \tfrac{1}{2} \int_0^t \sigma_s^2 [B_1^2(s, T) - B_1^2(s, t)]\, ds\right).$$

Note that this expression does not involve the parameter θ_t.

The case of constant λ and σ A simple special case of the extended Vasicek model assumes that $\lambda_t = \lambda$ and $\sigma_t = \sigma$ (but not θ_t) are constant. This leads to simpler formulas, but still allows for the initial zero curve to be matched exactly.

Since in this case $\Lambda_t = e^{\lambda t}$, we get

$$B_1(t_1, t_2) = \frac{1 - e^{-\lambda(t_2 - t_1)}}{\lambda}$$

and

$$B_2(t_1, t_2) = \frac{\sigma^2}{\lambda^2}\left(t_2 - t_1 + \frac{2}{\lambda}(1 - e^{-\lambda(t_2 - t_1)}) - \frac{1}{2\lambda}(1 - e^{-2\lambda(t_2 - t_1)})\right).$$

Calibrating the Hull and White model If we allow θ_t to vary with time, we can match today's zero curve (ie the numbers $P(0, T)$ or the numbers $f(0, T)$)

exactly. Indeed, there is a one-to-one relation between the numbers $P(0, T)$ and the curve θ_T, and one can in most cases completely avoid working with θ_t.

To get an explicit expression for θ_T, take the derivative of $f(0, T)$ with respect to T. This gives

$$\theta_T = \lambda_T f(0, T) + \frac{\partial}{\partial T} f(0, T) + \Lambda_T^{-2} \int_0^T \sigma_s^2 \Lambda_s^2 \, ds.$$

Note, however, that this relation is more of theoretical interest. In practical applications of the Hull and White model, eg when using it with trees, it is advantageous to use a calibration that is adapted to the discrete setting. An example is described by Hull and White [38].

Later in this book we will discuss the calibration of the parameters λ_t and σ_t.

The correlation of short rate and cash bond The processes r_t (or $\Lambda_t r_t$) and $\log B_t$ form a two-dimensional Gaussian process. If we know their correlation, then their joint distribution is also known. Conditional on $\mathcal{F}_{t_1}$, we have

$$\operatorname{Cov}\left[\Lambda_{t_2} r_{t_2} - \Lambda_{t_1} r_{t_1}, \log\left(\frac{B_{t_2}}{B_{t_1}}\right)\right] = \operatorname{Cov}\left[\int_{t_1}^{t_2} \Lambda_s \sigma_s \, dW_s, \int_{t_1}^{t_2} \sigma_s B_1(s, t_2) \, dW_s\right]$$

$$= \int_{t_1}^{t_2} \Lambda_s \sigma_s^2 B_1(s, t_2) \, ds.$$

The forward measure It is often very convenient to switch from $\mathbb{P}$ to another measure $\mathbb{P}_T$, called the *forward measure*, associated with a certain fixed maturity T.

The forward measure is chosen such that

$$\frac{d\mathbb{P}_T}{d\mathbb{P}} = \frac{1}{P(0, T) B_T},$$

which implies

$$\mathbb{E}\left[\frac{d\mathbb{P}_T}{d\mathbb{P}} \,\bigg|\, \mathcal{F}_t\right] = \frac{P(t, T)}{P(0, T) B_t}.$$

Now we can price a derivative paying X at time T by

$$B_t \mathbb{E}[B_T^{-1} X \mid \mathcal{F}_t] = P(t, T) \mathbb{E}_{\mathbb{P}_T}[X \mid \mathcal{F}_t],$$

which follows from Bayes' rule. Here $\mathbb{E}_{\mathbb{P}_T}[\cdot]$ denotes expectation with regard to the measure $\mathbb{P}_T$. So, in effect, we have replaced stochastic discounting by a more tractable deterministic discounting.

A computation using Girsanov's theorem shows that, under the forward measure $\mathbb{P}_T$, the process $\tilde{W}_t$ given by $\tilde{W}_0 = 0$ and

$$d\tilde{W}_t = dW_t + \sigma_t B_1(t, T) \, dt$$

is a standard Brownian motion. Effectively, in our stochastic differential equation for the short rate r_t, the term θ_t is replaced by $\theta_t - \sigma_t^2 B_1(t, T)$.

The short rate under the forward measure The difference $\Lambda_{t_2} r_{t_2} - \Lambda_{t_1} r_{t_1}$ is normally distributed under $\mathbb{P}_T$ with mean

$$\int_{t_1}^{t_2} \Lambda_s \theta_s \, ds - \int_{t_1}^{t_2} \sigma_s^2 \Lambda_s B_1(s, T) \, ds$$

and the same variance as under the measure $\mathbb{P}$. In particular,
$$\mathbb{E}_{\mathbb{P}_T}[r_T] = f(0, t).$$

The cash bond under the forward measure We have

$$\log\left(\frac{B_{t_2}}{B_{t_1}}\right) =$$
$$r_{t_1}B_1(t_1, t_2) + \int_{t_1}^{t_2} \theta_s B_1(s, t_2)\,ds - \int_{t_1}^{t_2} \sigma_s^2 B_1(s, T)B_1(s, t_2)\,ds + \int_{t_1}^{t_2} \sigma_s B_1(s, t_2)\,d\tilde{W}_s.$$

So

$$\mathbb{E}_{\mathbb{P}_T}\left[\log\left(\frac{B_{t_2}}{B_{t_1}}\right)\right] = r_{t_1}B_1(t_1, t_2) + \int_{t_1}^{t_2} \theta_s B_1(s, t_2)\,ds - \int_{t_1}^{t_2} \sigma_s^2 B_1(s, T)B_1(s, t_2)\,ds$$

and

$$\mathrm{Var}_{\mathbb{P}_T}\left[\log\left(\frac{B_{t_2}}{B_{t_1}}\right)\right] = B_2(t_1, t_2).$$

1.5.2 Modelling Equity with Stochastic Interest Rates

So far we have just considered the interest rate side of our hybrid model. We now will also consider the equity process.

The model Modelling the short rate process as before, we find that the stock price process satisfies

$$\frac{dS_t}{S_t} = (r_t - d_t)\,dt + \sigma_t^S\,dW_t^S,$$

with deterministic dividend yield d_t and volatility σ_t^S, and with a Brownian motion W_t^S which is correlated with the Brownian motion of the short rate by

$$dW_t^S \cdot dW_t = \rho_t\,dt.$$

Here ρ_t is the deterministic instantaneous correlation.

The stock price distribution From Itô's lemma, it follows that

$$d\log S_t = [r_t - d_t - \tfrac{1}{2}(\sigma_t^S)^2]\,dt + \sigma_t^S\,dW_t^S.$$

We can write

$$\log\left(\frac{S_{t_2}}{S_{t_1}}\right) = \log\left(\frac{B_{t_2}}{B_{t_1}}\right) - \int_{t_1}^{t_2} [d_s + \tfrac{1}{2}(\sigma_s^S)^2]\,ds + \int_{t_1}^{t_2} \sigma_s^S\,dW_s^S.$$

So $\log(S_{t_2}/S_{t_1})$ is normally distributed (conditional on $\mathcal{F}_{t_1}$) with mean

$$\mathbb{E}\left[\log\left(\frac{B_{t_2}}{B_{t_1}}\right)\right] - \int_{t_1}^{t_2} [d_s + \tfrac{1}{2}(\sigma_s^S)^2]\,ds$$

and variance

$$B_2(t_1, t_2) + 2\int_{t_1}^{t_2} \rho_s \sigma_s \sigma_s^S B_1(s, t_2)\,ds + \int_{t_1}^{t_2} (\sigma_s^S)^2\,ds.$$

Correlation of stock with short rate and cash bond The covariance between the stock process and the short rate is computed as

$$\mathrm{Cov}\left[\Lambda_{t_2} r_{t_2} - \Lambda_{t_1} r_{t_1}, \log\left(\frac{S_{t_2}}{S_{t_1}}\right)\right] = \int_{t_1}^{t_2} \Lambda_s \sigma_s^2 B_1(s, t_2)\, ds + \int_{t_1}^{t_2} \rho_s \sigma_s \sigma_s^S \Lambda_s\, ds,$$

and the covariance between the stock process and the cash bond is

$$\mathrm{Cov}\left[\log\left(\frac{B_{t_2}}{B_{t_1}}\right), \log\left(\frac{S_{t_2}}{S_{t_1}}\right)\right] = B_2(t_1, t_2) + \int_{t_1}^{t_2} \rho_s \sigma_s \sigma_s^S B_1(s, t_2)\, ds.$$

The stock price distribution under the forward measure Using the multi-dimensional Girsanov theorem, one sees that

$$dW_t^S + \rho_t \sigma_t B_1(t, T)\, dt$$

is a Brownian motion under the forward measure $\mathbb{P}_T$. Thus we get

$$\mathbb{E}_{\mathbb{P}_T}\left[\log\left(\frac{S_{t_2}}{S_{t_1}}\right)\right] = \mathbb{E}_{\mathbb{P}_T}\left[\log\left(\frac{B_{t_2}}{B_{t_1}}\right)\right] - \int_{t_1}^{t_2} \rho_s \sigma_s \sigma_s^S B_1(s, T)\, ds - \int_{t_1}^{t_2} [d_s + \tfrac{1}{2}(\sigma_s^S)^2]\, ds.$$

and

$$\mathrm{Var}_{\mathbb{P}_T}\left[\log\left(\frac{S_{t_2}}{S_{t_1}}\right)\right] = \mathrm{Var}_{\mathbb{P}}\left[\log\left(\frac{S_{t_2}}{S_{t_1}}\right)\right],$$

of course.

1.5.3 Piecewise Constant Parameters

To be able to work with the extended Vasicek model in practice, one can assume, for example, that the model parameters λ_t and σ_t are piecewise constant. Specifically, we assume that we have an increasing sequence τ_i of times and that λ_t and σ_t have the constant values λ_i and σ_i, respectively, on the interval $[\tau_i, \tau_{i+1})$.

In this case we can compute the integrals Λ_t, $B_1(t_1, t_2)$ and $B_2(t_1, t_2)$ explicitly, as well as all other integrals that are needed to compute any first or second moment one needs for the hybrid model, eg to do a Monte Carlo simulation. The computations are straightforward but occasionally cumbersome, and so we consider in the following only some special cases.

To compute Λ, we simply note that

$$\Lambda_{\tau_n} = \exp\left(\sum_{i=0}^{n-1} \lambda_i (\tau_{i+1} - \tau_i)\right).$$

For the computation of B_1, we observe that in the case of constant parameters we have

$$B_1(\tau_k, \tau_{k+1}) = \frac{1 - e^{-\lambda_k(\tau_{k+1} - \tau_k)}}{\lambda_k}.$$

So, in the more general case, we have

$$B_1(\tau_i, \tau_j) = \Lambda_{\tau_i} \sum_{k=i}^{j-1} \frac{B_1(\tau_k, \tau_{k+1})}{\Lambda_{\tau_k}}$$

$$= \sum_{k=i}^{j-1} \exp\left(-\sum_{l=i}^{k-1} \lambda_l(\tau_{l+1} - \tau_l)\right) B_1(\tau_k, \tau_{k+1})$$

Next, we want to compute B_2. We again start with the case of an interval with constant parameters. In this case we have

$$B_2(\tau_k, \tau_{k+1}) = \frac{\sigma_k^2}{\lambda_k^2}[\tau_{k+1} - \tau_k - 2B_1(\tau_k, \tau_{k+1}) + \tilde{B}_1(\tau_k, \tau_{k+1})],$$

where we write

$$\tilde{B}_1(\tau_k, \tau_{k+1}) = \frac{1 - e^{-2\lambda_k(\tau_{k+1} - \tau_k)}}{2\lambda_k}.$$

Further,

$$B_2(\tau_i, \tau_j) = \sum_{k=i}^{j-1} \sigma_k^2 \int_{\tau_k}^{\tau_{k+1}} B_1^2(s, \tau_j) \, ds,$$

where

$$\sigma_k^2 \int_{\tau_k}^{\tau_{k+1}} B_1^2(s, \tau_j) \, ds = \sigma_k^2 \int_{\tau_k}^{\tau_{k+1}} \left(B_1(s, \tau_{k+1}) + \frac{\Lambda_s}{\Lambda_{\tau_{k+1}}} B_1(\tau_{k+1}, \tau_j)\right)^2 ds$$

$$= B_2(\tau_k, \tau_{k+1}) + 2\sigma_k^2 \Lambda_{\tau_{k+1}}^{-1} B_1(\tau_{k+1}, \tau_j) \int_{\tau_k}^{\tau_{k+1}} \Lambda_s B_1(s, \tau_{k+1}) \, ds$$

$$+ \sigma_k^2 \Lambda_{\tau_{k+1}}^{-2} B_1^2(\tau_{k+1}, \tau_j) \int_{\tau_k}^{\tau_{k+1}} \Lambda_s^2 \, ds.$$

To evaluate these expressions, we need

$$\int_{\tau_k}^{\tau_{k+1}} \Lambda_s B_1(s, \tau_{k+1}) \, ds = \tfrac{1}{2} \Lambda_{\tau_{k+1}} B_1^2(\tau_k, \tau_{k+1})$$

and

$$\int_{\tau_k}^{\tau_{k+1}} \Lambda_s^2 \, ds = \frac{\Lambda_{\tau_k}^2}{2\lambda_k}(e^{2\lambda_k(\tau_{k+1} - \tau_k)} - 1) = \Lambda_{\tau_{k+1}}^2 \tilde{B}_1(\tau_k, \tau_{k+1}).$$

Thus

$$\sigma_k^2 \int_{\tau_k}^{\tau_{k+1}} B_1^2(s, \tau_j) \, ds$$

$$= B_2(\tau_k, \tau_{k+1}) + \sigma_k^2 B_1^2(\tau_k, \tau_{k+1}) B_1(\tau_{k+1}, \tau_j) + \sigma_k^2 \tilde{B}_1(\tau_k, \tau_{k+1}) B_1^2(\tau_{k+1}, \tau_j).$$

All other integrals can be evaluated in the same spirit, and we leave it to the reader to do the necessary computations.

1.6 The Multi-Currency Hybrid Model

So far we have assumed that we only have to model one stochastic interest rate process. In general, claims may depend on two currencies, the most common being quanto and composite options as introduced in Section 1.3.4. Options written on several underlyings and interest rates may involve even more than two currencies. In order to price multiple-currency products, we fix a probability measure, namely the probability measure of the payoff currency, and express the payoff in terms of this measure. This involves a new class of processes, namely the *exchange rate processes*.

1.6.1 *The Stochastic Differential Equations*

The general multi-asset set-up involves the following system of stochastic differential equations (SDEs):

$$\left.\begin{aligned} dr_t^i &= (\theta_t^i - \lambda_t^i r_t^i)\, dt + \sigma_t^{r^i}\, d\tilde{W}_t^{r^i} \\ \frac{dS_t^i}{S_t^i} &= (r_t^i - d_t^i)\, dt + \sigma_t^{S^i}\, d\tilde{W}_t^{S^i} \\ \frac{dX_t^i}{X_t^i} &= (r_t^i - r_t^p)\, dt + \sigma_t^{X^i}\, dW_t^{X^i} \end{aligned}\right\} \quad \text{for } i = 1, 2, \ldots, n.$$

Here, the parameters θ^i, λ^i, σ^{r^i}, σ^{S^i}, σ^{X^i} and d^i are supposed to be deterministic functions of time t, the processes $\tilde{W}^{r^i}$ and $\tilde{W}^{S^i}$ are standard Brownian motions with respect to $\mathbb{P}^i$, the domestic risk-neutral probability measure of asset i, and W^{X^i} is a Brownian motion with respect to $\mathbb{P}$, the risk-neutral probability measure of the payoff currency. We will denote expectations with respect to the payoff currency by $\mathbb{E}[\cdot]$ and with respect to $\mathbb{P}^i$ by $\mathbb{E}_i[\cdot]$.

Now, if H is a payoff in currency i to be made at time T, the fair value Π^H for H at time $t=0$, expressed in the payoff currency p, is given by

$$\Pi^H = X_0^i \mathbb{E}_i\Big[e^{-\int_0^T r_u^i\, du} H \Big] = \mathbb{E}\Big[e^{-\int_0^T r_u^p\, du} X_T^i H \Big],$$

where the latter equation follows from the no-arbitrage condition. This implies that all option pricing can be done with respect to the common measure $\mathbb{P}$.

The above system of SDEs for (r^i, S^i, X^i) can be expressed with respect to $\mathbb{P}$ as

$$dr_t^i = (\theta_t^i - \lambda_t^i r_t^i - \rho_t^{r^i, X^i} \sigma_t^{r^i} \sigma_t^{X^i})\, dt + \sigma_t^{r^i}\, dW_t^{r^i},$$

$$\frac{dS_t^i}{S_t^i} = (r_t^i - d_t^i - \rho_t^{S^i, X^i} \sigma_t^{S^i} \sigma_t^{X^i})\, dt + \sigma_t^{S^i}\, dW_t^{S^i},$$

$$\frac{dX_t^i}{X_t^i} = (r_t^i - r_t^p)\, dt + \sigma_t^{X^i}\, dW_t^{X^i},$$

where $\rho_t^{A,B}$ denotes the instantaneous correlation of the Brownian motions driving the processes A and B, eg

$$\rho_t^{r^i, X^i}\, dt = \langle dW_t^{r^i}, dW_t^{X^i} \rangle,$$

and where W^{r^i}, W^{S^i} and W^{X^i}, with

$$W_t^{r^i} = \tilde{W}_t^{r^i} + \int_0^t \rho_u^{r^i,X^i} \sigma_u^{r^i} \sigma_u^{X^i} \, du,$$

$$W_t^{S^i} = \tilde{W}_t^{S^i} + \int_0^t \rho_u^{S^i,X^i} \sigma_u^{S^i} \sigma_u^{X^i} \, du,$$

are Brownian motions with respect to $\mathbb{P}$.

The cash bond and the zero bond with expiry S in currency i are denoted by B^i and $P^i(S)$ respectively (see Section 1.5.1). Moreover, we use the obvious notation Λ^i and B_1^i (see (1.19) and (1.20)). Finally, we introduce B_2 terms related to the covariances of

$$\Lambda_t^i r_t^i - \Lambda_s^i r_s^i, \quad \log\frac{B_t^i}{B_s^i}, \quad \log\frac{S_t^i}{S_s^i}, \quad \log\frac{X_t^i}{X_s^i},$$

namely

$$B_2^{r^i,r^j}(s,t) = \int_s^t \rho_u^{r^i,r^j} \sigma_u^{r^i} \sigma_u^{r^j} \Lambda_u^i \Lambda_u^j \, du,$$

$$B_2^{r^i,B^j}(s,t,v) = \int_s^t \rho_u^{r^i,r^j} \sigma_u^{r^i} \sigma_u^{r^j} \Lambda_u^i B_1^j(u,v) \, du,$$

$$B_2^{r^i,X^j}(s,t) = \int_s^t \rho_u^{r^i,X^j} \sigma_u^{r^i} \sigma_u^{X^j} \Lambda_u^i \, du,$$

$$B_2^{r^i,S^j}(s,t) = \int_s^t \rho_u^{r^i,S^j} \sigma_u^{r^i} \sigma_u^{S^j} \Lambda_u^i \, du,$$

$$B_2^{B^i,B^j}(s,t,v,w) = \int_s^t \rho_u^{r^i,r^j} \sigma_u^{r^i} \sigma_u^{r^j} B_1^i(u,v) B_1^j(u,w) \, du,$$

$$B_2^{B^i,X^j}(s,t,v) = \int_s^t \rho_u^{r^i,X^j} \sigma_u^{r^i} \sigma_u^{X^j} B_1^i(u,v) \, du,$$

$$B_2^{B^i,S^j}(s,t,v) = \int_s^t \rho_u^{r^i,S^j} \sigma_u^{r^i} \sigma_u^{S^j} B_1^i(u,v) \, du,$$

$$B_2^{X^i,X^j}(s,t) = \int_s^t \rho_u^{X^i,X^j} \sigma_u^{X^i} \sigma_u^{X^j} \, du,$$

$$B_2^{X^i,S^j}(s,t) = \int_s^t \rho_u^{X^i,S^j} \sigma_u^{X^i} \sigma_u^{S^j} \, du,$$

$$B_2^{S^i,S^j}(s,t) = \int_s^t \rho_u^{S^i,S^j} \sigma_u^{S^i} \sigma_u^{S^j} \, du.$$

We remark that B_2, introduced in (1.21), reads $B_2^{B^i,B^i}(s,t)$ in the multi-currency setting of this section.

Solving the SDEs In this section we solve the above system of stochastic differential equations. This allows us to derive the joint distribution of

$$\log\frac{S_t^i}{S_s^i}, \quad \log\frac{X_t^i}{X_s^i}, \quad \log\frac{B_t^i}{B_s^i}, \quad \Lambda_t^i r_t^i - \Lambda_s^i r_s^i.$$

We begin with

$$\log \frac{B^i_t}{B^i_s} = r^i_s B^i_1(s,t) + \int_s^t \theta^i_u B^i_1(u,t)\,du - B^{B^i,X^i}_2(s,t) + \int_s^t \sigma^{r^i}_u B^i_1(u,t)\,dW^{r^i}_u.$$

For the asset and exchange rate processes, we have

$$\log \frac{S^i_t}{S^i_s} = \int_s^t [r^i_u - d^i_u - \rho^{S^i,X^i}_u \sigma^{S^i}_u \sigma^{X^i}_u - \tfrac{1}{2}(\sigma^{S^i}_u)^2]\,du + \int_s^t \sigma^{S^i}_u\,dW^{S^i}_u$$

$$= \log \frac{B^i_t}{B^i_s} - \int_s^t d^i_u\,du - B^{S^i,X^i}_2(s,t) - \tfrac{1}{2} B^{S^i,S^i}_2(s,t) + \int_s^t \sigma^{S^i}_u\,dW^{S^i}_u,$$

as well as

$$\log \frac{X^i_t}{X^i_s} = \int_s^t [r^p_u - r^i_u - \tfrac{1}{2}(\sigma^{X^i}_u)^2]\,du + \int_s^t \sigma^{X^i}_u\,dW^{X^i}_u$$

$$= \log \frac{B^p_t}{B^p_s} - \log \frac{B^i_t}{B^i_s} - \tfrac{1}{2} B^{X^i,X^i}_2(s,t) + \int_s^t \sigma^{X^i}_u\,dW^{X^i}_u.$$

The interest rate process is given by

$$\Lambda^i_t r^i_t - \Lambda^i_s r^i_s = \int_s^t \Lambda^i_u \theta^i_u\,du - B^{r^i,X^i}_2(s,t) + \int_s^t \Lambda_u \sigma^{r^i}_u\,dW^{r^i}_u.$$

1.6.2 Expectations, Variances and Correlations

In this section, we give the joint distribution of

$$\Lambda^i_t r^i_t - \Lambda^i_s r^i_s, \quad \log \frac{B^i_t}{B^i_s}, \quad \log \frac{S^i_t}{S^i_s}, \quad \log \frac{X^i_t}{X^i_s}, \qquad i = 1, 2, \ldots, n,$$

with respect to $\mathbb{P}$, the risk-neutral probability measure of the payoff currency. Since all random variables are normal, it suffices to give expectations, variances and correlations. Suppressing the argument (s,t) (respectively (s,t,t) and (s,t,t,t)) in all B_2 terms, we have

$$\mathbb{E}[\Lambda^i_t r^i_t - \Lambda^i_s r^i_s] = \int_s^t \Lambda^i_u \theta^i_u\,du - B^{r^i,X^i}_2,$$

$$\mathbb{E}\left[\log \frac{B^i_t}{B^i_s}\right] = r^i_s B^i_1 + \int_s^t \theta^i_u B^i_1(u,t)\,du - B^{B^i,X^i}_2,$$

$$\mathbb{E}\left[\log \frac{S^i_t}{S^i_s}\right] = r^i_s B^i_1 - \int_s^t d^i_u\,du + \int_s^t \theta^i_u B^i_1(u,t)\,du - B^{B^i,X^i}_2 - B^{S^i,X^i}_2 - \tfrac{1}{2} B^{S^i,S^i}_2,$$

$$\mathbb{E}\left[\log \frac{X^i_t}{X^i_s}\right] = r^p_s B^p_1 + \int_s^t \theta^p_u B^p_1(u,t)\,du - r^i_s B^i_1$$

$$- \int_s^t \theta^i_u B^i_1(u,t)\,du + B^{B^i,X^i}_2 - \tfrac{1}{2} B^{X^i,X^i}_2.$$

For the covariances (and hence variances) involving the short rate, we have

$$\text{Cov}_\mathbb{P}[\Lambda_t^i r_t^i - \Lambda_s^i r_s^i, \Lambda_t^j r_t^j - \Lambda_s^j r_s^j] = B_2^{r^i, r^j},$$

$$\text{Cov}_\mathbb{P}\left[\log \frac{B_t^i}{B_s^i}, \Lambda_t^j r_t^j - \Lambda_s^j r_s^j\right] = B_2^{B^i, r^j},$$

$$\text{Cov}_\mathbb{P}\left[\log \frac{S_t^i}{S_s^i}, \Lambda_t^j r_t^j - \Lambda_s^j r_s^j\right] = B_2^{B^i, r^j} + B_2^{S^i, r^j},$$

$$\text{Cov}_\mathbb{P}\left[\log \frac{X_t^i}{X_s^i}, \Lambda_t^j r_t^j - \Lambda_s^j r_s^j\right] = B_2^{B^p, r^j} - B_2^{B^i, r^j} + B_2^{X^i, r^j},$$

and, for covariances of cash bond, asset and exchange rate, we have

$$\text{Cov}_\mathbb{P}\left[\log \frac{B_t^i}{B_s^i}, \log \frac{B_t^j}{B_s^j}\right] = B_2^{B^i, B^j},$$

$$\text{Cov}_\mathbb{P}\left[\log \frac{S_t^i}{S_s^i}, \log \frac{B_t^j}{B_s^j}\right] = B_2^{B^i, B^j} + B_2^{S^i, B^j},$$

$$\text{Cov}_\mathbb{P}\left[\log \frac{X_t^i}{X_s^i}, \log \frac{B_t^j}{B_s^j}\right] = B_2^{B^p, B^j} - B_2^{B^i, B^j} + B_2^{X^i, B^j},$$

$$\text{Cov}_\mathbb{P}\left[\log \frac{S_t^i}{S_s^i}, \log \frac{S_t^j}{S_s^j}\right] = B_2^{B^i, B^j} + B_2^{S^i, B^j} + B_2^{B^i, S^j} + B_2^{S^i, S^j},$$

$$\text{Cov}_\mathbb{P}\left[\log \frac{S_t^i}{S_s^i}, \log \frac{X_t^j}{X_s^j}\right] = B_2^{B^i, B^p} + B_2^{S^i, B^p} - B_2^{B^i, B^j} - B_2^{S^i, B^j} + B_2^{B^i, X^j} + B_2^{S^i, X^j},$$

The covariance of two exchange rates is given by

$$\text{Cov}_\mathbb{P}\left[\log \frac{X_t^i}{X_s^i}, \log \frac{X_t^j}{X_s^j}\right] = B_2^{B^p, B^p} - B_2^{B^i, B^p} + B_2^{X^i, B^p} - B_2^{B^p, B^j}$$
$$+ B_2^{B^i, B^j} - B_2^{X^i, B^j} + B_2^{B^p, X^j} - B_2^{B^i, X^j} + B_2^{X^i, X^j}.$$

1.6.3 Forward Measure

The forward measure $\mathbb{P}_T$ is defined by

$$P_0(T)\mathbb{E}_{\mathbb{P}_T}[F] = \mathbb{E}[(B_T^p)^{-1} F], \quad F \in \mathcal{F}_T.$$

It is often convenient to use the forward measure rather than the risk-neutral measure $\mathbb{P}$ in order to avoid modelling the cash bond B^p. In this section, we detail the distribution of

$$\Lambda_t^i r_t^i - \Lambda_s^i r_s^i, \quad \log \frac{S_t^i}{S_s^i}, \quad \log \frac{X_t^i}{X_s^i}, \quad i = 1, 2, \ldots, n,$$

with respect to $\mathbb{P}_T$, assuming $s < t < T$. The variances and correlations are not affected by the change of measure. The expectations are given by

$$\mathbb{E}_{\mathbb{P}_T}[\Lambda_t^i r_t^i - \Lambda_s^i r_s^i] = \mathbb{E}[\Lambda_t^i r_t^i - \Lambda_s^i r_s^i] - \int_s^t \rho_u^{r^i, r^p} \sigma_u^{r^i} \sigma_u^{r^p} \Lambda_u^i B_1^p(u, T) \, du$$

$$= \mathbb{E}[\Lambda_t^i r_t^i - \Lambda_s^i r_s^i] - B_2^{r^i, B^p}(s, t, T),$$

Mathematical Fundamentals

as well as

$$\mathbb{E}_{\mathbb{P}_T}\left[\log\frac{S^i_t}{S^i_s}\right] = \mathbb{E}\left[\log\frac{S^i_t}{S^i_s}\right] - \int_s^t \rho_u^{r^i,r^p}\sigma_u^{r^i}\sigma_u^{r^p}B_1^i(u,t)B_1^p(u,T)\,du$$

$$- \int_s^t \rho_u^{S^i,r^p}\sigma_u^{S^i}\sigma_u^{r^p}B_1^p(u,T)\,du$$

$$= \mathbb{E}\left[\log\frac{S^i_t}{S^i_s}\right] - B_2^{B^i,B^p}(s,t,t,T) - B_2^{B^p,S^i}(s,t,T)$$

and

$$\mathbb{E}_{\mathbb{P}_T}\left[\log\frac{X^i_t}{X^i_s}\right] = \mathbb{E}\left[\log\frac{X^i_t}{X^i_s}\right] - \int_s^t \sigma_u^{r^p}\sigma_u^{r^p}B_1^p(u,t)B_1^p(u,T)\,du$$

$$+ \int_s^t \rho_u^{r^i,r^p}\sigma_u^{r^i}\sigma_u^{r^p}B_1^i(u,t)B_1^p(u,T)\,du - \int_s^t \rho_u^{X^i,r^p}\sigma_u^{X^i}\sigma_u^{r^p}B_1^p(u,T)\,du$$

$$= \mathbb{E}\left[\log\frac{X^i_t}{X^i_s}\right] - B_2^{B^p,B^p}(s,t,t,T) + B_2^{B^i,B^p}(s,t,t,T) - B_2^{B^p,X^i}(s,t,T).$$

1.6.4 Quantos and Composites

Let $f(x) = (x - K)^+$ be the call payoff, with K denoting the strike. As detailed in Section 1.3.4, the fair price Π^{qf} for a call option on asset S^i, maturing at time T and quantoed into currency p, is given by

$$\Pi^{qf} = \mathbb{E}\left[(B_T^p)^{-1}X_0^i f(S_T^i)\right] = P_0(T)\mathbb{E}_{\mathbb{P}_T}\left[X_0^i f(S_T^i)\right].$$

More generally, a payoff F in currency i depends on several underlyings and bonds S^j and P^j ($j = 1, 2, \ldots, n$), to be made at T and quantoed into currency p, is worth

$$\Pi^{qF} = \mathbb{E}\left[(B_T^p)^{-1}X_0^i F(S^j, P^j, j = 1, 2, \ldots, n)\right]$$

$$= P_0(T)\mathbb{E}_{\mathbb{P}_T}\left[X_0^i F(S^j, P^j, j = 1, 2, \ldots, n)\right].$$

As for composite options, the corresponding call option is given by

$$\Pi^{cf} = \mathbb{E}\left[(B_T^p)^{-1}f(X_T^i S_T^i)\right] = P_0(T)\mathbb{E}_{\mathbb{P}_T}\left[f(X_T^i S_T^i)\right]$$

and, in the more general case,

$$\Pi^{cF} = \mathbb{E}\left[(B_T^p)^{-1}F(X^j S^j, X^j P^j, j = 1, 2, \ldots, n)\right]$$

$$= P_0(T)\mathbb{E}_{\mathbb{P}_T}F(X^j S^j, X^j P^j, j = 1, 2, \ldots, n)\right].$$

1.6.5 Piecewise Constant Parameters

To be able to work with the extended Vasicek model in practice, we assume that the model parameters λ_t and σ_t are piecewise constant.

Specifically, we assume an increasing sequence τ_i of times to be given, such that λ_t and σ_t assume the constant values λ_i and σ_i in the interval $[\tau_i, \tau_{i+1}]$. In Section 1.5.1, the terms $B_2^{B_i,B_i}$, $B_2^{r_i,B_i}$ and $B_2^{r_i,r_i}$ have been calculated for this restricted model.

In order to calculate the terms in Section 1.6.3, we need the terms

$$B_2^{r_i,r_j}(\tau_k, \tau_{k+1}, \tau_l, \tau_m), \quad B_2^{r_i,B_j}(\tau_k, \tau_{k+1}, \tau_l, \tau_m), \quad B_2^{B_i,B_j}(\tau_k, \tau_{k+1}, \tau_l, \tau_m).$$

We first introduce the notation

$$\tilde{B}_1^{ij}(\tau_k, \tau_{k+1}) := \frac{1 - e^{-(\lambda_k^i + \lambda_k^j)(\tau_{k+1} - \tau_k)}}{\lambda_k^i + \lambda_k^j}.$$

Now the first two terms can be rewritten as

$$B_2^{r_i,r_j}(\tau_k, \tau_{k+1}) = \Lambda_{\tau_{k+1}}^i \Lambda_{\tau_{k+1}}^j \tilde{B}_1^{ij}(\tau_k, \tau_{k+1})$$

and

$$B_2^{r_i,B_j}(\tau_k, \tau_{k+1}, \tau_l) = \int_{\tau_k}^{\tau_{k+1}} \Lambda_s^i \Lambda_s^j \left(\frac{B_1^j(s, \tau_{k+1})}{\Lambda_s^j} + \frac{B_1^j(\tau_{k+1}, \tau_l)}{\Lambda_{\tau_{k+1}}^j} \right) ds$$

$$= \int_{\tau_k}^{\tau_{k+1}} \Lambda_s^i B_1^j(s, \tau_{k+1}) \, ds + B_2^{r_i,r_j}(\tau_k, \tau_{k+1}) \frac{B_1^j(\tau_{k+1}, \tau_l)}{\Lambda_{\tau_{k+1}}^j},$$

where the first term in the sum reads

$$\int_{\tau_k}^{\tau_{k+1}} \Lambda_s^i B_1^j(s, \tau_{k+1}) \, ds = \frac{\Lambda_{\tau_{k+1}}^i}{\Lambda_{\tau_k}^j} [B_1^i(\tau_k, \tau_{k+1}) - \tilde{B}_1^{ij}(\tau_k, \tau_{k+1})].$$

The third term is

$$B_2^{B_i,B_j}(\tau_k, \tau_{k+1}, \tau_l, \tau_m) = \sum_{k=i}^{j-1} \rho_k^{r^i,r^j} \sigma_k^i \sigma_k^j \int_{\tau_k}^{\tau_{k+1}} B_1^i(s, \tau_l) B_1^j(s, \tau_m) \, ds.$$

Now, since

$$B_1^i(s, \tau_l) = B_1^i(s, \tau_k) + \frac{\Lambda_s^i}{\Lambda_{\tau_{k+1}}^i} B_1^i(\tau_{k+1}, \tau_l),$$

the above integral can be evaluated using

$$\int_{\tau_k}^{\tau_{k+1}} B_1^i(s, \tau_l) B_1^j(s, \tau_m) \, ds = B_2^{B_i,B_j}(\tau_k, \tau_{k+1}, \tau_{k+1}, \tau_{k+1})$$

$$+ \Lambda_{\tau_{k+1}^i}^{-1} B_1^i(\tau_{k+1}, \tau_m) \int_{\tau_k}^{\tau_{k+1}} \Lambda_s^j B_1^j(s, \tau_{k+1}) \, ds$$

$$+ \Lambda_{\tau_{k+1}^j}^{-1} B_1^j(\tau_{k+1}, \tau_l) \int_{\tau_k}^{\tau_{k+1}} \Lambda_s^i B_1^i(s, \tau_{k+1}) \, ds$$

$$+ \Lambda_{\tau_{k+1}^i}^{-1} \Lambda_{\tau_{k+1}^j}^{-1} B_1^i(\tau_{k+1}, \tau_l) B_1^j(\tau_{k+1}, \tau_m)$$

$$= B_2^{B_i,B_j}(\tau_k, \tau_{k+1}, \tau_{k+1}, \tau_{k+1})$$

$$+ \frac{\Lambda_{\tau_{k+1}}^i}{2\Lambda_{\tau_{k+1}}^i} B_1^i(\tau_{k+1}, \tau_l) \tilde{B}_1^i(\tau_k, \tau_{k+1})$$

$$+ \frac{\Lambda_{\tau_{k+1}}^i}{2\Lambda_{\tau_{k+1}}^j} B_1^j(\tau_{k+1}, \tau_m) \tilde{B}_1^j(\tau_k, \tau_{k+1})$$

$$+ B_1^i(\tau_{k+1}, \tau_l) B_1^j(\tau_{k+1}, \tau_m) \tilde{B}_1^{ij}(\tau_k, \tau_{k+1}).$$

Chapter 2
Closed-Form Solutions for Standard Products

In this chapter we derive closed-form and semi-closed-form solutions for derivative equity products that are commonly found in the OTC market and, to an increasing extent, in the exchange-traded market.

These standard products are mainly European exercise-style options, ie options that can only be exercised at a single maturity date. However, for American-style options, which can be exercised throughout their life, we provide several analytical approximations that are commonly used in their valuation. Typically, though, American options are valued using tree methods (Chapter 6) or, more generally, partial differential equation methods (Chapter 8). We complete the chapter by detailing valuation solutions for a wide variety of digital, barrier and Asian options.

2.1 Basic Products

The simplest derivative equity products are those whose payoff depends in a linear fashion on the price of the underlying stock or index. Examples of such products are forward contracts and equity swaps. In some sense, linearity means that there is no optionality present. Mathematically, there is not very much to say on the pricing of those products.

On the next level of sophistication, we have plain vanilla (standard) European options. For completeness we shall present their pricing methodology, even though it can be found in any book on equity derivatives. It will also serve as a model case and as an introduction to some of the methods that will be used throughout the book.

2.1.1 Forward Contracts

A *forward contract* is a product that *requires* one party (which is said to be "long the contract") to buy the underlying equity or index at a specified price F at a specified maturity T. When such a contract is initiated, the price F is usually chosen in such a way that the value of the contract is zero. This price is simply called the *forward*.

Denoting the stock price process of the underlying stock (or index) by $S = (S_t, t \geq 0)$, the payoff X (which can be negative) of the contract is

$$X = S_T - F.$$

The fair value at time $t = 0$ is then given by

$$e^{-R_T}\mathbb{E}[S_T - F] = e^{-R_T}(\mathbb{E}[S_T] - F).$$

This is equal to zero precisely when $F = \mathbb{E}[S_T]$. But

$$\mathbb{E}[S_T] = S_0 e^{R_T - D_T}.$$

So the forward is given by

$$F = S_0 e^{R_T - D_T}.$$

2.1.2 Plain Vanilla European Options

A *plain vanilla* option simply gives the holder the right either to buy (in the case of a call) or to sell (in the case of a put) the underlying asset at a specified price, the strike, which is usually denoted by K, at the maturity date, usually denoted by T.

So the payoff X of a plain vanilla call can be written as

$$X = (S_T - K)^+,$$

where we use the convenient "plus" notation $x^+ = \max\{x, 0\}$. In addition, we have again assumed that $S = (S_t, t \geqslant 0)$ is the stock price process of the underlying stock (or index). Similarly, the put payoff X is given by

$$X = (K - S_T)^+.$$

From the theory presented in Section 1.2, we get the fair price of an option by integrating the expected payoff and then discounting. So, to price a plain vanilla call, we have to evaluate the integral

$$e^{-R_T}\mathbb{E}[(S_T - K)^+].$$

Recall that e^{-R_T} is the discount factor from maturity T to the evaluation date $t = 0$. Now note that we can split up the integral as

$$\mathbb{E}[(S_T - K)^+] = \mathbb{E}[S_T \cdot 1_{\{S_T > K\}}] - K \cdot \mathbb{P}[S_T > K].$$

The probability can now be evaluated as

$$\mathbb{P}[S_T > K] = \mathbb{P}[\log S_T > \log K] = N\left(\frac{\log(S_0/K) + \mathcal{N}_T}{\Sigma_T}\right),$$

since $\log S_T$ is normally distributed with mean $\log S_0 + \mathcal{N}_T$ and variance Σ_T^2. (See Section 1.2 for the notation.) Recall that $N(\cdot)$ denotes the cumulative normal distribution function, which can be computed very efficiently [61].

It remains to evaluate the integral $\mathbb{E}[S_T 1_{\{S_T > K\}}]$. An elegant way to do this is to use a Radon–Nikodym measure to transform the integral into a probability under this new measure, and then use Girsanov's theorem to determine the stock price distribution under this measure. So, we choose a measure $\tilde{\mathbb{P}}$, with

$$\frac{d\tilde{\mathbb{P}}}{d\mathbb{P}} = \exp\left(\int_0^T \sigma_t \, dW_t - \tfrac{1}{2}\Sigma_T^2\right) = e^{D_T - R_T}\frac{S_T}{S_0}.$$

Then
$$\mathbb{E}[S_T \cdot 1_{\{S_T > K\}}] = e^{R_T - D_T} S_0 \cdot \tilde{\mathbb{E}}[1_{\{S_T > K\}}],$$

where $\tilde{\mathbb{E}}$ denotes expectation with respect to $\tilde{\mathbb{P}}$. Of course,
$$\tilde{\mathbb{E}}[1_{\{S_T > K\}}] = \tilde{\mathbb{P}}[S_T > K].$$

However, Girsanov's theorem tells us that $\log S_t$ is a Brownian motion under $\tilde{\mathbb{P}}$ with drift $\mathcal{N}_t + \Sigma_t^2 = R_t - D_t + \frac{1}{2}\Sigma_t^2$. Thus

$$\tilde{\mathbb{P}}[S_T > K] = \mathbb{P}[\log S_T > \log K] = N\left(\frac{\log(S_0/K) + \mathcal{N}_T + \Sigma_t^2}{\Sigma_T}\right).$$

Putting all the pieces together, we arrive at the classical Black–Scholes formula
$$C = e^{-D_T} S_0 N(d_1) - e^{-R_T} K N(d_2)$$

for the price C of a European call option, with

$$d_1 = \frac{\log(S_0/K) + \mathcal{N}t_T}{\Sigma_T} \quad \text{and} \quad d_2 = \frac{\log(S_0/K) + \mathcal{N}_T}{\Sigma_T}.$$

To find the formula for the put price P, we have alternative ways to proceed. We can, for example, review and redo the computations and check where signs change, and arrive at

$$P = e^{-R_T} K N(-d_2) - e^{-D_T} S_0 N(-d_1).$$

Note that we can unify these two price formulas by introducing a call–put flag χ, with $\chi = 1$ for a call and $\chi = -1$ for a put. We then get a price

$$\chi\left(e^{-D_T} S_0 N(\chi d_1) - e^{-R_T} K N(\chi d_2)\right).$$

Another way is to use call–put parity. For this, we note that if we are long a call and short a put (with the same strike K and maturity T) then we get the payoff

$$(S_T - K)^+ - (K - S_T)^+ = S_T - K,$$

which, after taking expectations of the discounted payoff, gives

$$C - P = e^{-D_T} S_0 - e^{-R_T} K. \tag{2.1}$$

This is called *call–put parity* and can be used to price the put as

$$P = C - e^{-D_T} S_0 + e^{-R_T} K.$$

2.1.3 Equity Swaps

Another product that is often traded in the markets is the equity swap. An *equity swap* is the exchange of an equity-related payment stream against an interest-related payment stream. In practice, the payment currency and the currency of the underlying equity or index are often distinct. In this case, the equity swap is either of quanto or of composite type. How these features affect the pricing is covered in Section 1.3.4, and so we will restrict ourselves to the single-currency case, as we shall do for the pricing of all other products.

Let us for simplicity assume that the swap starts today and that payment periods for both streams are identical. We denote these periods by

$$[0, t_1], [t_1, t_2], \ldots, [t_{n-1}, t_n].$$

Payments are made at the ends of these intervals.

Floating notional Equity swaps can either have a fixed or a floating notional. Let us treat the floating notional case first. The *initial notional* is denoted by $N = N_1$. Fix the number $m = N/S_0$, which can be regarded as the number of contracts. The *notional for the period* $[t_{i-1}, t_i]$ is set at time t_{i-1} to $N_i = mS_{t_{i-1}}$. Then the equity-related payment for the interval $[t_{i-1}, t_i]$ made at time t_i is

$$N_i\left(\frac{S_{t_i}}{S_{t_{i-1}}} - 1\right) = m(S_{t_i} - S_{t_{i-1}}).$$

So the fair value of the equity-linked payment streams is now given by

$$m \sum_{i=1}^{n} e^{-R_{t_i}}(\mathbb{E}[S_{t_i}] - \mathbb{E}[S_{t_{i-1}}]) = m \sum_{i=1}^{n} e^{-R_{t_i}}(F_{t_i} - F_{t_{i-1}}).$$

Interest payments for the period $[t_{i-1}, t_i]$ are also made at time t_i and on the notional N_i. Therefore, we get a present value of the interest rate payments of

$$\mathbb{E}\left[\sum_{i=1}^{n} e^{-R_{t_i}}(N_i e^{R_{t_i} - R_{t_{i-1}}} - N_i)\right] = m \sum_{i=1}^{n} (e^{-R_{t_{i-1}}} F_{t_i} - e^{-R_{t_i}} F_{t_i}).$$

Recall that we assume interest rates to be deterministic.

Fixed notional In the fixed notional case, as the name implies, the notional N does not change. Therefore the equity-linked payment for the period $[t_{i-1}, t_i]$ is

$$N\left(\frac{S_{t_i}}{S_{t_{i-1}}} - 1\right),$$

and so the value of these streams is now given by

$$N \sum_{i=1}^{n} \left(\mathbb{E}\left[\frac{S_{t_i}}{S_{t_{i-1}}}\right] - 1\right) = N \sum_{i=1}^{n} \left(\frac{F_{t_i}}{F_{t_{i-1}}} - 1\right).$$

The interest side is now deterministic since the notional is fixed, and its value is

$$N \sum_{i=1}^{n} e^{-R_{t_i}}(e^{R_{t_i} - R_{t_{i-1}}} - 1) = N(1 - e^{-R_{t_n}}).$$

2.2 American Options

American options are extensions to the plain vanilla European options discussed in Section 2.1.2. They have all the properties of their European counterpart, as well as the additional property that they can be exercised at any time between the start of the contract and its maturity.

As we have seen, closed-form solutions for European options are easy to obtain, but this is not the case for American plain vanilla options. Therefore the most common approaches to pricing American options usually involve some numerical approximation scheme. Section 6.2 provides such a scheme for a binomial–trinomial tree method, and Section 8.5 deals with the partial differential equation approach. A commonly stated principle in financial mathematics is that American options cannot be priced using Monte Carlo methods. However, this is not true and in Section 7.5 we illustrate such an approach.

These numerical approximation schemes can be used to achieve arbitrary accuracy, but they are computationally expensive. In this section we provide analytical approximations to the pricing of American options which dramatically increase the calculational speed, and which provide additional insights into the behaviour of these options.

Pricing of American options is a classic example of an optimal hitting time problem of the form

$$P^A(S_t, t) = \sup_{\tau \in \Phi_{t,T}} \mathbb{E}[e^{-R_{t\tau}} h(S_\tau)],$$

where $\Phi_{t,T}$ is the set of all possible stopping times $\tau \in [t, T]$. The payoff function $h(S_\tau)$ is simply $(S_\tau - K)^+$ for an American call and $(K - S_\tau)^+$ for an American put.

The optimal stopping time is then given by

$$\tau^* = \inf\{u \in [t, T] : P(S_u, u) = h(S_u)\},$$

and is the first time the option value falls to the intrinsic (immediate exercise) price. For example, it will not be optimal to exercise an American put option at time t if $P(S_t, t) > K - S_t$, whereas exercise will occur when $P(S_t, t) = K - S_t$. This naturally leads to the definition of a continuation region $\mathcal{C}$ (where the American option stays alive) and a stopping region $\mathcal{S}$ (where the American option is exercised):

$$\mathcal{C} = \{(x, t) \in (0, \infty) \times [0, T] : P(x, t) > h(x)\},$$
$$\mathcal{S} = \{(x, t) \in (0, \infty) \times [0, T] : P(x, t) = h(x)\}.$$

The line separating these two regions defines the optimal exercise boundary $S^*(t)$. The value of the American option in the stopping region is simply the intrinsic value $h(S_\tau)$. The value inside the continuation region is then given by the solution to the standard Black–Scholes partial differential equation.

Before proceeding with the analytical approximations, it is worth looking at some general properties of American options and their relationship to European options. The *additional* privilege of early exercise means that the value of an American option will always be more than its European counterpart,

$$P^A(S_t, t) \geq P^E(S_t, t),$$

and, because they can exercise early, their values will also always be at least the intrinsic value of the option,

$$P^A(S_t, t) \geq h(S_t).$$

These results are true for both call and put American options. The call–put parity relationship (2.1) for European options does not hold for American options.

Although generally no exact solution exists for American options, there is one important exception, the American call option on an asset paying no dividends. This can be seen by considering the associated European call, which (with no dividends) has a value that is bounded from below by

$$C^E(S_t, t) > S_t - Ke^{-R_{T,t}}.$$

As the value of an American option is at least the value of the corresponding European option, we have

$$C^A(S_t, t) > S_t - Ke^{-R_{T,t}} > S_t - K,$$

and as such it will never be optimal to exercise the American option early. The value of an American call option on an asset paying no dividends is therefore the same as the corresponding European call,

$$C^A(S_t, t) = C^E(S_t, t).$$

2.2.1 The Barone-Adesi–Whaley Approximation

Proceeding from the Black–Scholes equation (1.17), Barone-Adesi and Whaley [2] derived an approximate ordinary differential equation for the price difference between an American option and the corresponding European option. This can be compared with the control variate technique in Monte Carlo calculations (see Section 7.2.2). The quantity

$$d(S, t) = C^A(S, t) - C^E(S, t),$$

being the difference between American and European option prices, is guaranteed to be non-negative and satisfies the Black–Scholes equation in that region of (S, t) space in which $C^A(S, t)$ and $C^E(S, t)$ also satisfy the equation.

Clearly, $d(S_0, 0)$ is the premium paid for the right to exercise the option before its maturity.

We proceed by defining

$$k_1 = \frac{2r}{\sigma^2} \quad \text{and} \quad k_2 = \frac{2(r-d)}{\sigma^2}$$

and introducing a quantity $h(t)$, which is, in essence, a new time coordinate,

$$h(t) = 1 - e^{-r(T-t)},$$

and then re-expressing $d(S, t)$ in terms of this variable as

$$d(S, t) = h(t) \cdot f(S, h),$$

Closed-Form Solutions for Standard Products

with the result that we obtain

$$S^2\frac{\partial^2 f}{\partial S^2} + k_2 S\frac{\partial f}{\partial S} - \frac{k_1}{h}\left(f + h(1-h)\frac{\partial f}{\partial h}\right) = 0.$$

The approximation consists of dropping the last term in the above equation. This is reasonable as the term is generally small and tends to zero in the limits $T \to \infty$ and $T \to t$, and results in

$$S^2\frac{\partial^2 f}{\partial S^2} + k_2 S\frac{\partial f}{\partial S} - \frac{k_1}{h}f = 0, \qquad (2.2)$$

an ordinary differential equation. Henceforth, we take $t = 0$ and drop all time dependence.

Equation (2.2) has the solution

$$f(S) = AS^{\alpha_1} + BS^{\alpha_2},$$

in which α_1 and α_2 are roots of the auxiliary equation

$$\alpha^2 + \alpha(k_2 - 1) - \frac{k_1}{h} = 0.$$

The two roots are

$$\alpha_{2,1} = \tfrac{1}{2}\left(-(k_2 - 1) \pm \sqrt{(k_2 - 1)^2 + 4\frac{k_1}{h}}\right),$$

of which the negative root must be discarded since we require $C^A(S) \to 0$ as $S \to 0$ as a boundary condition. Thus we have

$$C^A(S) \simeq C^E(S) + BhS^{\alpha_2}. \qquad (2.3)$$

All that remains is to determine the constant B, for which we require a further boundary condition.

The above treatment assumes that C^A satisfies the Black–Scholes equation (1.17). However, in that region of (S, t) space in which it is optimal to exercise the American option, its value is intrinsic and equation (1.17) is not satisfied. Denoting by S^* the spot price at which the early-exercise boundary lies at $t = 0$, we have

$$C^A(S^*) = S^* - K,$$

which is the required boundary condition.

However, S^* is itself not known and a second equation is required. This is obtained by imposing the additional condition that $\partial C^A/\partial S$ be continuous across the exercise boundary. Thus

$$S^* - K = C^E(S^*) + BhS^{*\alpha_2},$$

$$1 = e^{-dT}N(d_1(S^*)) + Bh\alpha_2 S^{*\alpha_2 - 1},$$

and, by elimination,

$$S^* - K = C^E(S^*) + \frac{S^*}{\alpha_2}\left[1 - e^{-dT}N(d_1(S^*))\right], \tag{2.4}$$

which may be solved iteratively for S^*.

Given S^*, the approximate American call price is obtained by substitution into equation (2.3), and we have, for $d > 0$,

$$C^A(S) \simeq C^E(S) + \frac{S^*}{\alpha_2}\left[1 - e^{-dT}N(d_1(S^*))\right]\left(\frac{S}{S^*}\right)^{\alpha_2}, \quad S < S^*. \tag{2.5}$$

For $d = 0$, we have $C^A = C^E$, as the call will not be exercised early.

The analogous procedure for American puts yields

$$P^A(S) \simeq P^E(S) - \frac{S^*}{\alpha_1}\left[1 - e^{-dT}N(-d_1(S^*))\right]\left(\frac{S}{S^*}\right)^{\alpha_1}, \quad S > S^*,$$

with S^* obtained from

$$K - S^* = P^E(S^*) - \frac{S^*}{\alpha_1}\left[1 - e^{-dT}N(-d_1(S^*))\right].$$

2.2.2 Perpetual American Options

A *perpetual American* option is a standard American option but with an infinite time to expiry, ie it can be exercised at any point from its start date. The useful thing about these options is that a closed-form solution for its price exists.

At first sight these options seem an unnatural instrument to investigate. However, they prove to be useful in establishing an upper bound for the standard American option, which complements the lower bound of the corresponding European option:

$$P^A(S_\infty, \infty) > P^A(S_t, t) > P^E(S_t, t).$$

We can value these options exactly by making use of the Barone-Adesi–Whaley approximation of the previous section. However, because the time to maturity is infinite, the quantity $h(t)$ is 1 and their approximation becomes an exact result. The value of the perpetual American call is then given by taking the limit as $T \to \infty$ of equation (2.5):

$$C^A(S, \infty) = \frac{S^*}{\alpha}\left(\frac{S}{S^*}\right)^\alpha, \quad S < S^*,$$

where

$$\alpha = \tfrac{1}{2}\left[-(k_2 - 1) + \sqrt{(k_2 - 1)^2 + 4k_1}\right]$$

and k_1 and k_2 are as given in Section 2.2.1. The optimal exercise boundary S^* can also be obtained exactly from equation (2.4):

$$S^* = \frac{\alpha K}{\alpha - 1}.$$

Note that $S^* = \lim_{t \to \infty} S^*(t)$.

2.2.3 The Geske–Johnson Technique

An American option can be exercised at any time during the lifetime $[0, T]$ of the option, while a European option is exercised at maturity only. A compromise between these two types of plain vanilla options is the so-called *Bermudan option* (note the geographical location of Bermuda with respect to Europe and America), which allows exercise at a subset $\mathcal{T} \subset [0, T]$ with $T \in \mathcal{T}$. If $\mathcal{T} = \{T\}$ the Bermudan option is in fact European, and if $\mathcal{T} = [0, T]$ it is an American option. It is obvious that, if $\mathcal{T}^1 \subset \mathcal{T}^2$, the corresponding Bermudan option values B^i satisfy $B^1 \leqslant B^2$. Moreover, if the mesh size $|\mathcal{T}^n|$ of a given sequence $(\mathcal{T}^n, n = 1, 2, \ldots)$ tends to 0, we have

$$\lim_{n \to \infty} B^n = P^A,$$

with P^A denoting the corresponding American option value. The Geske–Johnson approach [28] is based on these facts as well as on a closed-form formula for Bermudan options with finite $\mathcal{T}$.

We now show how to obtain the closed-form formula for a Bermudan put. For

$$\mathcal{T} = \{t_1, t_2, \ldots, t_n = T\},$$

the Bermudan put price at t_{n-1} is given by

$$P^B(S_{t_{n-1}}, t_{n-1}, T) = \max\{P^E(S_{t_{n-1}}, t_{n-1}, T), (K - S_{t_{n-1}})^+\},$$

where $P^E(S_t, t, T)$ denotes the European put value at time t with maturity T, if the asset at t has value S_t. Let $S^*_{t_{n-1}}$ denote the (uniquely defined) asset value such that

$$P^E(S^*_{t_{n-1}}, t_{n-1}, T) = (K - S^*_{t_{n-1}})^+,$$

and represents the optimal exercise boundary at time t_{n-1}. Then $P^B(S_{t_{n-2}}, t_{n-2}, T)$ is given by the greater of $(K - S_{t_{n-2}})^+$ and

$$P^B(S_{t_{n-2}}, t_{n-2}, T)$$
$$= e^{-R_T} \mathbb{E}\big[(K - S_T)^+ 1_{\{S_{t_{n-1}} \geqslant S^*_{t_{n-1}}\}}\big] + e^{-R_{t_{n-1}}} \mathbb{E}\big[(K - S_{t_{n-1}})^+ 1_{\{S_{t_{n-1}} < S^*_{t_{n-1}}\}}\big].$$

The closed-form value for this expression can be obtained by using an argument similar to the derivation of the fade-in call price in Section 3.2. Repeating this argument, we obtain a sequence of critical asset prices $S^*_{t_i}$

($i = 1, 2, \ldots, n$), with $S^*_{t_n} = S^*_T = K$, such that the Bermudan put value at $t = 0$ is given as

$$P^B(S_0, 0, T) = \sum_{i=1}^{n} e^{-R_{t_i}} \mathbb{E}\left[(K - S_{t_i})^+ 1_{\{S_{t_i} < S^*_{t_i}, S_{t_j} \geqslant S^*_{t_j}, j=1,2,\ldots,i-1\}} \right].$$

We remark that the closed-form value involves *n*-variate cumulative normal distribution functions. These distribution functions are accurate and reasonably fast to compute up to at least $n = 5$. In order to obtain a good approximation for the American put, Geske and Johnson suggest using extrapolation schemes based on the sequence $\left(P^B_n(S_0, 0, T), n = 1, 2, \ldots, N\right)$, where P^B_n denotes the Bermudan put price with exercise dates i/n ($i = 1, 2, \ldots, n$). Bunch and Johnson [10] improve these results by defining P^B_n to be the most expensive Bermudan put with not more than n exercise times in $[0, T]$. Another fast approximation technique can be found in the article by Ho, Stapleton and Subrahmanyam [35]

2.3 Digital Options

2.3.1 European Digital Options

A *European digital call (put)* option pays a fixed amount if the price of the underlying security at maturity is above (below) a previously agreed strike price K; otherwise it pays nothing.

Digital options can be used as part of a structured product to create a *pay late* or *contingent premium* option. Consider, for example, an investor who wants to have a plain vanilla put option with strike K and maturity day T, but who does not want to pay any premium if the option finishes out-of-the-money. This investor can buy an ordinary put option and simultaneously sell short enough digital put options with this strike and maturity to finance the costs of the plain vanilla put.

However, as with barrier options (see Section 2.4), digital options exhibit an exploding delta if the price of the underlying security gets close to the critical value K near maturity.

Pricing formulas The payoff X at maturity T of a *digital call* that is sold and paid for at time $t = 0$ is

$$X = 1_{\{S_T > K\}}.$$

The fair value is easily computed to be

$$D^{\text{in}} = e^{-R_T} \mathbb{P}[S_T > K] = e^{-R_T} N(d_2),$$

with

$$d_2 = \frac{\log(S_0/K) + \mathcal{N}_T}{\Sigma_T}.$$

Similarly, the price for the corresponding put with payoff

$$X = 1_{\{S_T < K\}} = 1 - 1_{\{S_T > K\}}$$

is given by
$$D_{in} = e^{-R_T}[1 - N(d_2)] = e^{-R_T} N(-d_2).$$

Hedging and volatility smile The most common approach to hedging digital calls is the usage of a *call spread*, which consists of a long and a short call position where the long call has a lower strike $K_L < K$ and the short call has the strike $K_U > K$. The notional of the two calls has to be adjusted by the factor $1/(K_U - K_L)$.

It turns out that, in the case of a strike-dependent volatility at maturity T, the call spread price may be substantially different from the digital price calculated above. We now quantify this difference.

Limit of call spreads The payoff of a digital call option with strike K and maturity T can be written as a limit of call spread payoffs in the form

$$\frac{1}{K_U - K}[(S_T - K)^+ - (S_T - K_U)^+] \xrightarrow{K_U \to K} 1_{\{S_T \geq K\}},$$

where we assume $K_U > K$. We assume the implied volatility Σ_T to be differentiable at K with derivative $\Sigma'_T(K)$.

Let $C(K)$ denote the price of a call with maturity T and strike K using a volatility of $\Sigma_T(K)$ for the underlying. With

$$d_1(K) = d_2(K) + \Sigma_T(K) \quad \text{and} \quad d_2(K) = \frac{\log(S_0/K) + \mathcal{N}_T}{\Sigma_T(K)},$$

we obtain, for the digital call price,

$$\begin{aligned}
\frac{C(K) - C(K_U)}{K_U - K} &= \frac{1}{K_U - K}\{e^{-R_T}[-KN(d_2(K)) + K_U N(d_2(K))] \\
&\quad + S_0 e^{-D_T}[N(d_1(K)) - N(d_1(K_U))] \\
&\quad + e^{-R_T}[-K_U N(d_2(K)) + K_U N(d_2(K_U))]\} \\
&\xrightarrow{K_U \to K} e^{-R_T} N(d_2(K)) \\
&\quad - S_0 e^{-D_T} \varphi(d_1(K)) d'_1(K) \\
&\quad + K e^{-R_T} \varphi(d_2(K)) d'_2(K) \\
&= e^{-R_T} N(d_2(K)) - S_0 e^{-D_T} \varphi(d_1(K)) \Sigma_T(K)'.
\end{aligned}$$

Here we used the relations

$$d'_{1/2}(K) = -\frac{1}{K \Sigma_T(K)} - d_{1/2}(K) \frac{\Sigma_T(K)'}{\Sigma_T(K)} \pm \Sigma_T(K)'$$

as well as

$$\varphi(d_2(K)) = \frac{S_0}{K} e^{R_T - D_T} \varphi(d_1(K)).$$

In other words, the call spread limit differs from the digital price by an amount Δ proportional to the slope of the volatility:

$$\Delta = S_0 e^{-D_T} \varphi(d_1(K)) \Sigma_T(K)'.$$

Limit of put spreads A similar reasoning leads to a skewed digital put price of

$$e^{-R_T} N(-d_2(K)) + \Delta.$$

Hence, as expected, the sum of the digital call and put gives the discount factor.

Remarks

- If we replace the function $\Sigma_T(K)$ by $\tilde{\Sigma}_T(\alpha) = \Sigma_T(\alpha S_0)$, which amounts to expressing the volatility smile as a function of "moneyness", the correction term reads

$$\Delta = e^{-D_T} \varphi(d_1(K)) \tilde{\Sigma}_T(K/S_0)'.$$

- There is a no-arbitrage condition to be imposed on the slope of Σ_T. A sufficient condition to ensure positive digital prices is

$$|\tilde{\Sigma}_T'| < \frac{S_0 N(-|d_2(K)|)}{K \varphi(d_2(K))}.$$

2.3.2 Asset-or-Nothing Options

An *asset-or-nothing call (put)* option pays one unit of the underlying security if the price of the underlying at maturity T is above (below) a previously agreed strike price K; otherwise it pays nothing.

There is a relation between asset-or-nothing calls and puts: a long position in both an asset-or-nothing call and an asset-or-nothing put is equivalent to receiving one unit of the underlying at time T and has thus the same price as a forward.

Furthermore, there is an arbitrage relationship between European options, digital options and asset-or-nothing options: a long position in an asset-or-nothing call with strike K combined with a short position of K digital calls with strike K is equivalent to a long position in a plain vanilla call option with strike K. A similar relationship holds for the asset-or-nothing put.

Like barrier and digital options, asset-or-nothing options exhibit an exploding delta if the price of the underlying security gets close to the critical value K near maturity.

Pricing formulas The payoff X at maturity T for an asset-or-nothing call is

$$X = 1_{\{S_T > K\}} S_T.$$

From this, the fair value of X at time $t = 0$ is easily computed to be

$$e^{-R_T} \mathbb{E}[X] = S_0 e^{-D_T} N(d_1),$$

with

$$d_1 = \frac{\log(S_0/K) + \mathcal{N} t_T}{\Sigma_T}.$$

The fair price for the corresponding put is

$$e^{-R_T} \mathbb{E}[1_{\{S_T < K\}} S_T] = S_0 e^{-D_T} N(-d_1).$$

2.3.3 American Digital Options

An *American digital* option pays a fixed amount if the price of the underlying security attains a certain threshold K before or at maturity T; otherwise it pays nothing. In the case of the standard American digital option, the payment is made right at the time K is reached. In a product variant called *American digital option with deferred payoff*, the amount will only be paid at maturity day T provided that K has been attained before or at that day.

The American digital option is named *call (put)* if the barrier level K has to be reached from below (above).

Usage of the option: contingent premiums, rebates, and ladders

Contingent premium options: By analogy with the digital options, American digital options can be used to create pay-later or contingent premium options.

Barrier options with rebate: In conjunction with the formulas for barrier options (see Section 2.4), this section allows the pricing of barrier options with rebate, because a rebate is nothing but the payment of a fixed amount if a certain barrier level is reached (or not reached) during a specified time period.

Ladder options: A *ladder* option is a structured product that allows one to "lock in" the performance of the underlying at certain stipulated levels. We assume that N levels $S_0 < c_1 < c_2 < \cdots < c_N$ are given and that the ladder option is set at-the-money. If c_i is the highest level that was reached by the underlying price during the lifetime of the ladder option, then the ladder option pays at maturity the amount $c_i - S_0 + (S_T - c_i)^+$.

The payoff profile of the ladder option can be constructed using a combination of American digital calls and barrier options. The following structure is equivalent to the above ladder option:

- Long one plain vanilla call option with strike S_0,
  short one up-and-in call with strike S_0 and in-barrier c_1,
  long $c_1 - S_0$ American digital calls with deferred payoff and barrier c_1.

- Long one up-and-in call with strike c_1 and in-barrier c_1,
  short one up-and-in call with strike c_1 and in-barrier c_2,
  long $c_2 - c_1$ American digital calls with deferred payoff and barrier c_2.

  $\vdots$

- Long one up-and-in call with strike c_{N-1} and in-barrier c_{N-1},
  short one up-and-in call with strike c_{N-1} and in-barrier c_N,
  long $c_N - c_{N-1}$ American digital calls with deferred payoff and barrier c_N.

- Long one up-and-in call with strike c_N and in-barrier c_N.

A simple modification leads to a ladder option that pays out locked-in profits immediately: just replace the American digital calls with deferred payoffs by standard American digital calls.

Practical hedging and pricing issues Similar to European digital options, American digital options display an exploding delta if the price of the underlying gets close to the barrier. Despite their construction from digital and barrier options, ladder options have a continuous payoff profile and are thus easier to hedge.

Owing to the path-dependence nature of these instruments, closed-form solutions are only available if it is assumed all interest rates, dividend yields and volatilities are constants.

Pricing formulas for deferred payoff The hitting time of level b of a drifted Brownian motion $(W_t + tv, \, t \in [0, T])$ is denoted by

$$T_b^v = \inf\{t \mid W_t + vt = b\}.$$

Assuming constant drift and volatility parameters, the payoff X at maturity T of an American digital call can now be expressed as

$$X = 1_{\{T_b^v < T\}},$$

with drift and level parameters given by

$$v = \frac{\mathcal{N}_T}{\Sigma_T} \quad \text{and} \quad b = \frac{\log(K/S_0)}{\Sigma_T}.$$

In Section 1.1.15 the density of T_b^v was derived as

$$\mathbb{P}[\,T_b^v \in dt\,] = \frac{b}{\sqrt{2\pi t^3}}\, e^{-\frac{(b-vt)^2}{2t}} dt$$

for $b > 0$ and $t > 0$. Assuming $K > S_0$ ensures $b > 0$, we obtain

$$\mathbb{P}[\,T_b \leqslant T\,] = \frac{1}{\sqrt{2\pi}} \left(\tfrac{1}{2}\int_0^T (b+vt) t^{-3/2} e^{-\frac{(b-vt)^2}{2t}} dt + \tfrac{1}{2} e^{2bv} \int_0^T (b-vt) t^{-3/2} e^{-\frac{(b+vt)^2}{2t}} dt \right).$$

Changing variables from t to $(b-vt)/\sqrt{t}$ in the first integral and to $(b+vt)/\sqrt{t}$ in the second integral, and defining

$$d_1 = \frac{b+vT}{\sqrt{T}} \quad \text{and} \quad d_2 = \frac{b-vT}{\sqrt{T}},$$

gives

$$\mathrm{D}^{\text{in, mat}} = e^{-R_T} \mathbb{E}[\,X\,] = e^{-R_T} \left[N(-d_2) + \left(\frac{K}{S_0}\right)^{\mathcal{N}_T/\Sigma_T^2} N(-d_1) \right] \qquad (2.6)$$

for the fair value of the American digital call option with deferred payoff.

We now assume $S_0 > K$, ie $b < 0$. The American digital put option with deferred payoff has payoff X, with

$$X = 1_{\{T_b^v < T\}}.$$

The symmetry of the distribution of W implies

$$\mathbb{P}[\,T_b^v\,] = \mathbb{P}[\,T_{-b}^{-v}\,].$$

Hence it suffices to replace b by $-b$ and v by $-v$ in (2.6). This leads to

$$D_{\text{in, mat}} = e^{-R_T}\mathbb{E}[X] = e^{-R_T}\left[N(d_2) + \left(\frac{K}{S_0}\right)^{\mathcal{N}_T/\Sigma_T^2}N(d_1)\right]$$

for the American digital put price.

Pricing formulas for payoff at hit The discounted payoff of the standard American digital call option is given by

$$X = 1_{\{T_b^v \leq T\}}e^{-R_{T_b^v}}.$$

With $\tilde{v} := \sqrt{v^2 + 2r}$ and writing $R_t = rt$, we find

$$\mathbb{E}[X] = \int_0^T e^{-rt}\frac{|b|}{\sqrt{2\pi t^3}}e^{-\frac{(b-vt)^2}{2t}}dt$$

$$= \frac{1}{\sqrt{2\pi}}e^{-b(\tilde{v}-v)}\int_0^T t^{-3/2}be^{-\frac{(b-\tilde{v}t)^2}{2t}}dt.$$

Here we can use the previous calculation of the integral, which, with

$$d_1 = \frac{b + \tilde{v}T}{\sqrt{T}} \quad \text{and} \quad d_2 = \frac{b - \tilde{v}T}{\sqrt{T}},$$

leads to

$$D^{\text{in, hit}} = \mathbb{E}[X] = e^{-b(\tilde{v}-v)}[N(-d_2) + e^{2b\tilde{v}}N(-d_1)]$$

for the fair value of the standard American digital call option at time $t = 0$. This last equation corrects a mistake published on p. 51 of [49].

Similarly, the fair value of the standard American digital put option is

$$D_{\text{in, hit}} = \mathbb{E}[X] = e^{-b(\tilde{v}-v)}[N(d_2) + e^{2b\tilde{v}}N(d_1)].$$

2.3.4 Partial American Digital Options

A *partial American digital* option is a generalisation of an American digital option.

As with an American digital call (put), a partial American digital call (put) with strike K pays one unit if the underlying attains the level K from below (above). However, in the case of a partial American digital option, this level has to be attained during a previously agreed time interval $[s, t]$, which may begin after the evaluation date and end before maturity T, ie $0 \leq s \leq t \leq T$. The unit may be agreed to be paid at hit of the threshold K or at maturity T.

As with the American digital options, all interest rates, dividend yields and volatilities are assumed to be constants over $[s, t]$.

Note that, for $s = 0$ and $t = T$, we obtain an American digital option and, for $s = t = T$, we obtain a European digital option.

Call with payoff at maturity Let $\text{PD}^{\text{in, mat}}(S_0, 0)$ denote the price at evaluation date of a partial American digital call with spot S, maturity T, time interval

$[s, t]$ and payoff at maturity. We denote the corresponding American digital call with payoff at maturity by $D^{\text{in, mat}}(S_s, s)$. Then

$$e^{R_T}\text{PD}^{\text{in, mat}}(S_0, 0) = e^{R_s T}\mathbb{E}[D^{\text{in, mat}}(S_s, s)]$$

$$= \mathbb{E}\left[1_{\{S_s \leq K\}}N\left(\frac{1}{\Sigma_{st}}\left\{\log\frac{S_s}{K} + \mathcal{N}_{st}\right\}\right)\right]$$

$$+ \mathbb{E}\left[1_{\{S_s \leq K\}}\left(\frac{K}{S_s}\right)^{2\mathcal{N}_{st}/\Sigma_{st}^2}N\left(\frac{1}{\Sigma_{st}}\left\{\log\frac{S_s}{K} - \mathcal{N}_{st}\right\}\right)\right]$$

$$+ \mathbb{P}[S_s > K]$$

$$= \int_{-\infty}^{\gamma} N(\alpha x + \beta_1)\varphi(x)\,dx$$

$$+ f\int_{-\infty}^{\gamma} e^{\delta x}N(\alpha x + \beta_2)\varphi(x)\,dx + \mathbb{P}[S_s > K],$$

with

$$\alpha = \frac{\Sigma_s}{\Sigma_{st}},$$

$$\beta_1 = \frac{1}{\Sigma_{st}}\left\{\log\frac{S_0}{K} + \mathcal{N}_t\right\},$$

$$\beta_2 = \frac{1}{\Sigma_{st}}\left\{\log\frac{S_0}{K} - \mathcal{N}_t + 2\mathcal{N}_s\right\},$$

$$\gamma = \frac{1}{\Sigma_s}\left\{\log\frac{K}{S_0} - \mathcal{N}_s\right\},$$

$$\delta = -2\mathcal{N}_{st}\frac{\Sigma_s}{\Sigma_{st}^2},$$

$$f = \exp\left(\frac{2\mathcal{N}_{st}}{\Sigma_{st}^2}\left\{\log\frac{K}{S_0} - \mathcal{N}_s\right\}\right).$$

This yields

$$e^{R_T}\text{PD}^{\text{in, mat}}(S_0, 0) = N_2\left(\frac{1}{\Sigma_s}\left\{\log\frac{K}{S_0} - \mathcal{N}_s\right\}, \frac{1}{\Sigma_t}\left\{\log\frac{S_0}{K} + \mathcal{N}_t\right\}; -\frac{\Sigma_s}{\Sigma_t}\right)$$

$$+ \exp\left\{2\frac{\mathcal{N}_{st}}{\Sigma_{st}^2}\left(\frac{\Sigma_s^2}{\Sigma_{st}^2}\mathcal{N}_{st} - \mathcal{N}_s\right)\right\}\left(\frac{K}{S_0}\right)^{2\mathcal{N}_{st}/\Sigma_{st}^2}$$

$$\cdot N_2\left(\frac{1}{\Sigma_s}\left\{\log\frac{K}{S_0} - \mathcal{N}_s + 2\frac{\Sigma_s^2}{\Sigma_{st}^2}\mathcal{N}_{st}\right\},\right.$$

$$\left.\frac{1}{\Sigma_t}\left\{\log\frac{S_0}{K} - \mathcal{N}_t + 2\mathcal{N}_s - 2\mathcal{N}_{st}\frac{\Sigma_s^2}{\Sigma_{st}^2}\right\}; -\frac{\Sigma_s}{\Sigma_t}\right)$$

$$+ N\left(\frac{1}{\Sigma_s}\left\{\log\frac{S_0}{K} + \mathcal{N}_s\right\}\right). \quad (2.7)$$

Call with payoff at hit Using the approach presented in the previous section on American digital options, we introduce

$$\mathcal{N}'_{st} = \sqrt{\mathcal{N}^2_{st} + 2r_{st}\Sigma^2_{st}}.$$

The price $\mathrm{PD}^{\mathrm{in,\,hit}}(S_0, 0)$ for the call with a payoff at hit is given in terms of the American digital price $\mathrm{D}^{\mathrm{in,\,hit}}$:

$$e^{R_s}\mathrm{PD}^{\mathrm{in,\,hit}}(S_0, 0) = \mathbb{E}[\,\mathrm{D}^{\mathrm{in,\,hit}}(S_s, s)\,]$$

$$= \mathbb{E}\left[\mathbf{1}_{\{S_s \leqslant K\}}\left(\frac{K}{S_s}\right)^{\frac{\mathcal{N}_{st}-\mathcal{N}'_{st}}{\Sigma^2_{st}}} N\left(\frac{1}{\Sigma_{st}}\left\{\log\frac{S_s}{K} + \mathcal{N}'_{st}\right\}\right)\right]$$

$$+ \mathbb{E}\left[\mathbf{1}_{\{S_s \leqslant K\}}\left(\frac{K}{S_s}\right)^{\frac{\mathcal{N}_{st}+\mathcal{N}'_{st}}{\Sigma^2_{st}}} N\left(\frac{1}{\Sigma_{st}}\left\{\log\frac{S_s}{K} - \mathcal{N}'_{st}\right\}\right)\right]$$

$$+ \mathbb{P}[S_s > K]$$

$$= f_1\int_{-\infty}^{\gamma} e^{\delta_1 x} N(\alpha x + \beta_1)\varphi(x)\,dx$$

$$+ f_2\int_{-\infty}^{\gamma} e^{\delta_2 x} N(\alpha x + \beta_2)\varphi(x)\,dx + \mathbb{P}[S_s > K],$$

with

$$\alpha = \frac{\Sigma_s}{\Sigma_{st}},$$

$$\beta_1 = \frac{1}{\Sigma_{st}}\left\{\log\frac{S_0}{K} + \mathcal{N}'_{st} + \mathcal{N}_s\right\},$$

$$\beta_2 = \frac{1}{\Sigma_{st}}\left\{\log\frac{S_0}{K} - \mathcal{N}'_{st} + \mathcal{N}_s\right\},$$

$$\gamma = \frac{1}{\Sigma_s}\left\{\log\frac{K}{S_0} - \mathcal{N}_s\right\},$$

$$\delta_1 = -\Sigma_s\frac{\mathcal{N}_{st} - \mathcal{N}'_{st}}{\Sigma^2_{st}},$$

$$\delta_2 = -\Sigma_s\frac{\mathcal{N}_{st} + \mathcal{N}'_{st}}{\Sigma^2_{st}},$$

$$f_1 = \exp\left(\frac{\mathcal{N}_{st} - \mathcal{N}'_{st}}{\Sigma^2_{st}}\left\{\log\frac{K}{S_0} - \mathcal{N}_s\right\}\right),$$

$$f_2 = \exp\left(\frac{\mathcal{N}_{st} + \mathcal{N}'_{st}}{\Sigma^2_{st}}\left\{\log\frac{K}{S_0} - \mathcal{N}_s\right\}\right).$$

Explicitly, this gives

$$e^{R_s}\text{PD}^{\text{in, hit}}(S_0, 0) = \exp\left(\frac{\mathcal{N}_{st} - \mathcal{N}'_{st}}{\Sigma_{st}^2}\left\{\log\frac{K}{S_0} - \mathcal{N}_s + \frac{\Sigma_s^2}{2\Sigma_{st}^2}(\mathcal{N}_{st} - \mathcal{N}'_{st})\right\}\right)$$
$$\cdot N_2\left(\frac{1}{\Sigma_s}\left\{\log\frac{K}{S_0} - \mathcal{N}_s + \frac{\Sigma_s^2}{\Sigma_{st}^2}(\mathcal{N}_{st} - \mathcal{N}'_{st})\right\},\right.$$
$$\left.\frac{1}{\Sigma_t}\left\{\log\frac{S_0}{K} + \mathcal{N}'_{st} + \mathcal{N}_s - \frac{\Sigma_s^2}{\Sigma_{st}^2}(\mathcal{N}_{st} - \mathcal{N}'_{st})\right\}; -\frac{\Sigma_s}{\Sigma_t}\right)$$
$$+ \exp\left(\frac{\mathcal{N}_{st} + \mathcal{N}'_{st}}{\Sigma_{st}^2}\left\{\log\frac{K}{S_0} - \mathcal{N}_s + \frac{\Sigma_s^2}{2\Sigma_{st}^2}(\mathcal{N}_{st} + \mathcal{N}'_{st})\right\}\right)$$
$$\cdot N_2\left(\frac{1}{\Sigma_s}\left\{\log\frac{K}{S_0} - \mathcal{N}_s + \frac{\Sigma_s^2}{\Sigma_{st}^2}(\mathcal{N}_{st} + \mathcal{N}'_{st})\right\},\right.$$
$$\left.\frac{1}{\Sigma_t}\left\{\log\frac{S_0}{K} + \mathcal{N}_s - \mathcal{N}'_{st} - \frac{\Sigma_s^2}{\Sigma_{st}^2}(\mathcal{N}_{st} + \mathcal{N}'_{st})\right\}; -\frac{\Sigma_s}{\Sigma_t}\right)$$
$$+ N\left(\frac{1}{\Sigma_s}\left\{\log\frac{S_0}{K} + \mathcal{N}_s\right\}\right). \quad (2.8)$$

Put options Observe that the prices of American digital put options are obtained from the corresponding formulas for call options by replacing the argument in the cumulative normal functions N by their negative value, ie

$$\text{D}^{\text{in, hit}}(S_0, 0) = f_1 N(a) + f_2 N(b)$$

would imply

$$\text{D}_{\text{in, hit}}(S_0, 0) = f_1 N(-a) + f_2 N(-b).$$

This leads to the following recipe for partial American digital put options. If

$$\text{PD}^{\text{in}}(S_0, 0) = g_1 N_2(a_1, b_1; \rho) + g_2 N_2(a_2, b_2; \rho) + g_3 N(a_3),$$

ie (2.7) for deferred payoff and (2.8) for payoff at hit, then

$$\text{PD}_{\text{in}}(S_0, 0) = g_1 N_2(-a_1, -b_1; \rho) + g_2 N_2(-a_2, -b_2; \rho) + g_3 N(-a_3).$$

Usage of the option Partial American options can be used for the same purposes as European and American digital options. In particular, they can be used to price rebates for partial barrier options.

2.3.5 American Double Digital Options

An *American double digital* option is the extension of an American digital option with threshold K to the case of two thresholds U and L, with $L < S_0 < U$. It pays a premium if the price of the underlying attains either U from below or L from above before or at maturity T; otherwise it pays nothing. In the case of the standard American double digital option, the payment is made right at the time the first threshold is reached. The amount of premium can depend on the threshold hit first. In a product variant called *American double digital option with deferred payoff*, the amount will only be

Closed-Form Solutions for Standard Products

paid at the maturity day T provided a threshold has been attained before or at that day.

The product described so far is the case of two "knock-in" thresholds. One or both thresholds can also be "knock-out" thresholds. In the presence of two knock-out thresholds, payment occurs if none of the thresholds is hit. One knock-in and one knock-out threshold imply that payment depends on never hitting the knock-out threshold or hitting the knock-in threshold at least once. We remark that, in contrast to the double barrier option, a knocked-out American digital can be knocked in later while a knocked-in option cannot be knocked out. This is due to the usage of this derivative as a rebate for the double barrier option.

Derivation of formulas Again, as with the American digital options, the results developed here are true only under the assumption of constant interest rates, dividend yields and volatilities. This is required so that we can extract closed-form solutions.

A straightforward deterministic transformation of S shows that $L \leqslant S_t \leqslant U$ if and only if $0 \leqslant X_t \leqslant a$, with $(X_t = x + W_t + tv, t \in [0, T])$ denoting a Brownian motion with drift v starting in x,

$$v = -\frac{\mu}{\sigma}, \qquad x = \frac{\log(S_0/L)}{\sigma}, \qquad a = \frac{\log(U/L)}{\sigma}.$$

We define the hitting time of the level a by X as

$$T_a := \inf\{t : X_t = a\}.$$

Let $\mathbb{P}_x^v$ denote the distribution of X. Given a premium of u for hitting U and a premium l for hitting L, the fair price of the option (with payment at maturity) is given by

$$\mathrm{D}_{\mathrm{in,mat}}^{\mathrm{in, mat}} = e^{-R_T}\{l\mathbb{P}_x^v[T_0 \leqslant T, T_0 \leqslant T_a] + u\mathbb{P}_x^v[T_a \leqslant T, T_a < T_0]\},$$

with T denoting time to maturity in years. Using Girsanov's theorem, we obtain

$$e^{R_T}\mathrm{D}_{\mathrm{in,mat}}^{\mathrm{in,mat}} = l\mathbb{E}_x\left[1_{\{T_0 \leqslant T, T_0 \leqslant T_a\}}\exp(-vx - \tfrac{1}{2}v^2 T_0)\right]$$

$$+ u\mathbb{E}_x\left[1_{\{T_a \leqslant T, T_a < T_0\}}\exp[v(a-x) - \tfrac{1}{2}v^2 T_a]\right]$$

$$= le^{-vx}\int_0^T \exp(-\tfrac{1}{2}v^2 s)\mathbb{P}_x[T_0 \in ds, T_0 \leqslant T_a]$$

$$+ ue^{v(a-x)}\int_0^T \exp(-\tfrac{1}{2}v^2 s)\mathbb{P}_x[T_a \in ds, T_a < T_0].$$

Now, the densities of $\mathbb{P}_x[T_0 \in ds, T_0 \leqslant T_a]$ and $\mathbb{P}_x[T_a \in ds, T_a < T_0]$ with respect to Lebesgue measure ds can be found in the literature [44] and are given by

$$\mathbb{P}_x[T_0 \in ds, T_0 \leqslant T_a] = \frac{1}{\sqrt{2\pi s^3}}\sum_{n=-\infty}^{\infty} a_n \exp\left(-\frac{a_n^2}{2s}\right)ds, \qquad (2.9)$$

$$\mathbb{P}_x[T_a \in ds, T_a < T_0] = \frac{1}{\sqrt{2\pi s^3}}\sum_{n=-\infty}^{\infty} b_n \exp\left(\frac{-b_n^2}{2s}\right)ds, \qquad (2.10)$$

with

$$a_n = 2na + x \quad \text{and} \quad b_n = (2n+1)a - x.$$

Recalling a formula derived in Section 2.3.3 on American digital options, we have, for all $b \in \mathbb{R}$,

$$\int_0^T \frac{|b|}{\sqrt{2\pi s^3}} \exp\left(-\frac{(b-vs)^2}{2s}\right) ds = N\left(\frac{-b+vT}{\sqrt{T}}\right) + e^{2bv} N\left(\frac{-b-vT}{\sqrt{T}}\right).$$

This formula implies

$$\int_0^T \frac{b}{\sqrt{2\pi s^3}} \exp\left(\frac{-b^2}{2s} - \frac{v^2 s}{2}\right) ds = e^{bv} N\left(\frac{-b-vT}{\sqrt{T}}\right) + e^{-bv} N\left(\frac{-b+vT}{\sqrt{T}}\right)$$

for all $b > 0$, as well as

$$\int_0^T \frac{b}{\sqrt{2\pi s^3}} \exp\left(-\frac{b^2}{2s} - \frac{v^2 s}{2}\right) ds = -e^{-bv} N\left(\frac{b-vT}{\sqrt{T}}\right) - e^{bv} N\left(\frac{b+vT}{\sqrt{T}}\right)$$

for all $b < 0$. Since $0 < x < a$, we have $a_n > 0$ and $b_n > 0$ if and only if $n \geqslant 0$. Hence, these results imply the following formula for the price $D_{\text{in, mat}}^{\text{in, mat}}$ of an American double digital option with deferred payoff:

$$\begin{aligned}
e^{R_T} D_{\text{in, mat}}^{\text{in, mat}} &= l e^{-vx} \int_0^T \exp\left(-\frac{v^2 s}{2}\right) \frac{1}{\sqrt{2\pi s^3}} \sum_{n=-\infty}^{\infty} a_n \exp\left(-\frac{a_n^2}{2s}\right) ds \\
&\quad + u e^{v(a-x)} \int_0^T \exp\left(-\frac{v^2 s}{2}\right) \frac{1}{\sqrt{2\pi s^3}} \sum_{n=-\infty}^{\infty} b_n \exp\left(-\frac{b_n^2}{2s}\right) ds \\
&= e^{-vx} \sum_{n=-\infty}^{\infty} \left[l \int_0^T \frac{a_n}{\sqrt{2\pi s^3}} \exp\left(-\frac{a_n^2}{2s} - \frac{v^2 s}{2}\right) ds \right. \\
&\qquad \left. + u e^{va} \int_0^T \frac{b_n}{\sqrt{2\pi s^3}} \exp\left(-\frac{b_n^2}{2s} - \frac{v^2 s}{2}\right) ds \right] \\
&= e^{-vx} \left[l \left(e^{vx} N\left(\frac{-x-vT}{\sqrt{T}}\right) + e^{-vx} N\left(\frac{-x+vT}{\sqrt{T}}\right) \right) \right. \\
&\qquad + u e^{va} \left(e^{v(a-x)} N\left(\frac{x-a-vT}{\sqrt{T}}\right) + e^{-v(a-x)} N\left(\frac{x-a+vT}{\sqrt{T}}\right) \right) \\
&\qquad + \sum_{n=1}^{\infty} \left\{ l \left(e^{a_n v} N\left(\frac{-a_n - vT}{\sqrt{T}}\right) + e^{-a_n v} N\left(\frac{-a_n + vT}{\sqrt{T}}\right) \right) \right. \\
&\qquad + u e^{va} \left(e^{b_n v} N\left(\frac{-b_n - vT}{\sqrt{T}}\right) + e^{-b_n v} N\left(\frac{-b_n + vT}{\sqrt{T}}\right) \right) \\
&\qquad - l \left(e^{-a_{-n} v} N\left(\frac{a_{-n} - vT}{\sqrt{T}}\right) + e^{a_{-n} v} N\left(\frac{a_{-n} + vT}{\sqrt{T}}\right) \right) \\
&\qquad \left. \left. - u e^{va} \left(e^{-b_{-n} v} N\left(\frac{b_{-n} - vT}{\sqrt{T}}\right) + e^{b_{-n} v} N\left(\frac{b_{-n} + vT}{\sqrt{T}}\right) \right) \right\} \right].
\end{aligned}$$

We now derive the price of an American double digital option with payoff of the upper premium at hit (ie if the upper threshold is hit before any hitting of the lower threshold) and the lower premium (ie if the lower threshold is hit before any hitting of the upper threshold) at maturity. Since the discounted payoff of this option is given by

$$le^{-R_T}1_{\{T_0 \leq T, T_0 \leq T_a\}} + ue^{-R_T^a}1_{\{T_a \leq T, T_a \leq T_0\}},$$

the fair price is given by

$$D_{\text{in, mat}}^{\text{in, hit}} = le^{-R_T - vx} \int_0^T \exp(-\tfrac{1}{2}v^2 s) \mathbb{P}_x[T_0 \in ds, T_0 \leq T_a]$$

$$+ ue^{v(a-x)} \int_0^T \exp(-\tfrac{1}{2}v^2 s - rs) \mathbb{P}_x[T_a \in ds, T_a < T_0].$$

We summarise the result in the next section.

Summary of American double digital option pricing formulas We first introduce a third premium n, which will be used as a "no-hit premium". The possible instrument combinations are as follows:

- *Two "knock-in" thresholds*
  The price of an American double digital option with two "knock-in" thresholds U and L and deferred payoff at maturity of premium u if threshold U is the first threshold to be hit and of l at hit if L is the first threshold to be hit (and no payoff if neither threshold is hit by maturity day) will be denoted by $D_{\text{in, hit}}^{\text{in, mat}}$.

- *One "knock-in" and one "knock-out" threshold*
  If, instead, L is a "knock-out" threshold, the product pays n at maturity if neither threshold is hit. The premium u is paid at first hit of the upper threshold U. If the upper threshold is never hit and the lower threshold L is hit at least once, nothing is paid. The fair price for this product will be denoted by $D_{\text{out}}^{\text{in, hit}}$.

- *Two "knock-out" thresholds*
  An American double digital option with two "knock-out" thresholds U and L pays the "no-hit premium" n at maturity if neither threshold is hit. The fair price of this product will be denoted by $D_{\text{out}}^{\text{out}}$.

- *One "knock-in" threshold*
  The fair prices of (single-threshold) American digital options are given in Section 2.3.3. We refer to the fair prices of an American digital call and put respectively with thresholds U and L respectively and payoffs u and l respectively by $D^{\text{in, hit}}$ and $D_{\text{in, hit}}$. The corresponding products with deferred payoff are denoted by $D^{\text{in, mat}}$ and $D_{\text{in, mat}}$.

We first list the formulas for two "knock-in" thresholds. The upper and lower premium can be paid (if hit first) either at hit or at maturity of the option.

There is a single formula covering the values of $D_{in,\,hit}^{in,\,hit}$, $D_{in,\,mat}^{in,\,hit}$, $D_{in,\,hit}^{in,\,mat}$, $D_{in,\,mat}^{in,\,mat}$.

$$D_{in}^{in} = e^{-vx}\Biggl[\tilde{l}\biggl(e^{v_l x}N\biggl(\frac{-x - v_l T}{\sqrt{T}}\biggr) + e^{-v_l x}N\biggl(\frac{-x + v_l T}{\sqrt{T}}\biggr)\biggr)$$

$$+ \tilde{u}e^{va}\biggl(e^{v_u(a-x)}N\biggl(\frac{x - a - v_u T}{\sqrt{T}}\biggr) + e^{-v_u(a-x)}N\biggl(\frac{x - a + v_u T}{\sqrt{T}}\biggr)\biggr)$$

$$+ \sum_{n=1}^{\infty}\biggl\{\tilde{l}\biggl(e^{a_n v_l}N\biggl(\frac{-a_n - v_l T}{\sqrt{T}}\biggr) + e^{-a_n v_l}N\biggl(\frac{-a_n + v_l T}{\sqrt{T}}\biggr)\biggr)$$

$$+ \tilde{u}e^{va}\biggl(e^{b_n v_u}N\biggl(\frac{-b_n - v_u T}{\sqrt{T}}\biggr) + e^{-b_n v_u}N\biggl(\frac{-b_n + v_u T}{\sqrt{T}}\biggr)\biggr)$$

$$- \tilde{l}\biggl(e^{-a_{-n} v_l}N\biggl(\frac{a_{-n} - v_l T}{\sqrt{T}}\biggr) + e^{a_{-n} v_l}N\biggl(\frac{a_{-n} + v_l T}{\sqrt{T}}\biggr)\biggr)$$

$$- \tilde{u}e^{va}\biggl(e^{-b_{-n} v_u}N\biggl(\frac{b_{-n} - v_u T}{\sqrt{T}}\biggr) + e^{b_{-n} v_u}N\biggl(\frac{b_{-n} + v_u T}{\sqrt{T}}\biggr)\biggr)\biggr\}\Biggr].$$

The four cases are distinguished through the definition of v_u, v_l, $\tilde{u}$, $\tilde{l}$:

$D_{in,\,hit}^{in,\,hit}$: $\quad v_l = v_u = \sqrt{v + 2r}, \quad \tilde{l} = l, \quad \tilde{u} = u,$

$D_{in,\,mat}^{in,\,hit}$: $\quad v_l = v, \quad v_u = \sqrt{v + 2r}, \quad \tilde{l} = e^{-R_T}l, \quad \tilde{u} = u,$

$D_{in,\,hit}^{in,\,mat}$: $\quad v_l = \sqrt{v + 2r}, \quad v_u = v, \quad \tilde{l} = l, \quad \tilde{u} = e^{-R_T}u,$

$D_{in,\,mat}^{in,\,mat}$: $\quad v_u = v_l = v, \quad \tilde{l} = e^{-R_T}l, \quad \tilde{u} = e^{-R_T}u.$

We now summarise the formulas for American double digital options with one or two "knock-out" thresholds. In order to price an option with two "knock-out" thresholds, we assume the upper, lower and no-hit premium (u, l and n, respectively) to be equal. Then, $D_{out}^{out} = ne^{-R_T} - D_{in,\,mat}^{in,\,mat}$. The prices of the remaining options are given in terms of a sum of an American digital option and an American double digital option with two "knock-out" thresholds:

$$D_{in,\,hit}^{out} = D_{in,\,hit} + D_{out}^{out},$$

$$D_{in,\,mat}^{out} = D_{in,\,mat} + D_{out}^{out},$$

$$D_{out}^{in,\,hit} = D^{in,\,hit} + D_{out}^{out},$$

$$D_{out}^{in,\,mat} = D^{in,\,mat} + D_{out}^{out}.$$

2.4 Barrier Options

An important and very common class of equity derivatives are the barrier options. For this type of product, the payoff depends on whether the underlying asset does or does not reach a certain barrier level during a specified time interval. The event of reaching this barrier level is called the *trigger* or *barrier* event.

The barrier can either be of *in* type or of *out* type:

- For an *in* barrier, the option payoff will only be paid if the barrier is reached, ie if the trigger event happens (this event is called a *knock-in*).

- For an *out* barrier, the option payoff will not be paid if the barrier is reached (a *knock-out* has happened).

One also distinguishes between *up* and *down* barriers, depending on whether the asset price has to be above (or at) the barrier level (up) or below it (down) for the trigger event to occur. One usually talks of *down-and-out* options, etc. Note that for our valuation purposes we do not have to care whether the trigger event involves a strict or a weak inequality, ie whether the case of the asset price being equal to the barrier is included or excluded in the trigger event.

As an illustrative example, consider a fund manager who is very bullish and wants to buy a one-year at-the-money call on a certain stock index, which he could buy as a simple plain vanilla product. However, he believes that the index will at no point during the next year drop below 90% of today's level. So he can instead buy an at-the-money down-and-out call, with a barrier level of 90%, which will be substantially cheaper than the corresponding plain vanilla call.

Usually barrier options pay a fixed amount, called a *rebate*, in the case of a knock-out or failed knock-in. The knock-out rebate comes usually in two variations. It can be paid either immediately at knock-out or at option maturity. This feature can make the hedging of the barrier option more manageable and therefore cheaper. Since the rebate feature amounts to a digital option of some sort, which we have treated in Section 2.3, we assume in the following that there is no rebate feature.

Note also that most barrier options are of European exercise style. American-style barriers can in most cases not be valued in closed form.

There are many variations of the basic barrier product. We shall consider the following:

- A *double barrier* option has two barrier levels, an up level and a down level. They can both be of out type, both of in type, or mixed.

- A *partial barrier* option has a barrier that ceases to be active before maturity, after evaluation date, or both.

- A *time-dependent barrier* option is a barrier option with a barrier level that is not fixed in time. We will treat the case of an exponentially growing barrier level.

- An *outside barrier* option has two underlying assets: one is used to compute the payoff at maturity; the other is used to determine the barrier event.

- A *compound barrier* option is an in barrier option on a barrier option, ie the option holder receives a standard barrier option if the initial barrier knocks in.

Owing to the path dependence of the barrier feature, we have to assume constant parameters (interest rate, dividend yield and volatility) during the time that the barrier is active in order to be able to derive closed-form solutions.

For most barrier options, the payoff, in the case that it is paid, is of plain vanilla type [65]. However, we shall present a barrier pricing formula allowing for a general payoff, by reducing the barrier case to the non-barrier case.

2.4.1 Single Barriers

The most basic barrier option type is the single barrier type. Usually, the barrier level is constant. However, it is not very difficult to allow for an exponential growth of the barrier level, which we will do in this section. So, we assume the barrier level is given by $H_t = H_0 e^{at}$. In addition, we will give pricing formulas for arbitrary European, path-independent payoffs in terms of the corresponding non-barrier formulas. In particular, we give standard barrier pricing formulas in terms of standard plain vanilla pricing formulas.

We shall concentrate on the case of *in* barrier options, since any *out* barrier option can be represented as a short in barrier and a long non-barrier option. We call this relation *in–out parity*.

In the following, it is convenient to use the auxiliary process

$$R_t = \frac{\log(S_t/S_0) - at}{\sigma}, \quad t \geq 0.$$

By Itô's lemma, this process follows the stochastic differential equation

$$dR_t = v\, dt + dW_t,$$

with $R_0 = 0$ and

$$v = \frac{r - d - a - \frac{1}{2}\sigma^2}{\sigma}.$$

General up-and-in pricing formula We want to price an up-and-in barrier option that pays an amount $f(S_T)$ at maturity T. We denote its fair value by $V_f^{\text{ui}}(S_0)$, given the spot price S_0. In the following, we will use $V_f(S_0)$ to denote the value of a European option with payoff f but with no barrier feature.

Now

$$V_f^{\text{ui}}(S_0) = e^{-R_T} \cdot \mathbb{E}[f(S_T) \cdot 1_A],$$

where A is the barrier event, ie

$$A = \left\{ \max_{0 \leq t \leq T}(S_t - H_t) \geq 0 \right\}.$$

Let us define

$$h = \frac{\log(H_0/S_0)}{\sigma}.$$

The condition $\max(S_t - H_t) > 0$ is equivalent to the condition $\max R_t > h$. Thus

$$A = \left\{ \max_{0 \leq t \leq T} R_t \geq h \right\}.$$

In the following, we can assume that the barrier is initially above the spot price (otherwise the option is already knocked in). So, $h > 0$.

Defining the two functions

$$f^*(S_T) = \begin{cases} f(S_T) & \text{if } S_T \leq H_T \\ 0 & \text{otherwise} \end{cases} \quad \text{and} \quad f^{**}(S_T) = \begin{cases} 0 & \text{if } S_T \leq H_T \\ f(S_T) & \text{otherwise} \end{cases}$$

means that we can decompose the function f into $f = f^* + f^{**}$. Since $f^{**} \cdot 1_A = f^{**}$, we have

$$V_f^{\text{ui}}(S_0) = e^{-R_T} \cdot \mathbb{E}[f^*(S_T) \cdot 1_A] + e^{-R_T}\mathbb{E}[f^{**}(S_T)]$$

$$= e^{-R_T} \cdot \mathbb{E}[f^*(S_T) \cdot 1_A] + V_{f^{**}}(S_0).$$

It remains to evaluate $e^{-R_T} \cdot \mathbb{E}[f^*(S_T) \cdot 1_A]$.

The process R_t is a Brownian motion with drift under the original risk-neutral measure $\mathbb{P}$. Now consider the measure $\hat{\mathbb{P}}$ with Radon–Nikodym derivative

$$\frac{d\hat{\mathbb{P}}}{d\mathbb{P}} = \exp(\nu R_T - \tfrac{1}{2}\nu^2 T).$$

We shall denote the expectation operator with respect to $\hat{\mathbb{P}}$ by $\hat{\mathbb{E}}$. By Girsanov's theorem, the process R_t is a driftless Brownian motion under the new measure $\hat{\mathbb{P}}$.

We can now use the reflection principle. We reflect each path of R on the interval $[0, \tau]$ along the barrier h, where τ is the first hitting time of the barrier h (or T, if there is no hit). This gives us a new driftless Brownian motion starting at $2h$, which we can therefore write as $\bar{R}_t + 2h$, with a standard Brownian motion $\bar{R}_t$ (under $\hat{\mathbb{P}}$) starting at 0.

The function that we have to integrate does not depend on the value of the paths before τ. Therefore we can replace R_t by $\bar{R}_t + 2h$ in the computation below.

Now

$$\mathbb{E}[f^*(S_T) \cdot 1_A] = \hat{\mathbb{E}}\left[f^*(S_T) \cdot 1_A \cdot \frac{d\mathbb{P}}{d\hat{\mathbb{P}}}\right]$$

$$= \hat{\mathbb{E}}\left[f^*(S_0 e^{\sigma R_T + aT}) \cdot 1_A \cdot e^{-\nu R_T + \nu^2 T/2}\right]$$

$$= \hat{\mathbb{E}}\left[f^*(S_0 e^{\sigma \bar{R}_T + aT + 2\sigma h}) \cdot e^{-\nu \bar{R}_T + \nu^2 T/2 + 2\nu h}\right]$$

$$= e^{2\nu h} \cdot \hat{\mathbb{E}}\left[f^*(S_T e^{2\sigma h}) \cdot \frac{d\mathbb{P}}{d\hat{\mathbb{P}}}\right]$$

$$= e^{2\nu h} \cdot \mathbb{E}[f^*(S_T e^{2\sigma h})]$$

$$= e^{R_T} e^{2\nu h} \cdot V_{f^*}(S_0 e^{2\sigma h})$$

$$= e^{R_T} \cdot \left(\frac{H_0}{S_0}\right)^{2\nu/\sigma} \cdot V_{f^*}\left(\frac{H_0^2}{S_0}\right).$$

To summarise, the value of the up-and-in barrier option is given by

$$V_f^{\text{ui}}(S_0) = \left(\frac{H_0}{S_0}\right)^{2v/\sigma} \cdot V_{f^*}\left(\frac{H_0^2}{S_0}\right) + V_{f^{**}}(S_0).$$

Note that the case of constant barrier can be recovered from this formula by simply setting $a = 0$, and thus

$$\frac{2v}{\sigma} = \frac{2\mathcal{N}_T}{\Sigma_T^2}.$$

General down-and-in pricing formula For a down-and-in barrier option, the same procedure gives the formula

$$V_f^{\text{di}}(S_0) = \left(\frac{H_0}{S_0}\right)^{2v/\sigma} \cdot V_{f^{**}}\left(\frac{H_0^2}{S_0}\right) + V_{f^*}(S_0),$$

where we again assume that the option is not already knocked in, ie we assume that $S_0 > H_0$.

Plain vanilla barrier formulas It is now easy to apply our formulas to the case of a plain vanilla call or put option.

- For an *up-and-in call*, the payoff is given by $f(S) = (S - K)^+$, where K denotes the strike.
  If $K \geq H_T$, then $f = f^*$, and so we simply get

  $$C^{\text{in}}(S_0, K) = C(S_0, K),$$

  where $C(S_0, K)$ and $C^{\text{in}}(S_0, K)$ denotes a plain vanilla and up-and-in call, respectively, with spot S_0 and strike K. This equality is evident since this option is only in-the-money if the spot is above the barrier, and thus knocked in.
  The more interesting case is $K < H_T$, where we get

  $$C^{\text{in}}(S_0, K)$$
  $$= \left(\frac{H_0}{S_0}\right)^{2v/\sigma} \cdot \left[C\left(\frac{H_0^2}{S_0}, K\right) - C\left(\frac{H_0^2}{S_0}, H_T\right) - (H_T - K) \cdot D^{\text{in}}\left(\frac{H_0^2}{S_0}, H_T\right)\right]$$
  $$+ C(S_0, H_T) + (H_T - K) \cdot D^{\text{in}}(S_0, H_T),$$

  where $D^{\text{in}}(S_0, K)$ denotes a European digital call with spot S_0 and strike K.

- Now consider a *down-and-in call*. If $K \geq H_T$, then

  $$C_{\text{in}}(S_0, K) = \left(\frac{H_0}{S_0}\right)^{2v/\sigma} \cdot C\left(\frac{H_0^2}{S_0}, K\right).$$

  In the case $K < H_T$, we get

  $$C_{\text{in}}(S_0, K) = \left(\frac{H_0}{S_0}\right)^{2v/\sigma} \cdot \left[C\left(\frac{H_0^2}{S_0}, H_T\right) + (H_T - K) \cdot D^{\text{in}}\left(\frac{H_0^2}{S_0}, H_T\right)\right]$$
  $$+ C(S_0, K) - C(S_0, H_T) - (H_T - K) \cdot D^{\text{in}}(S_0, H_T).$$

- In the case of an *up-and-in put*, we get, for $K < H_T$,

$$P^{\text{in}}(S_0, K) = \left(\frac{H_0}{S_0}\right)^{2v/\sigma} \cdot P\left(\frac{H_0^2}{S_0}, K\right),$$

and, for $K \geqslant H_T$,

$$P^{\text{in}}(S_0, K) = \left(\frac{H_0}{S_0}\right)^{2v/\sigma} \cdot \left[P\left(\frac{H_0^2}{S_0}, H_T\right) + (K - H_T) \cdot D_{\text{in}}\left(\frac{H_0^2}{S_0}, H_T\right)\right]$$
$$+ P(S_0, K) - P(S_0, H_T) - (K - H_T) \cdot D_{\text{in}}(S_0, H_T).$$

- Finally, let us consider an *down-and-in put*. We get, for $K < H_T$,

$$P_{\text{in}}(S_0, K) = P(S_0, K),$$

and, for $K \geqslant H_T$,

$$P_{\text{in}}(S_0, K)$$
$$= \left(\frac{H_0}{S_0}\right)^{2v/\sigma} \cdot \left[P\left(\frac{H_0^2}{S_0}, K\right) - P\left(\frac{H_0^2}{S_0}, H_T\right) - (K - H_T) \cdot D_{\text{in}}\left(\frac{H_0^2}{S_0}, H_T\right)\right]$$
$$+ P(S_0, H_T) + (K - H_T) \cdot D_{\text{in}}(S_0, H_T).$$

2.4.2 Double Barriers

An option can also have two barriers: one *up* barrier and one *down* barrier. Each of them can independently be an *in* or *out* barrier.

We first treat the case when both of these barriers are *out* barriers. We denote the upper barrier by U and the lower barrier by L. We can assume $L < S_0 < U$.

Using the density derived in Chapter 1, we can compute the price of a double knock-out call. We assume that $K < U$. The price of the call is then given by

$$C_{\text{out}}^{\text{out}} = S^\beta e^\alpha \sum_{n=-\infty}^{\infty} \left\{ U^{-\beta} K \left(\frac{L}{U}\right)^{2n\beta} e^{\beta^2 \Sigma_T^2/2} [g_n(\beta) - h_n(\beta)] \right.$$
$$\left. - U^{1-\beta} \left(\frac{L}{U}\right)^{2n(\beta-1)} e^{(\beta-1)^2 \Sigma_T^2/2} [g_n(\beta-1) - h_n(\beta-1)] \right\},$$

with

$$g_n(\beta) = \left(\frac{U}{S}\right)^\beta \left[N\left(\frac{1}{\Sigma_T}\left(\log\frac{\max(K,L)}{S} - 2n\log\frac{U}{L} + \beta\Sigma_T^2\right)\right) \right.$$
$$\left. - N\left(\frac{1}{\Sigma_T}\left(\log\frac{U}{S} - 2n\log\frac{U}{L} + \beta\Sigma_T^2\right)\right) \right],$$

$$h_n(\beta) = \left(\frac{U}{S}\right)^{-\beta} \left[N\left(\frac{1}{\Sigma_T}\left(\log\frac{S\max(K,L)}{U^2} - 2n\log\frac{U}{L} + \beta\Sigma_T^2\right)\right) \right.$$
$$\left. - N\left(\frac{1}{\Sigma_T}\left(\log\frac{S}{U} - 2n\log\frac{U}{L} + \beta\Sigma_T^2\right)\right) \right],$$

and

$$\beta = -\frac{\mathcal{N}_T}{\Sigma_T^2}, \qquad \alpha = -\tfrac{1}{2}\Sigma_T^2\beta^2 - R_T.$$

The prices of the other variants of double barrier call options can be derived from C_{out}^{out}, single barrier and plain vanilla prices. For example, being long an up-and-out call and short an up-and-out down-and-out call is equivalent to a long position in an up-and-out down-and-in call. We have

$$C_{in}^{out} = C^{out} - C_{out}^{out},$$
$$C_{out}^{in} = C_{out} - C_{out}^{out},$$
$$C_{in}^{in} = C - C_{out}^{out}.$$

Now consider the double knock-out put. Again we can assume $L < S_0 < U$ and $K > L$, since the put is worthless otherwise. In this case we get

$$P_{out}^{out} = S^\beta e^\alpha \sum_{n=-\infty}^{\infty} \left\{ U^{-\beta} K \left(\frac{L}{U}\right)^{2n\beta} e^{\beta^2 \Sigma_T^2/2} [g_n'(\beta) - h_n'(\beta)] \right.$$
$$\left. - U^{1-\beta} \left(\frac{L}{U}\right)^{2n(\beta-1)} e^{(\beta-1)^2 \Sigma_T^2/2} [g_n'(\beta-1) - h_n'(\beta-1)] \right\},$$

with

$$g_n'(\beta) = \left(\frac{U}{S}\right)^\beta \left[N\left(\frac{1}{\Sigma_T}\left(\log\frac{\min(K,U)}{S} - 2n\log\frac{U}{L} + \beta\Sigma_T^2\right)\right) \right.$$
$$\left. - N\left(\frac{1}{\Sigma_T}\left(\log\frac{L}{S} - 2n\log\frac{U}{L} + \beta\Sigma_T^2\right)\right) \right],$$

$$h_n'(\beta) = \left(\frac{U}{S}\right)^{-\beta} \left[N\left(\frac{1}{\Sigma_T}\left(\log\frac{S\min(K,U)}{U^2} - 2n\log\frac{U}{L} + \beta\Sigma_T^2\right)\right) \right.$$
$$\left. - N\left(\frac{1}{\Sigma_T}\left(\log\frac{LS}{U^2} - 2n\log\frac{U}{L} + \beta\Sigma_T^2\right)\right) \right],$$

and with α and β as before.

Again, one can derive prices for other variants of double barrier put options using the relations

$$P_{in}^{out} = P^{out} - P_{out}^{out},$$
$$P_{out}^{in} = P_{out} - P_{out}^{out},$$
$$P_{in}^{in} = P - P_{out}^{out}.$$

2.4.3 Partial Barriers

In contrast to an ordinary barrier, a partial barrier is only active on a previously agreed time window $[s, t]$. This means a knock-out or a knock-in can only occur between time s and time t.

Pricing technique To illustrate the pricing technique, we will focus on a call option with a knock-out barrier active on $[s, t]$ with $0 < s < t < T$. The payoff of this option (discounted to time 0) is given by

$$X = e^{-R_T}(S_T - K)^+ 1_{\{\max_{s \leq u \leq t} S_u < U\}}.$$

Closed-Form Solutions for Standard Products

The other partial barrier options can be valued using combinations of partial barrier options and plain vanilla options.

Our approach for pricing the partial barrier call is to calculate the price stepwise from maturity day back to evaluation day. At t, the price is given by the plain vanilla call formula with strike K and maturity T. The first step is to derive the price $C^{\text{out}}(S_s, s)$ of the option at s. This amounts to pricing a call with a knock-out barrier that is switched off at t. Finally, the price of the partial barrier option at evaluation date 0 is given by

$$e^{-R_s} \int_{-\infty}^{U} C^{\text{out}}(x, s) \, \mathbb{P}[\, S_s \in dx \mid S_0 \,].$$

Up-and-out call with semi-partial barrier We first derive the fair price $C^{\text{out}}(S_s, s)$ at s of a call option with up-and-out barrier on $[s, t]$ ($s < t < T$). So, the barrier is active immediately, but ends before maturity. We have, for $S_s < U$,

$$e^{R_{st}} C^{\text{out}}(S_s, s) = \mathbb{E}\big[\, C(S_t, t) \cdot 1_{\{S_u < U, \, s \leq u \leq t\}} \mid S_s \,\big]$$

$$= \mathbb{E}\big[\, [e^{-D_{tT}} S_t N(d_1) - K e^{-R_{tT}} N(d_2)] \cdot 1_{\{S_u < U, \, s \leq u \leq t\}} \mid S_s \,\big]$$

$$= \mathbb{E}\big[\, [e^{-D_{tT}} S_t N(d_1) - K e^{-R_{tT}} N(d_2)]$$

$$\cdot \mathbb{P}[\, S_u < U, \, u \in [s, t] \mid S_s, S_t \,] \mid S_s \,\big]$$

$$= \mathbb{E}\bigg[\, [e^{-D_{tT}} S_t N(d_1) - K e^{-R_{tT}} N(d_2)]$$

$$\cdot \left(1 - \exp\left(-\frac{2}{\Sigma_{st}^2} \log \frac{U}{S_s} \log \frac{U}{S_t}\right)\right) \cdot 1_{\{S_t < U\}} \,\bigg|\, S_s \,\bigg],$$

with

$$d_1 = \frac{\log(x/K) + \tilde{\mathcal{N}}_{st}}{\Sigma_{st}}, \qquad d_2 = \frac{\log(x/K) + \mathcal{N}_{st}}{\Sigma_{st}},$$

and $C(S_t, t)$ denoting the price of a plain vanilla call given a spot S_t at time t, and with strike K and maturity T. Here we have used formula (1.8) for the hitting time of a geometric Brownian bridge. We obtain, for $S_s < U$,

$$e^{R_{st}} C^{\text{out}}(S_s, s) = e^{-D_{tT}} \mathbb{E}\big[\, S_t N(d_1) \cdot 1_{\{S_t < U\}} \mid S_s \,\big] - K e^{-R_{tT}} \mathbb{E}\big[\, N(d_2) \cdot 1_{\{S_t < U\}} \mid S_s \,\big]$$

$$- e^{-D_{tT}} \mathbb{E}\bigg[\, S_t \exp\left(-\frac{2}{\Sigma_{st}^2} \log \frac{U}{S_s} \log \frac{U}{S_t}\right) N(d_1) \cdot 1_{\{S_t < U\}} \,\bigg|\, S_s \,\bigg]$$

$$+ K e^{-R_{tT}} \mathbb{E}\bigg[\, \exp\left(-\frac{2}{\Sigma_{st}^2} \log \frac{U}{S_s} \log \frac{U}{S_t}\right) N(d_2) \cdot 1_{\{S_t < U\}} \,\bigg|\, S_s \,\bigg],$$

$$= (\text{I}) + (\text{II}) + (\text{III}) + (\text{IV}). \tag{2.11}$$

We now calculate each of the four terms (I)–(IV) in (2.11) separately.

Term I Note that $S_t = U$ is equivalent to

$$B_t - B_s = \frac{1}{\Sigma_{st}} \left(\log \frac{U}{S_s} - \mathcal{N}_{st}\right).$$

Hence,

$$\mathbb{E}\big[S_t N(d_1) \cdot 1_{\{S_t < U\}} \mid S_s \big]$$

$$= S_s \exp(\mathcal{N}_{st}) \cdot \int_{-\infty}^{\gamma} e^{\delta z} N(\alpha z + \beta) \varphi(z)\, dz$$

$$= S_s e^{\mathcal{M}_{st}} \cdot N_2\left(\gamma - \delta, \frac{\beta + \alpha\delta}{\sqrt{1+\alpha^2}}; \frac{-\alpha}{\sqrt{1+\alpha^2}}\right)$$

$$= S_s e^{\mathcal{M}_{st}} N_2\left(\frac{1}{\Sigma_{st}}\left(\log\frac{U}{S_s} - \tilde{\mathcal{N}}_{st}\right), \frac{1}{\Sigma_{sT}}\left(\log\frac{S_s}{K} + \tilde{\mathcal{N}}_{sT}\right); -\frac{\sigma_{st}}{\sigma_{sT}}\right),$$

with

$$\alpha = \frac{\sigma_{st}}{\sigma_{tT}},$$

$$\beta = \frac{1}{\Sigma_{tT}}\left(\log\frac{S_s}{K} + \tilde{\mathcal{N}}_{sT} - \Sigma_{st}^2\right),$$

$$\gamma = \frac{1}{\Sigma_{st}}\left(\log\frac{U}{S_s} - \mathcal{N}_{st}\right),$$

$$\delta = \Sigma_{st}.$$

Term II The second term of (2.11) is obtained in much the same way.

$$\mathbb{E}\big[N(d_2) \cdot 1_{\{S_t < U\}} \mid S_s \big] = \int_{-\infty}^{\gamma} N(\alpha z + \beta)\varphi(z)\, dz$$

$$= N_2\left(\gamma, \frac{\beta}{\sqrt{1+\alpha^2}}; \frac{-\alpha}{\sqrt{1+\alpha^2}}\right)$$

$$= N_2\left(\frac{1}{\Sigma_{st}}\left(\log\frac{U}{S_s} - \mathcal{N}_{st}\right), \frac{1}{\Sigma_{sT}}\left(\log\frac{S_s}{K} + \mathcal{N}_{sT}\right); -\frac{\sigma_{st}}{\sigma_{sT}}\right),$$

with α and γ as for Term I,

$$\beta = \frac{1}{\Sigma_{sT}}\left(\log\frac{S_s}{K} + \mathcal{N}_{sT}\right)$$

and $\delta = 0$.

Term III The third term in (2.11) is given by

$$\mathbb{E}\left[S_t \exp\left(-\frac{2}{\Sigma_{st}^2}\log\frac{U}{S_s}\log\frac{U}{S_t}\right) \cdot N(d_1) \cdot 1_{\{S_t < U\}} \,\bigg|\, S_s \right]$$

$$= S_s \exp\left[\left(\mathcal{N}_{st} - \frac{2}{\Sigma_{st}^2}\log\frac{U}{S_s}\left(\log\frac{U}{S_s} - \mathcal{N}_{st}\right)\right)\right] \int_{-\infty}^{\gamma} e^{\delta z} N(\alpha z + \beta)\varphi(z)\, dz$$

$$= S_s\left(\frac{U}{S_s}\right)^{2\tilde{\mathcal{N}}_{st}/\Sigma_{st}^2} e^{\mathcal{M}_{st}} N_2\left(\frac{1}{\Sigma_{st}}\left(\log\frac{S_s}{U} - \tilde{\mathcal{N}}_{st}\right), \frac{1}{\Sigma_{sT}}\left(\log\frac{U^2}{S_s K} + \tilde{\mathcal{N}}_{sT}\right); -\frac{\sigma_{st}}{\sigma_{sT}}\right),$$

with α, β and γ as for Term I, and with

$$\delta = \frac{2}{\Sigma_{st}} \log \frac{U}{S_s} + \Sigma_{st}.$$

Term IV Finally, we have

$$\mathbb{E}\left[\exp\left(-\frac{2}{\Sigma_{st}^2} \log \frac{U}{S_s} \log \frac{U}{S_t}\right) N(d_2) \cdot 1_{\{S_t < U\}} \,\bigg|\, S_s\right]$$

$$= \exp\left[-\frac{2}{\Sigma_{st}^2} \log \frac{U}{S_s}\left(\log \frac{U}{S_s} - \tilde{\mathcal{N}}_{st}\right)\right] \int_{-\infty}^{\gamma} e^{\delta z} N(\alpha z + \beta) \varphi(z) \, dz$$

$$= \left(\frac{U}{S_s}\right)^{2\tilde{\mathcal{N}}_{st}/\Sigma_{st}^2} N_2\left(\frac{1}{\Sigma_{st}}\left(\log \frac{S_s}{U} - \tilde{\mathcal{N}}_{st}\right), \frac{1}{\Sigma_{sT}}\left(\log \frac{U^2}{S_s K} + \tilde{\mathcal{N}}_{sT}\right); -\frac{\sigma_{st}}{\sigma_{sT}}\right),$$

with α and γ as for Term I, β as in Term II and

$$\delta = -\frac{2}{\Sigma_{st}} \log \frac{U}{S_s}.$$

The sum of the four terms (I)–(IV) yields the price of the partial up-and-out call at time s with the barrier period ending at t. It is given by

$$\Pi(S, s) =$$

$$S e^{-D_{sT}} N_2\left(\frac{1}{\Sigma_{st}}\left(\log \frac{U}{S} - \tilde{\mathcal{N}}_{st}\right), \frac{1}{\Sigma_{sT}}\left(\log \frac{S}{K} + \tilde{\mathcal{N}}_{sT}\right); -\frac{\sigma_{st}}{\sigma_{sT}}\right)$$

$$- K e^{-R_{sT}} N_2\left(\frac{1}{\Sigma_{st}}\left(\log \frac{U}{S} - \mathcal{N}_{st}\right), \frac{1}{\Sigma_{sT}}\left(\log \frac{S}{K} + \mathcal{N}_{sT}\right); -\frac{\sigma_{st}}{\sigma_{sT}}\right)$$

$$- S e^{-D_{sT}} \left(\frac{U}{S}\right)^{2\tilde{\mathcal{N}}_{st}/\Sigma_{st}^2} N_2\left(\frac{1}{\Sigma_{st}}\left(\log \frac{S}{U} - \tilde{\mathcal{N}}_{st}\right), \frac{1}{\Sigma_{sT}}\left(\log \frac{U^2}{SK} + \tilde{\mathcal{N}}_{sT}\right); -\frac{\sigma_{st}}{\sigma_{sT}}\right)$$

$$+ K e^{-R_{sT}} \left(\frac{U}{S}\right)^{\frac{2\mathcal{N}_{st}}{\Sigma_{st}^2}} N_2\left(\frac{1}{\Sigma_{st}}\left(\log \frac{S}{U} - \mathcal{N}_{st}\right), \frac{1}{\Sigma_{sT}}\left(\log \frac{U^2}{SK} + \mathcal{N}_{sT}\right); -\frac{\sigma_{st}}{\sigma_{sT}}\right).$$

(2.12)

Up-and-out call with full partial barrier Now we consider the case of a partial barrier call which starts after today and ends before maturity. We assume an up-and-out barrier active on $[s, t]$ ($0 < s < t < T$) and evaluate the option at time $t = 0$. In terms of the price $C^{\text{out}}(S_s, s)$ at s given in (2.12), the price $C^{\text{out}}(S_0, 0)$ of this option at 0 is given by

$$C^{\text{out}}(S_0, 0) = e^{-R_s} \mathbb{E}\left[C^{\text{out}}(S_s, s) \cdot 1_{\{S_s \leq U\}}\right]. \quad (2.13)$$

The calculation of this expression involves four terms (see (2.12)).

Term I We have

$$e^{-R_s}\mathbb{E}\left[S_s e^{-D_{sT}} N_2\left(\frac{1}{\Sigma_{st}}\left(\log\frac{U}{S_s}-\tilde{\mathcal{N}}_{st}\right), \frac{1}{\Sigma_{sT}}\left(\log\frac{S_s}{K}+\tilde{\mathcal{N}}_{sT}\right); -\frac{\sigma_{st}}{\sigma_{sT}}\right)\cdot 1_{\{S_s\leq U\}}\right]$$

$$= S_0 e^{-D_T-\Sigma_s^2 s/2}\mathbb{E}\left[e^{\delta X} N_2(\alpha_1 X+\beta_1, \alpha_2 X+\beta_2; \rho)\cdot 1_{\{X\leq\gamma\}}\right]$$

$$= S_0 e^{-D_T} N_3\left(\frac{1}{\Sigma_s}\left(\log\frac{U}{S_0}-\tilde{\mathcal{N}}_s\right), \frac{1}{\Sigma_t}\left(\log\frac{U}{S_0}-\tilde{\mathcal{N}}_t\right),\right.$$
$$\left.\frac{1}{\Sigma_T}\left(\log\frac{S_0}{K}+\tilde{\mathcal{N}}_T\right), \frac{\sigma_s}{\sigma_t}, -\frac{\sigma_s}{\sigma_T}, -\frac{\sigma_t}{\sigma_T}\right),$$

with X denoting a normally distributed random variable with mean 0 and variance 1, as well as

$$\alpha_1 = -\frac{\sigma_s}{\sigma_{st}},$$
$$\alpha_2 = \frac{\sigma_s}{\sigma_{sT}},$$
$$\beta_1 = \frac{1}{\Sigma_{st}}\left(\log\left(\frac{U}{S_0}\right) - \mathcal{N}_s - \tilde{\mathcal{N}}_{st}\right),$$
$$\beta_2 = \frac{1}{\Sigma_{sT}}\left(\log\left(\frac{S_0}{K}\right) + \mathcal{N}_s + \tilde{\mathcal{N}}_{sT}\right),$$
$$\gamma = \frac{1}{\Sigma_s}\left(\log\frac{U}{S_0} - \mathcal{N}_s\right),$$
$$\rho = -\frac{\sigma_{st}}{\sigma_{sT}}$$
$$\delta = \Sigma_s.$$

Term II The second term of (2.13), with $C^{\text{out}}(S_s, s)$ given by (2.12), is of the form

$$T_{\text{II}} = -K e^{-R_T}\mathbb{E}\left[N_2\left(\frac{1}{\Sigma_{st}}\left(\log\frac{U}{S_s}-\mathcal{N}_{st}\right), \frac{1}{\Sigma_{sT}}\left(\log\frac{S_s}{K}+\mathcal{N}_{sT}\right); -\frac{\sigma_{st}}{\sigma_{sT}}\right)\right]$$

$$= -K e^{-R_T}\mathbb{E}\left[N_2(\alpha_1 X+\beta_1, \alpha_2 X+\beta_2; \rho)\cdot 1_{\{X\leq\gamma\}}\right],$$

with α_1, α_2, γ and ρ as for Term I, as well as

$$\beta_1 = \frac{1}{\Sigma_{st}}\left(\log\left(\frac{U}{S_0}\right) - \mathcal{N}_s - \mathcal{N}_{st}\right),$$
$$\beta_2 = \frac{1}{\Sigma_{sT}}\left(\log\left(\frac{S_0}{K}\right) + \mathcal{N}_s + \mathcal{N}_{sT}\right).$$

Hence,

$$T_{\text{II}} = -K e^{-R_T} N_3\left(\frac{1}{\Sigma_s}\left(\log\frac{U}{S_0}-\mathcal{N}_s\right), \frac{1}{\Sigma_t}\left(\log\frac{U}{S_0}-\mathcal{N}_t\right),\right.$$
$$\left.\frac{1}{\Sigma_T}\left(\log\frac{S_0}{K}+\mathcal{N}_T\right), \frac{\sigma_s}{\sigma_t}, -\frac{\sigma_s}{\sigma_T}, -\frac{\sigma_t}{\sigma_T}\right).$$

Closed-Form Solutions for Standard Products

Term III The third term reads

$$T_{\text{III}} = e^{-D_sT-R_s}\mathbb{E}\left[S_s\left(\frac{U}{S_s}\right)^{2\tilde{\mathcal{N}}_{st}/\Sigma_{st}^2} \right.$$

$$\left. \cdot N_2\left(\frac{1}{\Sigma_{st}}\left(\log\frac{S_s}{U} - \tilde{\mathcal{N}}_{st}\right), \frac{1}{\Sigma_{sT}}\left(\log\frac{U^2}{S_sK} + \tilde{\mathcal{N}}_{sT}\right); -\frac{\sigma_{st}}{\sigma_{sT}}\right) \right]$$

$$= -S_0 e^{-D_T}\left(\frac{U}{S_0}\right)^{2\tilde{\mathcal{N}}_{st}/\Sigma_{st}^2} \mathbb{E}\left[e^{\delta X} N_2(\alpha_1 X + \beta_1, \alpha_2 X + \beta_2; \rho) \cdot 1_{\{X \leqslant \gamma\}}\right],$$

with

$$\alpha_1 = \frac{\sigma_s}{\sigma_{st}},$$

$$\alpha_2 = -\frac{\sigma_s}{\sigma_{sT}},$$

$$\beta_1 = \frac{1}{\Sigma_{st}}\left(\log\left(\frac{S_0}{U}\right) + \mathcal{N}_s - \tilde{\mathcal{N}}_{st}\right),$$

$$\beta_2 = \frac{1}{\Sigma_{sT}}\left(\log\left(\frac{U^2}{KS_0}\right) - \mathcal{N}_s + \tilde{\mathcal{N}}_{sT}\right),$$

$$\delta = -2\mathcal{M}_s/\Sigma_s,$$

and γ and ρ as for Term I. This yields

$$T_{\text{III}} = -S_0 e^{-D_T}\left(\frac{U}{S_0}\right)^{2\tilde{\mathcal{N}}_{st}/\Sigma_{st}^2} N_3\left(\frac{1}{\Sigma_s}\left(\log\frac{U}{S_0} + \tilde{\mathcal{N}}_s\right), \frac{1}{\Sigma_t}\left(\log\frac{S_0}{U} - \tilde{\mathcal{N}}_t\right),\right.$$

$$\left.\frac{1}{\Sigma_T}\left(\log\frac{U^2}{KS_0} + \tilde{\mathcal{N}}_T\right), -\frac{\sigma_s}{\sigma_t}, \frac{\sigma_s}{\sigma_T}, -\frac{\sigma_t}{\sigma_T}\right).$$

Term IV The last term is given by

$$T_{\text{IV}} = Ke^{-R_T}\mathbb{E}\left[\left(\frac{U}{S_s}\right)^{2\mathcal{N}_{st}/\Sigma_{st}^2} \right.$$

$$\left. \cdot N_2\left(\frac{1}{\Sigma_{st}}\left(\log\frac{S_s}{U} - \mathcal{N}_{st}\right), \frac{1}{\Sigma_{sT}}\left(\log\frac{U^2}{S_sK} + \mathcal{N}_{sT}\right); -\frac{\sigma_{st}}{\sigma_{sT}}\right) \right]$$

$$= Ke^{-R_T}\left(\frac{U}{S_0}\right)^{2\mathcal{N}_{st}/\Sigma_{st}^2} \mathbb{E}\left[e^{\delta X} N_2(\alpha_1 X + \beta_1, \alpha_2 X + \beta_2; \rho) \cdot 1_{\{X \leqslant \gamma\}}\right],$$

with α_1 and α_2 as for Term III and γ and ρ as for Term I, and with

$$\beta_1 = \frac{1}{\Sigma_{st}}\left(-\log\left(\frac{U}{S_0}\right) + \mathcal{N}_s - \mathcal{N}_{st}\right),$$

$$\beta_2 = \frac{1}{\Sigma_{sT}}\left(\log\left(\frac{U^2}{KS_0}\right) + -\mathcal{N}_s + \mathcal{N}_{sT}\right),$$

$$\delta = -\left(\frac{2\mathcal{N}_{st}\Sigma_s}{\Sigma_{st}^2}\right).$$

Hence, finally,

$$T_{IV} = Ke^{-R_T}\left(\frac{U}{S_0}\right)^{2\mathcal{N}_{st}/\Sigma_{st}^2} N_3\left(\frac{1}{\Sigma_s}\left(\log\frac{U}{S_0}+\mathcal{N}_s\right), \frac{1}{\Sigma_t}\left(\log\frac{S_0}{U}-\mathcal{N}_t\right),\right.$$
$$\left.\frac{1}{\Sigma_T}\left(\log\frac{U^2}{KS_0}+\mathcal{N}_T\right); -\frac{\sigma_s}{\sigma_t}, \frac{\sigma_s}{\sigma_T}, -\frac{\sigma_t}{\sigma_T}\right).$$

The sum of all four terms gives the fair price $C^{\text{out}}(S_0, 0)$ at $t = 0$ of a European call option with maturity T and an up-and-out barrier at level U, active from date s to date t ($0 < s < t < T$).

Put options with partial barriers In order to price put options, we remark that each of the four terms for the call $C^{\text{out}}(S_0, 0)$ (with the same set of parameters) – see (2.13) – can be written in the form

$$f\,\mathbb{E}\left[e^{\delta X} N_2(\alpha_1 X + \beta_1, \alpha_2 X + \beta_2; \rho) \cdot 1_{\{X<\gamma\}}\right], \qquad (2.14)$$

with some f, δ, α_1, α_2, β_1, β_2, ρ, γ, and with X denoting a standard normally distributed random variable. It is easy to see, following the lines of our earlier computation, that the corresponding term for the put is given by

$$-f\,\mathbb{E}\left[e^{\delta X} N_2(\alpha_1 X + \beta_1, -\alpha_2 X - \beta_2; -\rho) \cdot 1_{\{X<\gamma\}}\right]. \qquad (2.15)$$

Using formula (A.2) from Appendix A.3, we can rewrite (2.14) to give a term of the form

$$gN_3(a, b, c; \rho_{12}, \rho_{13}, \rho_{23}). \qquad (2.16)$$

Accordingly, (2.15) is equal to

$$-gN_3(a, b, -c; \rho_{12}, -\rho_{13}, -\rho_{23}).$$

Partial down barriers Using the results obtained so far, we now want to price options with a partial down-and-out barrier instead of an up-and-out barrier.

An inspection of the computation done so far shows that we have to replace $1_{\{S_t<U\}}$ by $1_{\{S_t>U\}}$ in formula (2.11). Hence, to obtain the formula $C_{\text{out}}(S_s, s)$ from the formula (2.12) for $C^{\text{out}}(S_s, s)$, we have to change the sign of the first and third argument of each of the four $N_2(\cdot)$ terms.

Accordingly, using (2.13) with $C_{\text{out}}(S_s, s)$ instead of $C^{\text{out}}(S_s, s)$, the price of the partial down-and-out call at $t = 0$ is obtained from the corresponding up-and-out call by changing the sign of the first, second, fifth and sixth argument in each of the four $N_3(\cdot)$ terms.

Options with partial in barriers The price for a partial up (down) and in call (put) can be obtained from the corresponding option with out feature using the fact that the sum of an option with in feature and the corresponding option with out feature is equivalent to a plain vanilla option.

Summary of partial barrier pricing formulas We now summarise the pricing formulas for call and put options with partial barriers derived in the previous sections. Throughout, we assume the barrier to be active only on $[s, t]$, with $0 < s < t < T$. We assume a barrier level U. We remark that we do not impose anything on the position of U relative to S_0.

Closed-Form Solutions for Standard Products

We use the notation

$$C^{\text{out}}, \quad C_{\text{out}}, \quad C^{\text{in}}, \quad C_{\text{in}}$$

for the prices of up-and-out, down-and-out, up-and-in, and down-and-in calls, respectively. Corresponding notation will be applied to put options. Plain vanilla call and put prices will be denoted by C and P.

We now list the formulas, starting with the up-and-out call:

$$
\begin{aligned}
C^{\text{out}} = & \, S_0 e^{-D_T} N_3\Bigg(\frac{1}{\Sigma_s}\left(\log\frac{U}{S_0} - \tilde{\mathcal{N}}_s\right), \frac{1}{\Sigma_t}\left(\log\frac{U}{S_0} - \tilde{\mathcal{N}}_t\right), \\
& \frac{1}{\Sigma_T}\left(\log\frac{S_0}{K} + \tilde{\mathcal{N}}_T\right); \frac{\sigma_s}{\sigma_t}, -\frac{\sigma_s}{\sigma_T}, -\frac{\sigma_t}{\sigma_T}\Bigg) \\
& - Ke^{-R_T} N_3\Bigg(\frac{1}{\Sigma_s}\left(\log\frac{U}{S_0} - \mathcal{N}_s\right), \frac{1}{\Sigma_t}\left(\log\frac{U}{S_0} - \mathcal{N}_t\right), \\
& \frac{1}{\Sigma_T}\left(\log\frac{S_0}{K} + \mathcal{N}_T\right); \frac{\sigma_s}{\sigma_t}, -\frac{\sigma_s}{\sigma_T}, -\frac{\sigma_t}{\sigma_T}\Bigg) \\
& - S_0 e^{-D_T}\left(\frac{U}{S_0}\right)^{2\tilde{\mathcal{N}}_{st}/\Sigma_{st}^2} N_3\Bigg(\frac{1}{\Sigma_s}\left(\log\frac{U}{S_0} + \tilde{\mathcal{N}}_s\right), \frac{1}{\Sigma_t}\left(\log\frac{S_0}{U} - \tilde{\mathcal{N}}_t\right), \\
& \frac{1}{\Sigma_T}\left(\log\frac{U^2}{KS_0} + \tilde{\mathcal{N}}_T\right); -\frac{\sigma_s}{\sigma_t}, \frac{\sigma_s}{\sigma_T}, -\frac{\sigma_t}{\sigma_T}\Bigg) \\
& + Ke^{-R_T}\left(\frac{U}{S_0}\right)^{2\mathcal{N}_{st}/\Sigma_{st}^2} N_3\Bigg(\frac{1}{\Sigma_s}\left(\log\frac{U}{S_0} + \mathcal{N}_s\right), \frac{1}{\Sigma_t}\left(\log\frac{S_0}{U} - \mathcal{N}_t\right), \\
& \frac{1}{\Sigma_T}\left(\log\frac{U^2}{KS_0} + \mathcal{N}_T\right); -\frac{\sigma_s}{\sigma_t}, \frac{\sigma_s}{\sigma_T}, -\frac{\sigma_t}{\sigma_T}\Bigg) \\
= & \sum_{i=1}^{4} f_i N_3(a_i, b_i, c_i; \alpha_i, \beta_i, \gamma_i).
\end{aligned}
$$

Next comes the down-and-out call:

$$C_{\text{out}} = \sum_{i=1}^{4} f_i N_3(-a_i, -b_i, c_i; \alpha_i, -\beta_i, -\gamma_i);$$

the up-and-out put:

$$P^{\text{out}} = \sum_{i=1}^{4} -f_i N_3(a_i, b_i, -c_i; \alpha_i, -\beta_i, -\gamma_i);$$

and the down-and-out put:

$$P_{\text{out}} = \sum_{i=1}^{4} -f_i N_3(-a_i, -b_i, -c_i; \alpha_i, \beta_i, \gamma_i).$$

The corresponding in options can be valued using the following relations:

$$C^{\text{in}} = C - C^{\text{out}},$$

$$C_{\text{in}} = C - C_{\text{out}},$$

$$P^{\text{in}} = P - P^{\text{out}},$$

$$P_{\text{in}} = P - P_{\text{out}}.$$

2.4.4 Compound Barriers

A *compound barrier* option is a barrier option that has an additional knock-in feature. The option holder receives a barrier option contingent that the underlying stock price reaches a certain knock-in level. Both the barrier level of the underlying option (the "second" barrier N_t) and the initial knock-in barrier level (the "first" barrier M_t) are allowed to grow exponentially with time.

Up-and-in on a down-and-in option This means that if the stock process S_t hits the upper barrier M_t at $0 \leqslant t \leqslant T$ we receive at this time the down-and-in option with maturity T, lower barrier N_t and payoff function f, which is assumed to depend only on S_T.

The two barriers follow the functions

$$M_t = M_0 e^{at},$$

$$N_t = N_0 e^{bt},$$

where the numbers are chosen in such a way that $N_t < M_t$ for all $t \in [0, T]$. This condition is equivalent to $N_0 < M_0$ and $N_T < M_T$. Note that the case of constant barriers can be recovered by setting $a = 0$ and $b = 0$.

In addition, we assume that $S_0 \leqslant M_0$. Note that S_0 can be greater or smaller than N_0.

To price our option, we have to evaluate

$$e^{-R_T} \cdot \mathbb{E}[f(S_T) \cdot 1_{A_2}],$$

with the event

$$A_2 = \big\{ \min_{\tau \leqslant t \leqslant T}(S_t - N_t) \leqslant 0 \big\},$$

where τ is the first hitting time for the upper barrier M_t.

Now consider the process

$$R_t = \frac{\log(S_t/S_0) - at}{\sigma}, \quad t \geqslant 0.$$

By Itô's lemma, it satisfies the stochastic differential equation

$$dR_t = v\,dt + dW_t,$$

with $R_0 = 0$ and

$$v = \frac{r - d - a - \tfrac{1}{2}\sigma^2}{\sigma}.$$

Closed-Form Solutions for Standard Products

Now

$$A_2 = \left\{\min_{\tau \leqslant t \leqslant T} R_t \leqslant \bar{d}\right\},$$

where τ is the first hitting time of R_t for the constant barrier $\bar{b} = \log(M_0/S_0)/\sigma$ and where $\bar{d} = \log(N_0/S_0)/\sigma$.

The process R_t is a Brownian motion with drift under the original measure $\mathbb{P}$. Now consider the measure $\hat{\mathbb{P}}$ with Radon–Nikodym derivative

$$\frac{d\hat{\mathbb{P}}}{d\mathbb{P}} = \exp(\nu R_T - \tfrac{1}{2}\nu^2 T).$$

By Girsanov's theorem, the process R_t is a driftless Brownian motion under the new measure $\hat{\mathbb{P}}$. We shall denote the expectation operator with respect to $\hat{\mathbb{P}}$ by $\hat{\mathbb{E}}$.

We can now use the reflection principle. We reflect each path of R on the interval $[0, \tau]$ along the barrier $\bar{b}$. (We set $\tau = \infty$ if there is no hit.) This gives us a new driftless Brownian motion starting at $2\bar{b}$, which we can therefore write as $\bar{R}_t + 2\bar{b}$ with a standard Brownian motion $\bar{R}_t$ (under $\hat{\mathbb{P}}$) starting at 0.

So, we can rewrite

$$A_2 = \left\{\min_{\tau \leqslant t \leqslant T} \bar{R}_t + 2\bar{b} \leqslant \bar{d}\right\}.$$

However, each path of $R_t + 2\bar{b}$ which hits $\bar{d}$ has to hit $\bar{b}$ before, since $\bar{d} < \bar{b} < 2\bar{b}$. Thus

$$A_2 = \left\{\min_{0 \leqslant t \leqslant T} \bar{R}_t + 2\bar{b} \leqslant \bar{d}\right\}.$$

We will also use the notation

$$A_1 = \left\{\min_{0 \leqslant t \leqslant T}(S_t - M_t) \geqslant 0\right\} = \left\{\min_{0 \leqslant t \leqslant T} R_t \geqslant \bar{b}\right\}.$$

Now

$$\mathbb{E}\big[f(S_T) \cdot 1_{A_1} \cdot 1_{A_2}\big] = \hat{\mathbb{E}}\left[f(S_T) \cdot 1_{A_1} \cdot 1_{A_2} \cdot \frac{d\mathbb{P}}{d\hat{\mathbb{P}}}\right]$$

$$= \hat{\mathbb{E}}\big[f(S_0 e^{\sigma R_t + at}) \cdot 1_{A_1} \cdot 1_{A_2} \cdot e^{-\nu R_t + \nu^2 T/2}\big]$$

$$= \hat{\mathbb{E}}\big[f(S_0 e^{\sigma \bar{R}_t + at + 2\sigma \bar{b}}) \cdot 1_{A_2} \cdot e^{-\nu \bar{R}_t + \nu^2 T/2 + 2\nu\bar{b}}\big]$$

$$= e^{2\nu\bar{d}} \cdot \hat{\mathbb{E}}\left[f(S_T e^{2\sigma\bar{b}}) \cdot 1_{A_2} \cdot \frac{d\mathbb{P}}{d\hat{\mathbb{P}}}\right]$$

$$= e^{2\nu\bar{d}} \cdot \mathbb{E}\big[f(S_T e^{2\sigma\bar{b}}) \cdot 1_{A_2}\big]$$

$$= e^{R_T} e^{2\nu\bar{b}} \cdot V_f^{\text{di}}(S_0 e^{2\sigma\bar{b}}, N_t)$$

$$= e^{R_T} \cdot \left(\frac{M_0}{S_0}\right)^{2\nu/\sigma} \cdot V_f^{\text{di}}\left(\frac{M_0^2}{S_0}, N_t\right),$$

where $V_f^{\text{di}}(S_0, N_t)$ denotes the value of a down-and-in option with payoff f, spot S_0 and barrier N_t.

Thus,
$$V_f^{ui/di}(S_0, M_t, N_t) = \left(\frac{M_0}{S_0}\right)^{2\nu/\sigma} \cdot V_f^{di}\left(\frac{M_0^2}{S_0}, N_t\right).$$

As an illustration, we compute the value $C^{ui/di}$ of an up-and-in option on a down-and-in call with strike $K \geq N_T$.

We recall from Section 2.4.1 that
$$C^{di}(S_0, N_t) = C_{in}(S_0, N_t) = \left(\frac{N_0}{S_0}\right)^{2\beta/\sigma} \cdot C\left(\frac{N_0^2}{S_0}\right),$$

with $\beta = (r - d - b - \frac{1}{2}\sigma^2)/\sigma$.

So,
$$C^{ui/di}(S_0, M_t, N_t) = \left(\frac{M_0}{S_0}\right)^{2\nu/\sigma} \cdot \left(\frac{N_0 S_0}{M_0^2}\right)^{2\beta/\sigma} \cdot C\left(\frac{N_0^2}{M_0^2} \cdot S_0\right).$$

Up-and-in on a down-and-out option To price an up-and-in option on a down-and-out option (where we still assume $N_t < M_t$), one can use the relation
$$V_f^{ui/do}(S_0, M_t, N_t) = V_f^{ui}(S_0, M_t) - V_f^{ui/di}(S_0, M_t, N_t).$$

Up-and-in on an up-and-in option An up-and-in option on an up-and-in option, where $M_t < N_t$ for all $t \in [0, T]$, is simply an up-and-in option with barrier N_t.

In our notation this becomes
$$V_f^{ui/ui}(S_0, M_t, N_t) = V_f^{ui}(S_0, N_t).$$

Up-and-in on an up-and-out option This option (where $M_t < N_t$) can be priced by
$$V_f^{ui/uo}(S_0, M_t, N_t) = V_f^{ui}(S_0, M_t) - V_f^{ui/ui}(S_0, M_t, N_t)$$
$$= V_f^{ui}(S_0, M_t) - V_f^{ui}(S_0, N_t).$$

Down-and-in on an up-and-in option As with the case of an up-and-in option, on a down-and-in option one gets
$$V_f^{di/ui}(S_0, M_t, N_t) = \left(\frac{M_0}{S_0}\right)^{2\nu/\sigma} \cdot V_f^{ui}\left(\frac{M_0^2}{S_0}, N_t\right),$$

where we assume that $M_t < N_t$.

Down-and-in on an up-and-out option The price a down-and-in option on a up-and-out option ($M_t < N_t$) is given by
$$V_f^{di/uo}(S_0, M_t, N_t) = V_f^{di}(S_0, M_t) - V_f^{di/ui}(S_0, M_t, N_t).$$

Down-and-in on a down-and-in option Here we have ($N_t < M_t$)
$$V_f^{di/di}(S_0, M_t, N_t) = V_f^{di}(S_0, N_t).$$

Down-and-in on a down-and-out option In this last case ($N_t < M_t$), we get

$$V_f^{\text{di/do}}(S_0, M_t, N_t) = V_f^{\text{di}}(S_0, M_t) - V_f^{\text{di/di}}(S_0, M_t, N_t)$$
$$= V_f^{\text{di}}(S_0, M_t) - V_f^{\text{di}}(S_0, N_t).$$

2.4.5 Rebates for Compound Barriers

For compound barrier options, there are two types of rebate. The first rebate is paid if the first barrier is never hit, and the second rebate is paid if the first barrier is hit and then the second barrier is never hit (for an in barrier) or if it is hit (for an out barrier).

The first type of rebate has the same pricing formula as the rebate in the single-barrier case. So, in the following, we only look at rebates of the second type, all of which can be expressed in terms of American digital calls or puts.

For notational simplicity, we assume the rebate amount to be 1.

Up-and-in on a down-and-out option The value of the rebate in this case is

$$G^{\text{ui/do}}(S_0, M_t, N_t) = V_1^{\text{ui/di}}(S_0, M_t, N_t) = \left(\frac{M_0}{S_0}\right)^{2\nu/\sigma} \cdot \text{D}_{\text{in}}\left(\frac{M_0^2}{S_0}, N_t\right),$$

where the American digital put pays out at hit or at maturity, corresponding to whether our rebate is paid out at hit or paid out at maturity. (The latter case is a slight generalisation of our earlier result.)

Up-and-in on an up-and-out option In this case, we get

$$G^{\text{ui/uo}}(S_0, M_t, N_t) = \text{D}^{\text{in}}\left(S_0, N_t\right).$$

Down-and-in on a up-and-out option Here,

$$G^{\text{di/uo}}(S_0, M_t, N_t) = V_1^{\text{di/ui}}(S_0, M_t, N_t) = \left(\frac{M_0}{S_0}\right)^{2\nu/\sigma} \cdot \text{D}^{\text{in}}\left(\frac{M_0^2}{S_0}, N_t\right).$$

Down-and-in on a down-and-out option In this case, we get

$$G^{\text{di/do}}(S_0, M_t, N_t) = \text{D}_{\text{in}}\left(S_0, N_t\right).$$

In on an in option These cases can be reduced to the in on out option cases by

$$G^{\text{i/i}}(S_0, M_t, N_t) = G^{\text{o}}(S_0, M_t) - G^{\text{i/o}}(S_0, M_t, N_t).$$

2.4.6 Range Options

If a customer believes that an asset will trade in a certain range, ie that its price will not fall below a lower barrier L and not rise above a upper barrier U, then he or she can purchase a *range* option. Such an option pays an amount that increases with the number of days the price of the asset stays in the range between L and U.

The range barriers L and U are defined as either *exploding* or *non-exploding*. Once an exploding barrier has been hit, the amount to be paid ceases to increase. More specifically, a range option pays at maturity an amount proportional to the number of days the price of the underlying stayed within the range before hitting an exploding range barrier.

Range option with no exploding barriers Let the dates on which it is checked whether the asset is in the range be denoted by $0 < t_1 < t_2 < \cdots < t_n$. So the payoff at maturity T of a range option with two non-exploding barriers is given by

$$\sum_{i=1}^{n} 1_{\{L \leqslant S_{t_i} \leqslant U\}}.$$

Hence, the fair value C^{range} is given by

$$C^{\text{range}} = e^{-R_T} \sum_{i=1}^{n} \mathbb{P}[L \leqslant S_{t_i} \leqslant U]$$

$$= e^{-R_T} \sum_{i=1}^{n} \left[N\left(\frac{\log(U/S_0) - \mathcal{N}_{t_i}}{\Sigma_{t_i}}\right) - N\left(\frac{\log(L/S_0) - \mathcal{N}_{t_i}}{\Sigma_{t_i}}\right) \right].$$

Of course, we can regard a range option with no exploding barrier as a series of digital call spreads with payoff delayed to T. Therefore we can also write

$$C^{\text{range}} = e^{-R_T} \sum_{i=1}^{n} [D^{\text{in}}(U, t_i) - D^{\text{in}}(L, t_i)],$$

where $D^{\text{in}}(K, t)$ denotes the price of a digital call with strike K and maturity t.

Range option with one exploding barrier Let us now assume that the upper barrier U is exploding, while the lower barrier L is non-exploding, so we have a payoff of

$$\sum_{i=1}^{n} 1_{\{S_{t_i} \geqslant L,\, S_u \leqslant b \text{ for all } 0 \leqslant u \leqslant t_i\}}$$

and a fair value of

$$e^{-R_T} \sum_{i=1}^{n} \mathbb{P}[S_{t_i} \geqslant L,\, S_u \leqslant U \text{ for all } 0 \leqslant u \leqslant t_i],$$

but this latter is just

$$e^{-R_T} \sum_{i=1}^{n} e^{R_{t_i}} V^{\text{uo}}_{D^{\text{in}}(L,U,t_i)},$$

with $V^{\text{uo}}_{D^{\text{in}}(K,U,t)}$ denoting an up-and-out digital barrier option with strike K, barrier U and maturity t. This can be priced with the general barrier pricing formula presented in Section 2.4.1.

In the converse case, where the lower barrier L is exploding and the upper barrier U is non-exploding, the value is

$$e^{-R_T} \sum_{i=1}^{n} e^{R_{t_i}} V^{\text{do}}_{D_{\text{in}}(U,L,t_i)}.$$

Range option with two exploding barriers Finally, we consider the case that both barriers L and U are exploding. For this structure, the payoff is

$$\sum_{i=1}^{n} 1_{\{L \leqslant S_u \leqslant H \text{ for all } 0 \leqslant u \leqslant t_i\}},$$

which gives us a fair price of

$$e^{-R_T} \sum_{i=1}^{n} \mathbb{P}[L \leqslant S_u \leqslant H \text{ for all } 0 \leqslant u \leqslant t_i].$$

So again we can reduce the pricing problem to something we already know, since the probability on the right-hand side is the (undiscounted) value of an American double knock-out digital option.

2.4.7 Long Barriers

Sometimes the payoff of a barrier option can be delayed, and the barrier is still active until the payoff date. So, the payoff is determined at some time $t > 0$, while the barrier is active until a time $T > t$. We call such an option a *long barrier option*. In the case of an up-and-out call, the payoff at maturity T is given by

$$(S_t - K)^+ 1_{\{S_u < H, 0 \leqslant u \leqslant T\}}.$$

Here K denotes the strike and H denotes the barrier, as usual. Note that in the extreme case of $t = T$ an up-and-out long barrier call is just a standard up-and-out call, while for $t = 0$ it is an American digital call with notional $(S_0 - K)^+$.

Application: protected barrier Asian option Let K denote a strike level and let a sequence of fixing dates $t_1 < t_2 < \cdots < t_n \leqslant T$ be given. A protected Asian call is an option to purchase an average S_F^A of the form

$$S_F^A = \sum_{i=1}^{n} \max(S_{t_i}, F)$$

at maturity T for the strike price K. In other words, every fixing is protected against falling below F. If the floor level F is equal to the strike, the payoff can be decomposed as

$$(S_K^A - K)^+ = \sum_{i=1}^{n} (S_{t_i} - K)^+.$$

Let H denote the barrier level of an up-and-out barrier. Then the payoff

$$(S_K^A - K)^+ 1_{\{S_u < H, 0 \leqslant u \leqslant T\}}$$

can be priced using long barrier options.

Pricing formulas for long up-and-out call For our pricing, we assume the interest rate, dividend yield and volatility to be constant on $[0, t)$ and $(t, T]$ separately.

At time t, the long barrier becomes an American digital, which we priced in Section 2.3.3. This price is

$$\mathbb{P}[S_u < H, t \leqslant u \leqslant T \mid S_t]$$
$$= N\left(\frac{\log(H/S_t) - \mathcal{N}_{tT}}{\Sigma_{tT}}\right) - \left(\frac{H}{S_t}\right)^{2\mathcal{N}_{tT}/\Sigma_{tT}^2} N\left(\frac{\log(S_t/H) - \mathcal{N}_{tT}}{\Sigma_{tT}}\right)$$

multiplied by the inner value $(S_t - K)^+$ of the call. To obtain the price today, we integrate this price at time t, multiplied by the indicator of the event of not hitting the barrier on $[0, t]$. Note that

$$\mathbb{P}[S_u < H, 0 \leqslant u \leqslant t \mid S_t] = 1 - \exp\left(-\frac{2\log(H/S_0)\log(H/S_t)}{\Sigma_t^2}\right).$$

We obtain

$$\mathbb{E}\left[(S_t - K)^+ 1_{\{S_u < H, 0 \leqslant u \leqslant T\}}\right]$$
$$= \mathbb{E}\left[(S_t - K)^+ 1_{\{S_u < H, 0 \leqslant u \leqslant t\}} \cdot \mathbb{P}[S_u < H, t \leqslant u \leqslant T \mid S_t]\right]$$
$$= \int_k^b \left[1 - \exp\left(-\frac{2\log(H/S_0)[\log(H/S_0) - \mathcal{N}_t]}{\Sigma_t^2}\right)\right][S_0 \exp(\Sigma_t x + \mathcal{N}_t) - K]$$
$$\cdot \left[N\left(\frac{\log(H/S_0) - \Sigma_t x - \mathcal{N}_{0T}}{\Sigma_{tT}}\right) - \exp\left(\frac{2\mathcal{N}_{tT}(\log(H/S_0) - \Sigma_t x - \mathcal{N}_t)}{\Sigma_{tT}^2}\right)\right.$$
$$\left. \cdot N\left(\frac{\log(S_0/H) + \Sigma_t x + \mathcal{N}_t - \mathcal{N}_{tT}}{\Sigma_{tT}}\right)\right]\varphi(x)\,dx$$
$$= \sum_{i=1}^{8} f_i \int_k^b e^{\delta_i x} N\left(\alpha_i x + \beta_i\right)\varphi(x)\,dx,$$

where k and b are defined by

$$k = \frac{\log(K/S_0) - \mathcal{N}_t}{\Sigma_t},$$
$$b = \frac{\log(H/S_0) - \mathcal{N}_t}{\Sigma_t}.$$

The coefficients $f_i, \alpha_i, \beta_i, \delta_i$ ($i = 1, 2, \ldots, 8$) are detailed in the next section.

Summary of up-and-out formulas The price of an up-and-out call with strike K, barrier level H, maturity T and payment date t is given by

$$e^{-R_T} \sum_{i=1}^{8} f_i \int_{\min(k,b)}^{b} e^{\delta_i x} N\left(\alpha_i x + \beta_i\right)\varphi(x)\,dx.$$

Closed-Form Solutions for Standard Products

The factors f_i are

$$f_1 = S_0 e^{\mathcal{N}_t},$$

$$f_2 = -\exp\left(\frac{-2\log(H/S_0)(\log(H/S_0) - \mathcal{N}_t)}{\Sigma_t^2}\right) S_0 e^{\mathcal{N}_t},$$

$$f_3 = -K,$$

$$f_4 = \exp\left(\frac{-2\log(H/S_0)(\log(H/S_0) - \mathcal{N}_t)}{\Sigma_t^2}\right) K,$$

$$f_{i+4} = -\exp\left(\frac{2\mathcal{N}_{tT}(\log(H/S_0) - \mathcal{N}_t)}{\Sigma_{tT}^2}\right) f_i, \quad i = 1, 2, 3, 4.$$

The other parameters are

$$\alpha_1 = \cdots = \alpha_4 = -\frac{\Sigma_t}{\Sigma_{tT}},$$

$$\alpha_5 = \cdots = \alpha_8 = \frac{\Sigma_t}{\Sigma_{tT}},$$

$$\beta_1 = \cdots = \beta_4 = \frac{\log(H/S_0) - \mathcal{N}_T}{\Sigma_{tT}^2},$$

$$\beta_5 = \cdots = \beta_8 = \frac{\log(S_0/H) + \mathcal{N}_t - \mathcal{N}_{tT}}{\Sigma_{tT}^2},$$

and

$$\delta_1 = \Sigma_t,$$

$$\delta_2 = \frac{2\log(H/S_0)}{\Sigma_t} + \Sigma_t,$$

$$\delta_3 = 0,$$

$$\delta_4 = \frac{2\log(H/S_0)}{\Sigma_t},$$

$$\delta_{i+4} = \delta_i - 2\mathcal{N}_{tT}\frac{\Sigma_t}{\Sigma_{tT}^2}, \quad i = 1, 2, 3, 4.$$

The price of the corresponding up-and-out put is given by

$$-e^{-R_T} \sum_{i=1}^{8} f_i \int_{-\infty}^{\min(k,b)} e^{\delta_i x} N(\alpha_i x + \beta_i) \varphi(x) \, dx.$$

We remark further that

$$\int_{-\infty}^{\gamma} e^{\delta x} N(\alpha x + \beta) \varphi(x) \, dx = e^{\delta^2/2} N_2\left(\gamma - \delta, \frac{\beta + \alpha\delta}{\sqrt{1+\alpha^2}}; \frac{-\alpha}{\sqrt{1+\alpha^2}}\right),$$

with N_2 denoting the cumulative bivariate normal function.

Down-and-out options The price of a down-and-out call with strike K, barrier level H, maturity T and payment date t is given by

$$e^{-R_T} \sum_{i=1}^{8} f_i \int_{-\infty}^{-\max(k,b)} e^{-\delta_i x} N(\alpha_i x - \beta_i) \varphi(x) \, dx.$$

The corresponding put has the value

$$-e^{-R_T} \sum_{i=1}^{8} f_i \int_{-\max(k,b)}^{b} e^{-\delta_i x} N(\alpha_i x - \beta_i) \varphi(x) \, dx.$$

In options A long in option is obtained as the difference of a plain vanilla option with maturity t and with payment deferred to time T and the corresponding long out option.

2.5 Asian Options

The payoff of a European option depends entirely on the price of the underlying asset at a date in the future. In volatile markets this payoff is very uncertain. A drop in the share price on the last day before maturity may affect the payoff drastically. This leverage may be a desired effect since, on the upside, huge gains may be made with relatively small investment.

If, however, investors have a view of the movement of an asset, this view will only tell something about the "average behaviour" of the asset movement. Therefore, if they want to bet on this behaviour only, rather than on the price of the asset at maturity, they may want to enter into an Asian option contract. An Asian option is written on the average performance of the underlying asset during the lifetime of the option. Generally, an option is called *Asian* if its payoff contains one or several arithmetic or geometric averages.

The forward of the average is easily computed as the average of the forwards. It is more difficult to capture the volatility of the average. We will discuss this issue in the following sections. Generally, it can be said that the volatility of the average is lower than the volatility of the asset. Therefore, the price of Asian options is lower than the price of their European counterparts.

In this section we discuss different types of pricing methods for Asian options. We begin with an exposition of approximation methods, followed by a variety of commonly used Asian products, including standard Asian options and Asian strike products.

2.5.1 *Approximations of the Average*

The distribution of the discrete arithmetic average

$$S_A = \frac{1}{n} \left(\sum_{i=1}^{n} S_{t_i} \right)$$

of our lognormal asset S cannot easily be given in explicit form. Even in the continuous limit $S_A = (1/T) \int_0^T S_t \, dt$, only the Laplace transform of the distribution is known (see [26]). A first attempt to make the problem tractable is to replace S_A by the lognormal geometric average, given by

$$S_G = \left(\prod_{i=1}^{n} S_{t_i}\right)^{1/n}.$$

Using the notation introduced in Section 1.2.1, we have

$$\mathbb{E}[\log S_G] = \frac{1}{n}\mathbb{E}\left[\sum_{i=1}^{n} \log S_{t_i}\right]$$

$$= \log S_0 + \sum_{i=1}^{n} \frac{n-i+1}{n} \mathcal{M}_{t_{i-1},t_i},$$

as well as

$$\mathrm{Var}[\log S_G] = \frac{1}{n^2} \mathrm{Var}\left[\sum_{i=1}^{n}\sum_{k=1}^{i} \int_{t_{k-1}}^{t_k} \sigma_t \, dW_t\right]$$

$$= \frac{1}{n^2} \mathrm{Var}\left[\sum_{i=1}^{n}(n-i+1)\int_{t_{i-1}}^{t_i} \sigma_t \, dW_t\right]$$

$$= \sum_{i=1}^{n}\left(\frac{n-i+1}{n}\right)^2 \Sigma_{t_{i-1},t_i}^2.$$

Just as the sum of normally distributed variables is itself normally distributed, the product of lognormally distributed variables remains lognormal (a direct consequence of the lognormal being the exponential of a normally distributed value). As such, the geometric average follows a lognormal process with the forward given by

$$F_G = \mathbb{E}[S_G] = S_0 \exp \sum_{i=1}^{n}\left[\frac{n-i+1}{n}\mathcal{M}_{t_{i-1},t_i} - \frac{1}{2}\left(\frac{n-i+1}{n}\Sigma_{t_{i-1},t_i}\right)^2\right]$$

and a volatility of

$$\Sigma_G^2 = \mathrm{Var}[\log S_G] = \sum_{i=1}^{n}\left(\frac{n-i+1}{n}\right)^2 \Sigma_{t_{i-1},t_i}^2.$$

As already stated, the case of arithmetic average is not as simple. This is because the sum of lognormally distributed variables is not a lognormal variable. There is a well-known inequality between arithmetic and geometric means

$$S_A \geqslant S_G,$$

which follows directly from Jensen's inequality applied to the convex log function. In some cases the geometric and arithmetic averages, S_A and S_G, are close enough to stop at this stage.

A useful refinement of this method is to use the approximation

$$\tilde{S}_A = S_G - \mathbb{E}[S_G] + \mathbb{E}[S_A],$$

where we can calculate $\mathbb{E}[S_G]$ and $\mathbb{E}[S_A]$. As an example, the value of a standard arithmetic Asian call with strike K can be approximated by

$$\mathbb{E}[(\tilde{S}_A - K)^+] = \mathbb{E}[(S_G - K')^+],$$

a standard geometric Asian call with strike $K' = K + \mathbb{E}[S_G] - \mathbb{E}[S_A]$, which can easily be valued.

The most common approach, however, is to assume that the arithmetic average process can be approximated by a lognormal distribution

$$\tilde{S}_A = F_A e^{\Sigma_A W_1 - \Sigma_A^2/2}, \tag{2.17}$$

with

$$F_A = \mathbb{E}[S_A] = \frac{1}{n}\sum_{i=1}^{n} F_{t_i}, \tag{2.18}$$

$$\Sigma_A^2 = \log\left(1 + \frac{\operatorname{Var}[S_A]}{\mathbb{E}[S_A^2]}\right). \tag{2.19}$$

Here the forward F_A and volatility Σ_A are defined such that the first two moments of S_A and $\tilde{S}_A$, ie expectation and variance, agree. This follows from

$$\mathbb{E}[\tilde{S}_A^m] = (F_A)^m e^{(m^2-m)\Sigma_A^2/2}.$$

A recursive method to calculate $\mathbb{E}[S_A^m]$ explicitly is presented in Section 9.2.1.

There are attempts to match the distribution of S_A to higher moments (see [74]). Let

$$a(y) = \tilde{a}(y) + \frac{c_2}{2!}\tilde{a}''(y) - \frac{c_3}{3!}\tilde{a}'''(y) + \frac{c_4}{4!}\tilde{a}''''(y) - \cdots \tag{2.20}$$

be an expansion of the density a of S_A around the lognormal density $\tilde{a}$ of $\tilde{S}_A$. Following [70], good values for the coefficients c_2, c_3 and c_4 are

$$c_2 = \operatorname{Var}[S_A] - \operatorname{Var}[\tilde{S}_A],$$

$$c_3 = \mathbb{E}[(S_A - \mathbb{E}[S_A])^3] - \mathbb{E}[(\tilde{S}_A - \mathbb{E}[\tilde{S}_A])^3],$$

$$c_4 = \mathbb{E}[(S_A - \mathbb{E}[S_A])^4] - \mathbb{E}[(\tilde{S}_A - \mathbb{E}[\tilde{S}_A])^4] + 3c_2^2.$$

We remark that, by construction, $c_2 = 0$. Section 9.3 elaborates on this approach and provides the first three derivatives of $\tilde{a}$.

2.5.2 Standard Asian Options

For standard Asian options with call and put payoffs

$$(S_A - K)^+ \quad \text{and} \quad (K - S_A)^+$$

at maturity T, we favour the two-moment approach as described in (2.17). The call and put prices are given by

$$C^{\text{Asian}} = e^{-R_T}[F_A N(d_1) - K N(d_2)],$$

$$P^{\text{Asian}} = e^{-R_T}[K N(-d_2) - F_A N(-d_1)],$$

Closed-Form Solutions for Standard Products

with
$$d_1 = \frac{\log(F_A/K) + \frac{1}{2}\Sigma_A^2}{\Sigma_A} \quad \text{and} \quad d_2 = \frac{\log(F_A/K) - \frac{1}{2}\Sigma_A^2}{\Sigma_A}.$$

Obviously, past fixings can be incorporated in the strike.

2.5.3 Standard Asian Options With Higher Moments

Using the expansion (2.20) up to fourth order, we obtain, for the call price C^{Asian},

$$e^{R_T} C^{\text{Asian}} = \int_k^\infty (x-k)\tilde{a}(x)\,dx - \frac{c_3}{3!}\tilde{a}'(k) + \frac{c_4}{4!}\tilde{a}''(k)$$

$$= F_A N(d_1) - K N(d_2) + \frac{c_3 \tilde{a}(k)}{k}\left(1 - \frac{\log(F_A/K)}{\Sigma_A^2}\right)$$

$$+ \frac{c_4 \tilde{a}(k)}{k^2}\left(2 - 3\frac{\log(F_A/K)}{\Sigma_A^2} + \frac{[\log(F_A/K)]^2}{\Sigma_A^4} - \frac{1}{\Sigma_A^2}\right),$$

where the higher moment terms are obtained by an integration by parts.

For the price P^{Asian} of the corresponding Asian put option, we use call–put parity
$$C^{\text{Asian}} - P^{\text{Asian}} = e^{-R_T}\big(\mathbb{E}[S_A] - K\big).$$

Turnbull and Wakeman [74] argue that three and four moments provide a better fit than two moments. However, we do not see that their results show a significant improvement.

2.5.4 Asian Strike Options

In the case of European forward start options, as described in Section 1.3.3, the strike is set at a fraction λS_t of the asset price at the start date t of the option. Since the strike dramatically affects the possible payoff, a risk-averse investor might want to set it at an average level

$$K_A = \frac{1}{m}\left(\sum_{j=1}^m S_{s_j}\right),$$

where $s_1 < s_2 < \cdots < s_m < t_1 < t_2 < \cdots < t_n \leqslant T$. This gives rise to Asian strike options with payoff
$$X = (S_A - K_A)^+.$$

Using our lognormal approximations

$$S_A = F_A e^{\Sigma_A W_1^1 - \Sigma_A^2/2},$$
$$K_A = F_K e^{\Sigma_K W_1^2 - \Sigma_K^2/2},$$

as in (2.17), the option pricing problem is equivalent to an outperformance option problem (which lognormal process performs better?), which is calculated in Section 4.4. The "correct" volatilities and correlation assuring that the first two moments of the joint distribution of (K_A, S_A) are matched are calculated in Section 9.2.1.

A variant of these options are the relative Asian options, which use the "in" – or "strike" – average as a notional:

$$X = \left(\frac{S_A}{K_A} - \lambda\right)^+.$$

The forward F_R and volatility Σ_R of the ratio S_A/K_A are given by

$$F_R = \frac{F_A}{F_K}\exp\bigl((\Sigma_K - \rho_A \Sigma_A)\Sigma_A\bigr),$$

$$\Sigma_R^2 = \Sigma_K^2 - 2\rho_A \Sigma_A \Sigma_K + \Sigma_A^2,$$

with F_A, Σ_A and F_K, Σ_K denoting the forward and volatility of S_A and K_A, respectively. Hence, the call and put prices, $C^{\text{rel Asian}}$ and $P^{\text{rel Asian}}$ are

$$C^{\text{rel Asian}} = e^{-R_T}[F_R N(d_1) - K N(d_2)],$$

$$P^{\text{rel Asian}} = e^{-R_T}[K N(-d_2) - F_R N(-d_1)],$$

with

$$d_1 = \frac{\log(F_R/K) + \tfrac{1}{2}\Sigma_R^2}{\Sigma_R} \quad \text{and} \quad d_2 = \frac{\log(F_R/K) - \tfrac{1}{2}\Sigma_R^2}{\Sigma_R},$$

which again can be compared with the results of the relative outperformance option (Section 4.3).

Chapter 3
Closed-Form Solutions for Non-Standard Products

3.1 Lookback Options

One feature of European options is that the overall performance of the underlying asset may be favourable for the investor but at maturity the asset may move in an undesired direction. As mentioned earlier, one way of avoiding this risk is the use of Asian options, which dampen the effect.

An alternative approach would be to use American options. However, they have the troublesome feature that the holder has to decide whether to exercise or hold on to the option. This involves regular monitoring of asset movements as well as market parameters such as volatility and interest rates.

An option that pays out the optimal highest value attained by the asset during a stipulated time interval is called *lookback* option. More formally, it is an option involving the running maximum

$$M_s^t = \max\{S_u : s \leqslant u \leqslant t\}$$

or minimum

$$m_s^t = \min\{S_u : s \leqslant u \leqslant t\}$$

of an underlying S on a previously agreed time interval $[s, t]$, called the lookback period. In this section we present closed-form formulas for various payoffs depending on S_T, m_s^t and M_s^t, with $T \geqslant t$ denoting maturity.

There are different flavours of lookback options. Products where the asset is replaced by a maximum or minimum over a given lookback period are called *fixed strike lookback* options. Their call and put payoffs at maturity T are defined as

$$\max(M_s^t - K, 0) \quad \text{and} \quad \max(K - m_s^t, 0),$$

respectively. Note that at-the-money options (ie $K = S_0$) of this type never expire worthless and are therefore relatively easy to price.

Products where the strike is replaced by an extremum are called *floating strike lookback* options. Their respective payoffs are

$$\max(S_T - m_s^t, 0) \quad \text{and} \quad \max(M_s^t - S_T, 0)$$

for call and put. The extrema m and M may be multiplied by a constant λ that plays the role of a strike. Finally, in order to lock in the performance of the asset relative to an extremum, this last payoff may be divided by the strike to obtain variable notional options with payoff

$$\frac{\max(S_T - m_s^t, 0)}{m_s^t} \quad \text{and} \quad \frac{\max(M_s^t - S_T, 0)}{M_s^t}.$$

Throughout, S is assumed to be a lognormal process as described in Section 1.2.1. The only restriction on the interest rate, dividend yield and volatility processes is that they are constant during the lookback period $[s, t]$. Our results generalise the formulas presented in [30, 34].

In practice, contracts involve extrema based on discrete fixings. For the difference between continuous and discrete sampling, we refer the reader to [11, 47, 48].

Floating strike lookback call The fair price C_{float} of a floating strike lookback call is

$$C_{\text{float}} = e^{-R_T} \mathbb{E}\left[(S_T - \lambda m_s^t)^+ \right].$$

We assume $s \leq 0$ and hence $m_s^0 \leq S_0$ to be a known constant. Recall that, assuming flat parameters on $[0, t]$, the lognormal asset process can be written as

$$\log \frac{S_u}{S_0} = \sigma_u W_u + \mathcal{N}_u, \quad u \in [0, t],$$

$$\log \frac{S_T}{S_t} = \sigma_{tT}(W_T - W_t) + \mathcal{N}_{tT},$$

with W denoting a Wiener process with respect to $\mathbb{P}$. Furthermore, interpreting S_T/F_T as a density and applying Girsanov's theorem from Section 1.1.12, we obtain the relation

$$e^{-R_T} \mathbb{E}\left[S_T F(S_t, 0 \leq t \leq T) \right] = S_0 e^{-D_T} \tilde{\mathbb{E}}\left[F(S_t, 0 \leq t \leq T) \right] \quad (3.1)$$

for any bounded function F on $(S_t, 0 \leq t \leq T)$. Here the process S with respect to $\tilde{\mathbb{P}}$ is distributed as

$$\log \frac{S_u}{S_0} = \sigma_u W_u + \tilde{\mathcal{N}}_u, \quad u \in [0, t],$$

$$\log \frac{S_T}{S_t} = \sigma_{tT}(W_T - W_t) + \tilde{\mathcal{N}}_{tT},$$

where W denotes a Wiener process with respect to $\tilde{\mathbb{P}}$. Introducing the notation

$$X = \log \frac{S_T}{S_0}, \qquad X_* = \log \frac{m_0^t}{S_0}, \qquad x_* = \log \frac{m_s^0}{S_0}$$

and splitting the minimum at time 0, in other words using the relations

$$m_s^t 1_{\{m_s^0 < m_0^t\}} = m_s^0 1_{\{m_s^0 < m_0^t\}}$$

$$m_s^t 1_{\{m_s^0 > m_0^t\}} = m_0^t 1_{\{m_s^0 > m_0^t\}},$$

we obtain

$$C_{\text{float}} = S_0 e^{-D_T} \{\tilde{\mathbb{P}}[X > \log\lambda + X_*, X_* \leq x_*] + \tilde{\mathbb{P}}[X > \log\lambda + x_*, X_* > x_*]\}$$
$$-\lambda e^{-R_T}\{S_0 \mathbb{E}[e^{X_*} 1_{\{X > \log\lambda + X_*, X_* \leq x_*\}}] + m_s^0 \mathbb{P}[X > \log\lambda + x_*, X_* > x_*]\}.$$

We now express X as a sum of two terms

$$X_1 = \log\frac{S_t}{S_0} \quad \text{and} \quad X_2 = \log\frac{S_T}{S_t}$$

and remark that (X_1, X_*) and X_2 are independent with respect to $\mathbb{P}$ (and hence with respect to $\tilde{\mathbb{P}}$). Moreover, by definition, $X_1 \geq X_*$ holds. We rewrite the integral as

$$C_{\text{float}} = S_0 e^{-D_T} \{\tilde{\mathbb{P}}[X_2 \leq \log\lambda, X_* \leq x_*, X > \log\lambda + X_*]$$
$$+ \tilde{\mathbb{P}}[X_2 \leq \log\lambda, X_* > x_*, X > \log\lambda + x_*]$$
$$+ \tilde{\mathbb{P}}[X_2 > \log\lambda]\}$$
$$- \lambda e^{-R_T}\{S_0 \mathbb{E}[e^{X_*} 1_{\{X_2 \leq \log\lambda, X_* \leq x_*, X > \log\lambda + X_*\}}]$$
$$+ S_0 \mathbb{P}[X_2 > \log\lambda] \mathbb{E}[e^{X_*} 1_{\{X_* \leq x_*\}}]$$
$$+ m_s^0 \mathbb{P}[X_2 \leq \log\lambda, X_* > x_*, X > \log\lambda + x_*]$$
$$+ m_s^0 \mathbb{P}[X_2 > \log\lambda] \mathbb{P}[X_* > x_*]\}.$$

This expression simplifies in the special case $t = T$ of a full lookback period $[0, T]$. We obtain, using $X_2 = 0$,

$$C_{\text{float}} = S_0 e^{-D_T} \{\tilde{\mathbb{P}}[X_* \leq x_*, X_1 > \log\lambda + X_*]$$
$$+ \tilde{\mathbb{P}}[X_* > x_*, X_1 > \log\lambda + x_*]\}$$
$$- \lambda e^{-R_T}\{S_0 \mathbb{E}[e^{X_*} 1_{\{X_* \leq x_*, X_1 > \log\lambda + X_*\}}]$$
$$+ m_s^0 \mathbb{P}[X_* > x_*, X_1 > \log\lambda + x_*]\}$$

if $\lambda \geq 1$. For $\lambda < 1$, we have

$$C_{\text{float}} = S_0 e^{-D_T} - \lambda e^{-R_T} S_0 \mathbb{E}[e^{X_*} 1_{\{X_* \leq x_*\}}] + m_s^0 \mathbb{P}[X_* > x_*]\}.$$

The probabilities and expectations in these expressions are calculated in Sections 1.1.17 and 1.1.18.

Floating strike lookback put The fair price of a fractional floating strike lookback put is

$$P_{\text{float}} = e^{-R_T} \mathbb{E}\left[\left(\frac{1}{\lambda} M_s^t - S_T\right)^+\right].$$

We reuse the notation

$$X_1 = \log\frac{S_t}{S_0}, \qquad X_2 = \log\frac{S_T}{S_t}, \qquad X = X_1 + X_2$$

from earlier in this chapter and introduce

$$X^* = \log\frac{M_0^t}{S_0} \quad \text{and} \quad x^* = \log\frac{M_s^0}{S_0},$$

with $M_s^0 \geqslant S_0$. It is easy to show that

$$\begin{aligned}
P_{\text{float}} = &-S_0 e^{-D_T}\{\tilde{\mathbb{P}}[-X_2 \leqslant \log\lambda, -X^* \leqslant -x^*, -X > \log\lambda - X^*] \\
&+ \tilde{\mathbb{P}}[-X_2 \leqslant \log\lambda, -x^* < -X^*, -X > \log\lambda - x^*] \\
&+ \tilde{\mathbb{P}}[-X_2 > \log\lambda]\} \\
&+ \frac{1}{\lambda}e^{-R_T}\{S_0\mathbb{E}[e^{-(-X^*)}1_{\{-X_2 \leqslant \log\lambda, -X^* \leqslant -x^*, -X > \log\lambda - X^*\}}] \\
&+ S_0\mathbb{P}[-X_2 > \log\lambda]\mathbb{E}[e^{-(-X^*)}1_{-X^* \leqslant -x^*}] \\
&+ M_s^0\mathbb{P}[-X_2 \leqslant \log\lambda, -X^* > -x^*, -X > \log\lambda - x^*] \\
&+ M_s^0\mathbb{P}[-X_2 > \log\lambda]\mathbb{P}[-X^* > -x^*]\}
\end{aligned}$$

where $\tilde{\mathbb{P}}$ is defined via (3.1). In the case of a full lookback period $[0, T]$ or $t = T$, we obtain

$$\begin{aligned}
P_{\text{float}} = &-S_0 e^{-D_T}\{\tilde{\mathbb{P}}[-X^* \leqslant -x^*, -X > \log\lambda - X^*] \\
&+ \tilde{\mathbb{P}}[-X^* > -x^*, -X > \log\lambda - x^*]\} \\
&+ \frac{1}{\lambda}e^{-R_T}\{S_0\mathbb{E}[e^{-(-X^*)}1_{\{-X^* \leqslant -x^*, -X > \log\lambda - X^*\}}] \\
&+ M_s^0\mathbb{P}[-X^* > -x^*, -X > \log\lambda - x^*]\}
\end{aligned}$$

if $\lambda \geqslant 1$. For $\lambda < 1$, we have

$$P_{\text{float}} = -S_0 e^{-D_T} + \frac{e^{-R_T}}{\lambda}\{S_0\mathbb{E}[e^{-(-X^*)}1_{\{-X^* \leqslant -x^*\}}] + M_s^0\mathbb{P}[-X^* > -x^*]\}.$$

A comparison with the call formulas in the last section shows that the put formula can be obtained from the call formula by:

- replacing (X, X_1, X_2, X_*) by $(-X, -X_1, -X_2, -X^*)$;

- using $-x^*$ rather than x_*;

- replacing the exponentials $\exp(x)$ by $\exp(-x)$;

- multiplying the result by -1.

The first statement translates into a replacement of the drifts μ_i by $-\mu_i$ ($i = 1, 2$). Again, all the probabilities and expectations in the above expressions can be found in Sections 1.1.17 and 1.1.18.

Fixed strike lookback put For a fixed strike lookback put with strike K, the fair price is given by

$$P_{\text{fix}} = e^{-R_T}\mathbb{E}\big[\,(K - m_t^T)^+\,\big].$$

We first assume a lookback period that has not yet started, ie $0 < t \leqslant T$. With X_*, X_2 and γ defined by

$$X_* = \log\frac{m_t^T}{S_t}, \qquad X_2 = \log\frac{S_t}{S_0}, \qquad \gamma = \log\frac{K}{S_0},$$

the price of the put can be written as

$$\begin{aligned}P_{\text{fix}} &= S_0 e^{-R_T}\big\{\mathbb{E}[\,(e^\gamma - e^{X_2+X_*})1_{\{X_*+X_2\leqslant\gamma,\,X_2\geqslant\gamma\}}\,] + \mathbb{E}[\,(e^\gamma - e^{X_2+X_*})1_{\{X_2<\gamma\}}\,]\big\} \\ &= K e^{-R_T}\mathbb{P}[\,X_* + X_2 \leqslant \gamma,\, X_2 \geqslant \gamma\,] - S_0 e^{-R_T}\mathbb{E}[\,e^{X_2+X_*}1_{\{X_*+X_2\leqslant\gamma,\,X_2\geqslant\gamma\}}\,] \\ &\quad + K e^{-R_T}\mathbb{P}[\,X_2 < \gamma\,] - S_0 e^{-R_t T + D_t}\tilde{\mathbb{P}}[\,X_2 < \gamma\,]\mathbb{E}[\,e^{X_*}\,].\end{aligned}$$

In the case when the lookback period has already started, in other words $t \leqslant 0 < T$, it is easy to see that

$$P_{\text{fix}} = e^{-R_T}\big\{\mathbb{E}[\,(\min(K, m_t^0) - m_0^T)^+\,] + (K - m_t^0)^+\big\}.$$

Hence, for $K > m_t^0$, the price is the sum of a lookback put with strike m_t^0 and a locked-in amount discounted. Introducing

$$\tilde{K} = \min(K, m_t^0) \quad\text{and}\quad \tilde{\gamma} = \log\frac{\tilde{K}}{S_0} \leqslant 0,$$

we obtain

$$P_{\text{fix}} = \tilde{K}\,e^{-R_T}\mathbb{P}[\,X_* \leqslant \tilde{\gamma}\,] - S_0 e^{-R_T}\mathbb{E}[\,e^{X_*}1_{\{X_*\leqslant\tilde{\gamma}\}}\,] + e^{-R_T}(K - m_t^0)^+.$$

Expressions for the integrals in this section can be found in Sections 1.1.17 and 1.1.18.

Fixed strike lookback call The price of a fixed strike lookback call is given by

$$\begin{aligned}C_{\text{fix}} &= e^{-R_T}\mathbb{E}[\,(M_t^T - K)^+\,] \\ &= S_0 e^{-R_T}\big\{\mathbb{E}[\,(e^{X_2+X^*} - e^\gamma)1_{\{X_2\geqslant\gamma\}}\,] + \mathbb{E}[\,(e^{X_2+X_*} - e^\gamma)1_{\{X^*+X_2\geqslant\gamma,\,X_2\leqslant\gamma\}}\,]\big\} \\ &\stackrel{(3.1)}{=} S_0 e^{-R_t T + D_t}\tilde{\mathbb{P}}[\,X_2 > \gamma\,]\mathbb{E}[\,e^{X^*}\,] - K\,e^{-R_T}\mathbb{P}[\,X_2 > \gamma\,] \\ &\quad + S_0 e^{-R_T}\mathbb{E}[\,e^{X_2+X^*}1_{\{X^*+X_2\geqslant\gamma,\,X_2\leqslant\gamma\}}\,] - K\,e^{-R_T}\mathbb{P}[\,X^* + X_2 \geqslant \gamma,\,X_2 \leqslant \gamma\,]\end{aligned}$$

if the lookback period starts in the future, ie $t > 0$. If, however, the lookback period started in the past, ie $t \leqslant 0$, setting

$$\tilde{K} = \max(K, M_t^0) \quad\text{and}\quad \tilde{\gamma} = \log\frac{\tilde{K}}{S_0} \geqslant 0$$

yields

$$C_{\text{fix}} = S_0 e^{-R_T}\mathbb{E}[\,e^{X^*}1_{\{X^*\leqslant\tilde{\gamma}\}}\,] - \tilde{K}e^{-R_T}\mathbb{P}[\,X^* \geqslant \tilde{\gamma}\,] + e^{-R_T}(M_t^0 - K)^+.$$

Variable notional lookback call Let $s \leqslant 0 < t < T$ and $\lambda \geqslant 1$. The fair price C_{vn} of a *variable notional lookback call* is given by

$$C_{vn} = e^{-R_T} \mathbb{E}\left[\frac{(S_T - \lambda m_s^t)^+}{m_s^t}\right] = e^{-R_T} \mathbb{E}\left[\left(\frac{S_T}{m_s^t} - \lambda\right)^+\right].$$

In the following, we assume $m_s^0 \leqslant S_0$ to be a known constant. The expression for C_{vn} can be written as

$$C_{vn} = e^{-R_T} \mathbb{E}\left[\frac{S_T}{m_s^t} 1_{\{S_T > \lambda m_s^t\}}\right] - \lambda e^{-R_T} \mathbb{P}[S_T > \lambda m_s^t]$$

$$= e^{-D_T}\left\{\frac{S_0}{m_s^0} \tilde{\mathbb{P}}[X > \log \lambda + \gamma, X_* > x_*] + \tilde{\mathbb{E}}[e^{-X_*} 1_{\{X > \log \lambda + X_*, X_* \leqslant x_*\}}]\right\}$$

$$- \lambda e^{-R_T}\left\{\mathbb{P}[X > \log \lambda + X_*, X_* \leqslant x_*] + \mathbb{P}[X > \log \lambda + x_*, X_* > x_*]\right\},$$

with

$$X = \log \frac{S_T}{S_0}, \qquad X_* = \log \frac{m_0^t}{S_0}, \qquad x_* = \log \frac{m_s^0}{S_0}.$$

The measures $\mathbb{P}$ and $\tilde{\mathbb{P}}$ are related via (3.1). We now introduce

$$X_1 = \log \frac{S_t}{S_0}, \qquad X_2 = \log \frac{S_T}{S_t}$$

and condition on the event $\{X_2 \leqslant \log \lambda\}$, so that we have

$$C_{vn} = \frac{S_0}{m_s^0} e^{-D_T}\left\{\tilde{\mathbb{P}}[X_2 \leqslant \log \lambda, X_* > x_*, X > \log \lambda + x_*]\right.$$
$$\left. + \tilde{\mathbb{P}}[X_2 > \log \lambda] \tilde{\mathbb{P}}[X_* > x_*]\right\}$$
$$+ e^{-D_T}\left\{\tilde{\mathbb{E}}[e^{-X_*} 1_{\{X_2 \leqslant \log \lambda, X_* \leqslant x_*, X > \log \lambda + X_*\}}]\right.$$
$$\left. + \tilde{\mathbb{P}}[X_2 > \log \lambda] \tilde{\mathbb{E}}[e^{-X_*} 1_{\{X_* \leqslant x_*\}}]\right\}$$
$$- \lambda e^{-R_T}\left\{\mathbb{P}[X_2 \leqslant \log \lambda, X_* \leqslant x_*, X > \log \lambda + X_*]\right.$$
$$+ \mathbb{P}[X_2 > \log \lambda]$$
$$\left. + \mathbb{P}[X_2 \leqslant \log \lambda, X_* > x_*, X > \log \lambda + x_*]\right\}.$$

We now consider the full period lookback $t = T$. For $\lambda \geqslant 1$ we obtain

$$C_{vn} = \frac{S_0}{m_s^0} e^{-D_T} \tilde{\mathbb{P}}[X_* > x_*, X > \log \lambda + x_*]$$
$$+ e^{-D_T} \tilde{\mathbb{E}}[e^{-X_*} 1_{\{X_* \leqslant x_*, X > \log \lambda + X_*\}}]$$
$$- \lambda e^{-R_T} \mathbb{P}[X_* \leqslant x_*, X > \log \lambda + X_*]$$
$$- \lambda e^{-R_T} \mathbb{P}[X_* > x_*, X > \log \lambda + x_*],$$

and for $\lambda < 1$ we have

$$C_{vn} = e^{-D_T}\left\{\frac{S_0}{m_s^0} \tilde{\mathbb{P}}[X_* > x_*] + \tilde{\mathbb{E}}[e^{-X_*} 1_{\{X_* \leqslant x_*\}}]\right\} - \lambda e^{-R_T}.$$

Closed-Form Solutions for Non-Standard Products

The probabilities and expectations in this expression are calculated in Sections 1.1.17 and 1.1.18.

Variable notional lookback put The fair price of the variable notional lookback put can be written as

$$P_{vn} = e^{-R_T}\mathbb{E}\left[\frac{((1/\lambda)M_s^t - S_T)^+}{M_s^t}\right] = e^{-R_T}\mathbb{E}\left[\left(\frac{1}{\lambda} - \frac{S_T}{M_s^t}\right)^+\right],$$

with $\lambda \geq 1$. We use the notation

$$X = \log\frac{S_T}{S_0}, \qquad X_* = \log\frac{m_0^t}{S_0}, \qquad x_* = \log\frac{m_s^0}{S_0}$$

and split the minimum at time 0, as in the floating strike lookback section,

$$m_s^t \mathbf{1}_{\{m_s^0 < m_0^t\}} = m_s^0 \mathbf{1}_{\{m_s^0 < m_0^t\}},$$

$$m_s^t \mathbf{1}_{\{m_s^0 > m_0^t\}} = m_0^t \mathbf{1}_{\{m_s^0 > m_0^t\}},$$

obtaining

$$P_{vn} = \frac{e^{-R_T}}{\lambda}\left\{\mathbb{P}[\log\lambda + X < X^*,\ X^* \geq \gamma] + \mathbb{P}[X + \log\lambda < \gamma,\ X^* < \gamma]\right\}$$

$$- e^{-D_T}\left\{\frac{S_0}{M_s^0}\tilde{\mathbb{P}}[X + \log\lambda < \gamma,\ X^* < \gamma] + \tilde{\mathbb{E}}[e^{-X^*}\mathbf{1}_{\{X + \log\lambda < X^*,\ X^* \geq \gamma\}}]\right\}.$$

Now we set a condition on the event $\{X_2 \geq -\log\lambda\}$ and conclude that

$$P_{vn} = \frac{e^{-R_T}}{\lambda}\left\{\mathbb{P}[X_2 \geq -\log\lambda,\ X^* \geq \gamma,\ X < -\log\lambda + X^*]\right.$$

$$+ \mathbb{P}[X_2 < -\log\lambda]$$

$$\left. + \mathbb{P}[X_2 \geq -\log\lambda,\ X^* < \gamma,\ X < -\log\lambda + \gamma]\right\}$$

$$- \frac{S_0}{M_s^0}e^{-D_T}\left\{\tilde{\mathbb{P}}[X_2 \geq -\log\lambda,\ X^* < \gamma,\ X < -\log\lambda + \gamma]\right.$$

$$\left. + \tilde{\mathbb{P}}[X_2 < -\log\lambda]\tilde{\mathbb{P}}[X^* < \gamma]\right\}$$

$$- e^{-D_T}\left\{\tilde{\mathbb{E}}[e^{-X^*}\mathbf{1}_{\{X_2 \geq -\log\lambda,\ X^* \geq \gamma,\ X < -\log\lambda + X^*\}}]\right.$$

$$\left. + \tilde{\mathbb{P}}[X_2 < -\log\lambda]\tilde{\mathbb{E}}[e^{-X^*}\mathbf{1}_{\{X^* \geq \gamma\}}]\right\}.$$

The full period ($t = T$) put prices are obtained by setting $X_2 = 0$. We obtain

$$P_{vn} = \frac{e^{-R_T}}{\lambda}\left\{\mathbb{P}[X < -\log\lambda + X^*,\ X^* \geq \gamma] + \mathbb{P}[X < -\log\lambda + \gamma,\ X^* < \gamma]\right\}$$

$$- e^{-D_T}\left\{\frac{S_0}{M_s^0}\tilde{\mathbb{P}}[X^* < \gamma,\ X < -\log\lambda + \gamma] + \tilde{\mathbb{E}}[e^{-X^*}\mathbf{1}_{\{X^* \geq \gamma,\ X < -\log\lambda + X^*\}}]\right\}$$

for $\lambda \geqslant 1$ and

$$P_{vn} = \frac{e^{-R_T}}{\lambda} - e^{-D_T}\left\{\frac{S_0}{M_s^0}\tilde{\mathbb{P}}[X^* < \gamma] + \tilde{\mathbb{E}}[e^{-X^*}1_{\{X^* \geqslant \gamma\}}]\right\}$$

for $\lambda < 1$.

3.2 Fade-In Options

A *fade-in* option is the product of a standard plain vanilla option and a fade-in factor λ. This factor λ increases with the time spent by the asset within a given range $[L, H]$. If the asset never leaves the range, the payoff is a plain vanilla payoff. More formally, the payoff of the fade-in call at maturity T is given by

$$\lambda(S_T - K)^+,$$

with

$$\lambda = \frac{1}{N}\sum_{i=1}^{N} 1_{\{S_{t_i} \in [L,H]\}},$$

where $(t_i, i = 1, 2, \ldots, N)$ is a set of fade-in dates within the lifetime of the option.

For an investor, the advantage of a fade-in option is that it is cheaper than the corresponding plain vanilla product. Moreover, it allows incorporation of a market view on the whole path of the asset into the product. This market view may either be positive in that λ is expected to be close to 1, thereby not affecting the payoff. In this case the cheaper price is the prevailing argument for the choice of this product. If the investor bets on the path not staying within the range, ie if a small λ factor is expected, this product may be of value as a hedge against the risk of adverse market movements. Figure 3.1 provides an example of a fade-in call option that has accrued approximately two-thirds of the call payoff (solid circles represent fade-in condition satisfied, while solid squares represent no-fade-in condition satisfied).

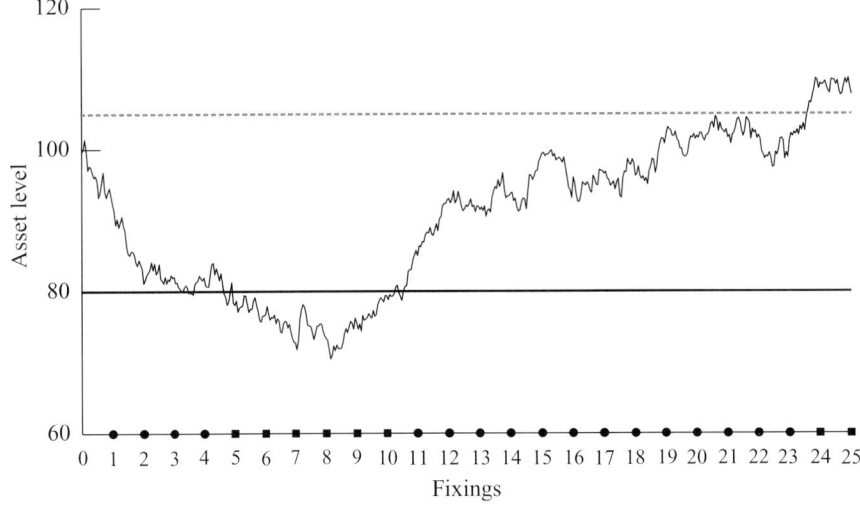

3.1 Fade-in call.

Similarly, a *fade-in barrier* option is the product of a barrier option, as described in Section 2.4, and λ. More specifically, the payoff of the fade-in up-and-in barrier call at maturity is given by

$$\lambda(S_T - K)^+ 1_{\{S_t \neq B, \, t \in [0,T]\}},$$

with the λ given above. Unlike the fade-in condition, the barrier condition is continuous and, as such, we represent this by $S_t \neq B$, where B represents the barrier level.

The corresponding *fade-out* options are defined with $1 - \lambda$ instead of λ. These types of options are also known as *accrual* options.

Finally, we remark that the formulas derived in this section allow us to price barrier options with two different sets of parameters, as well as two different barrier levels for the two time intervals $[0, t]$ and $(t, T]$.

Fade-in call We calculate the fair price of a payoff X given by

$$X = 1_{\{S_t \in [L,H]\}}(S_T - K)^+.$$

Using the notation

$$d_1(K, t) = d_2(K, t) + \Sigma_t \quad \text{and} \quad d_2(K, t) = \frac{1}{\Sigma_t}\left(\log\frac{S_0}{K} + \mathcal{M}_t\right),$$

we obtain the fair price FIC(t) for X as

$$\begin{aligned}
\text{FIC}(t) &= e^{-R_T}\,\mathbb{E}[\,1_{\{S_t \in [L,H]\}}(S_T - K)^+\,] \\
&= e^{-R_t} \int_{-d_2(L,t)}^{-d_2(H,t)} C(S_0 e^{\Sigma_t z + \mathcal{M}_t}, K, t, T)\varphi(z)\,dz \\
&= S_0 e^{-D_T}\{N_2(-d_1(H, t), d_1(K, T);\, -\Sigma_t/\Sigma_T) \\
&\qquad\qquad - N_2(-d_1(L, t), d_1(K, T);\, -\Sigma_t/\Sigma_T)\} \\
&\quad - Ke^{-R_T}\{N_2(-d_2(H, t), d_2(K, T);\, -\Sigma_t/\Sigma_T) \\
&\qquad\qquad - N_2(-d_2(L, t), d_2(K, T);\, -\Sigma_t/\Sigma_T)\}.
\end{aligned}$$

Here we have used the well-known Black–Scholes formula (see Sections 1.2 and 1.3) for a plain vanilla call $C(S_t, K, t, T)$ at time t with strike K and expiry T, as well as an identity (A.2) for $\int e^{\delta z} N(\alpha z + \beta)\varphi(z)\,dz$ given in Appendix A.3. Accordingly, the fair price FIC of a fade-in call based on a sequence $0 \leqslant t_1 < t_2 < \cdots < t_N \leqslant T$ is given by

$$\text{FIC} = \frac{1}{N}\sum_{i=1}^{N}\text{FIC}(t_i).$$

Fade-in put The price FIP(t) of the put subject to $S_t \in [L, H]$ is

$$\begin{aligned}
\text{FIP}(t) &= Ke^{-R_T}\{N_2(-d_2(H, t), -d_2(K, T);\, \Sigma_t/\Sigma_T) \\
&\qquad\qquad - N_2(-d_2(L, t), -d_2(K, T);\, \Sigma_t/\Sigma_T)\} \\
&\quad - S_0 e^{-D_T}\{N_2(-d_1(H, t), -d_1(K, T);\, \Sigma_t/\Sigma_T) \\
&\qquad\qquad - N_2(-d_1(L, t), -d_1(K, T);\, \Sigma_t/\Sigma_T)\}.
\end{aligned}$$

The price of a fade-in put based on $0 \leq t_1 < t_2 < \cdots < t_N \leq T$ is

$$\text{FIP} = \frac{1}{N} \sum_{i=1}^{N} \text{FIP}(t_i).$$

3.3 Fade-In Barrier Options

Our closed-form pricing formulas for fade-in barrier options will assume that the drift and volatility parameters r, d and σ are constant on $[0, t]$ as well as on $(t, T]$, where t denotes a fade-in date. In practice, one wants to compare the fade-in barrier prices with standard barrier options. The analytic pricing formula for standard barrier options, however, assumes constant parameters on $[0, T]$ (see Section 2.4). For consistency, it might be advisable to make this assumption for the fade-in barrier options as well.

Here we use the slightly more general set-up which, in turn, can be used to price barrier options with two different sets of parameters for the two time intervals $[0, t]$ and $(t, T]$ as well as two barrier levels.

As a building block for the pricing formula, we define a payoff X by

$$X = S_T^\alpha \, 1_{\{S_T \leq P_2\}} \, 1_{\{S_u \neq B_2, \, u \in [t, T]\}}.$$

We first calculate the expectation of not hitting an upper barrier and maturing below a strike, multiplied by the asset to the power α. Using equation (1.6), we obtain

$$\mathbb{E}[X \mid \mathcal{F}_t] = S_t^\alpha \exp\left(\alpha \mathcal{M}_{tT} + \frac{\alpha^2 \Sigma_{tT}^2}{2}\right)$$

$$\cdot \left\{ N\left(\frac{1}{\Sigma_{tT}} \log \frac{P_2}{S_t} - \mathcal{M}_{tT} - \alpha \Sigma_{tT}^2\right)\right.$$

$$\left. - \exp\left[2 \log \frac{B_2}{S_t}\left(\frac{\mathcal{M}_{tT}}{\Sigma_{tT}^2} + \alpha\right)\right] N\left(\frac{1}{\Sigma_{tT}} \log \frac{P_2 S_t}{B_2 B_2} - \mathcal{M}_{tT} - \alpha \Sigma_{tT}^2\right)\right\}$$

if $S_t \leq B_2$ and $P_2 < B_2$. The next step is to multiply X by a function Y depending on the path up to t, namely

$$Y = 1_{\{S_t \leq P_1\}} \, 1_{\{S_u \neq B_1, \, u \in [0, t]\}},$$

and calculate $\mathbb{E}[XY]$. We assume that the barrier levels B_1 and B_2 are higher than the adjoining point barrier levels P_1 and P_2, ie

$$P_1 \leq \min(B_1, B_2), \qquad P_2 \leq B_2, \qquad S_0 < B_1,$$

as shown in Figure 3.2.

We introduce the notation

$$D^{\text{out}}(\alpha, t, T, B_1, B_2, P_1, P_2)$$

$$= e^{-R_T} \mathbb{E}[S_T^\alpha \, 1_{\{S_t \leq P_1\}} \, 1_{\{S_T \leq P_2\}} \, 1_{\{S_u \neq B_1, \, u \in [0, t]\}} \, 1_{\{S_u \neq B_2, \, u \in [t, T]\}}],$$

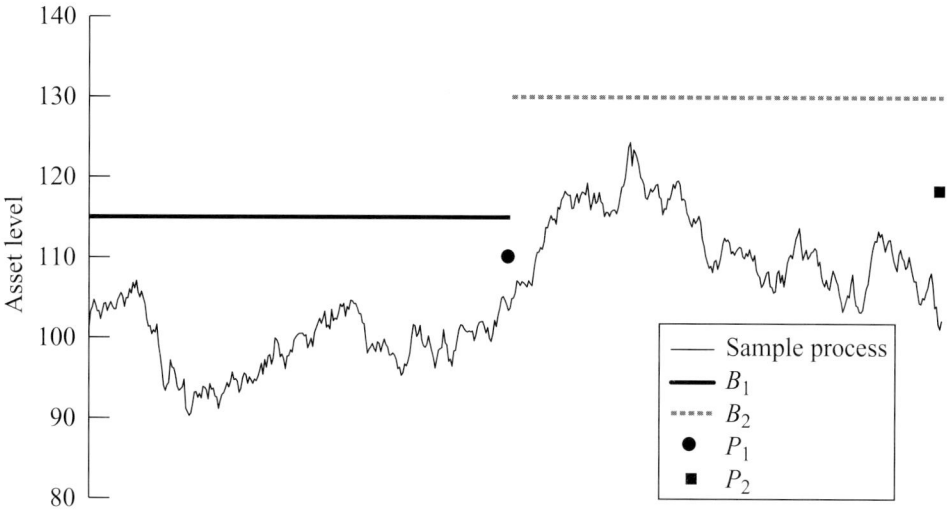

3.2 Digital barrier and fade-in event.

where D^{out} stands for the digital up-and-out barrier. We calculate

$$e^{R_T}D^{out}(\alpha, t, T, B_1, B_2, P_1, P_2)$$
$$= \mathbb{E}[\, X\, \mathbb{E}[\, Y \mid \mathcal{F}_t\,]\,]$$
$$= S_0^\alpha e^{\alpha \mathcal{M}_T + \alpha^2 \Sigma_{tT}^2/2} \int_{-\infty}^{-d_2(P_1, t)} \varphi(z) e^{\alpha \Sigma_t z}$$
$$\cdot \left\{ 1 - \exp\left[\frac{2}{\Sigma_t^2}\log\frac{B_1}{S_0}\left(\Sigma_t z + \mathcal{M}_t - \log\frac{B_1}{S_0}\right)\right]\right\}$$
$$\cdot \left\{ N\left(\frac{1}{\Sigma_{tT}}\left(\log\frac{P_2}{S_0} - \mathcal{M}_T - \Sigma_t z - \alpha\Sigma_{tT}^2\right)\right)\right.$$
$$- \exp\left[2\left(\log\frac{B_2}{S_0} - \mathcal{M}_t - \Sigma_t z\right)\left(\frac{\mathcal{M}_{tT}}{\Sigma_{tT}^2} + \alpha\right)\right]$$
$$\left. \cdot N\left(\frac{1}{\Sigma_{tT}}\left(\log\frac{P_2 S_0}{B_2 B_2} + \mathcal{M}_t + \Sigma_t z - \mathcal{M}_{tT} - \alpha\Sigma_{tT}^2\right)\right)\right\} dz$$
$$= \sum_{i=1}^{4} f_i N_2(a_i, b_i, \rho_i),$$

with factors f given by

$$f_1 = S_0^\alpha e^{\alpha \mathcal{M}_T + \alpha^2 \Sigma_T^2/2},$$
$$f_2 = -f_1 \left(\frac{B_1}{S_0}\right)^{2(\mathcal{M}_t/\Sigma_t^2 + \alpha)},$$
$$f_3 = -f_1 \exp\left[2\left(\frac{\mathcal{M}_{tT}}{\Sigma_{tT}^2} + \alpha\right)\left(\log\frac{B_2 S_0}{B_1 B_1} - \mathcal{M}_t + \frac{\Sigma_t^2}{\Sigma_{tT}^2}\mathcal{M}_{tT}\right)\right],$$
$$f_4 = -f_3 \left(\frac{B_1}{S_0}\right)^{2(\mathcal{M}_t/\Sigma_t^2 + \alpha)},$$

and N_2 arguments given by

$$a_1 = -d_2(P_1, t) - \alpha\Sigma_t, \qquad a_2 = a_1 - \frac{2}{\Sigma_t}\log\frac{B_1}{S_0},$$

$$a_3 = a_1 + 2\Sigma_t\left(\frac{\mathcal{M}_{tT}}{\Sigma_{tT}^2} + \alpha\right), \qquad a_4 = a_3 - \frac{2}{\Sigma_t}\log\frac{B_1}{S_0},$$

$$b_1 = \frac{1}{\Sigma_T}\left(\log\frac{P_2}{S_0} - \mathcal{M}_T - \alpha\Sigma_T^2\right), \qquad b_2 = b_1 - \frac{2}{\Sigma_T}\log\frac{B_1}{S_0},$$

$$b_3 = \frac{2}{\Sigma_T}\left(\log\frac{S_0}{B_2} - \alpha\Sigma_T^2\right) - \frac{2\mathcal{M}_{tT}}{\Sigma_{tT}^2}, \qquad b_4 = b_3 - \frac{2}{\Sigma_T}\log\frac{B_1}{S_0}.$$

The correlations are given by

$$\rho_1 = \rho_2 = -\rho_3 = -\rho_4 = \frac{\Sigma_t}{\Sigma_T}.$$

For down-and-out options, we shall need a formula for

$$\mathrm{D}_{\mathrm{out}}(\alpha, t, T, B_1, B_2, P_1, P_2)$$
$$= e^{-R_T}\mathbb{E}\big[S_T^\alpha\,1_{\{S_t \geqslant P_1\}}\,1_{\{S_T \geqslant P_2\}}\,1_{\{S_u \neq B_1,\,u\in[0,t]\}}\,1_{\{S_u \neq B_2,\,u\in[t,T]\}}\big].$$

Following the lines of the above calculation, it is easy to see that

$$\mathrm{D}_{\mathrm{out}}(\alpha, t, T, B_1, B_2, P_1, P_2) = e^{-R_T}\sum_{i=1}^{4} f_i N_2(-a_i, -b_i, \rho_i),$$

where we assume

$$P_1 \geqslant \max(B_1, B_2), \qquad P_2 \geqslant B_2, \qquad S_0 > B_1.$$

We will express the fade-in barrier prices in terms of the quantities

$$\mathrm{D}_{\mathrm{out}}(\alpha, t, P_1, P_2) = \mathrm{D}_{\mathrm{out}}(\alpha, t, T, B, B, P_1, P_2)$$

and

$$\mathrm{D}^{\mathrm{out}}(\alpha, t, P_1, P_2) = \mathrm{D}^{\mathrm{out}}(\alpha, t, T, B, B, P_1, P_2)$$

given above, where T denotes maturity, K the strike price and B the single barrier level.

Fade-in up-and-out call Assuming a barrier level $B > S_0$ as well as $0 < L \leqslant H \leqslant B$ and $K < B$, we have

$$\mathrm{FIC}^{\mathrm{out}} = e^{-R_T}\frac{1}{N}\sum_{i=1}^{N}\{\mathrm{D}^{\mathrm{out}}(1, t_i, H, B) - \mathrm{D}^{\mathrm{out}}(1, t_i, L, B)$$
$$- \mathrm{D}^{\mathrm{out}}(1, t_i, H, K) + \mathrm{D}^{\mathrm{out}}(1, t_i, L, K)$$
$$- K\mathrm{D}^{\mathrm{out}}(0, t_i, H, B) + K\mathrm{D}^{\mathrm{out}}(0, t_i, L, B)$$
$$+ K\mathrm{D}^{\mathrm{out}}(0, t_i, H, K) - K\mathrm{D}^{\mathrm{out}}(0, t_i, L, K)\}.$$

Fade-in up-and-out put Assuming a barrier level $B > S_0$ as well as $0 < L \leqslant H \leqslant B$, we have

$$\text{FIP}^{\text{out}} = e^{-R_T} \frac{1}{N} \sum_{i=1}^{N} \{ K \text{D}^{\text{out}}(0, t_i, H, \min(K, B)) \\ - K \text{D}^{\text{out}}(0, t_i, L, \min(K, B)) \\ - \text{D}^{\text{out}}(1, t_i, H, \min(K, B)) \\ + \text{D}^{\text{out}}(1, t_i, L, \min(K, B)) \}.$$

Fade-in down-and-out put Assuming a barrier level $0 < B < S_0$ as well as $B \leqslant L \leqslant H$ and $K > B$, we have

$$\text{FIP}_{\text{out}} = -e^{-R_T} \frac{1}{N} \sum_{i=1}^{N} \{ \text{D}_{\text{out}}(1, t_i, H, B) - \text{D}_{\text{out}}(1, t_i, L, B) \\ - \text{D}_{\text{out}}(1, t_i, H, K) + \text{D}_{\text{out}}(1, t_i, L, K) \\ - K \text{D}_{\text{out}}(0, t_i, H, B) + K \text{D}_{\text{out}}(0, t_i, L, B) \\ + K \text{D}_{\text{out}}(0, t_i, H, K) - K \text{D}_{\text{out}}(0, t_i, L, K) \}.$$

Fade-in down-and-out call Assuming a barrier level $0 < B < S_0$ as well as $B \leqslant L \leqslant H$, we have

$$\text{FIC}_{\text{out}} = -e^{-R_T} \frac{1}{N} \sum_{i=1}^{N} \{ K \text{D}_{\text{out}}(0, t_i, H, \max(K, B)) \\ - K \text{D}_{\text{out}}(0, t_i, L, \max(K, B)) \\ - \text{D}_{\text{out}}(1, t_i, H, \max(K, B)) \\ + \text{D}_{\text{out}}(1, t_i, L, \max(K, B)) \}.$$

Fade-out barrier options Fade-out barrier options are defined (and priced) as the difference between the underlying barrier option and the corresponding fade-in option.

In barriers An option with an in barrier can be written as the difference between an option without a barrier and the corresponding option with an out barrier.

3.4 Chooser Options

Let two plain vanilla options be given, one being a call and the other a put. Assume that the prices of these two options are similar. As the market moves, one option will eventually become worth more than the other. This holds, in particular, if the strikes and maturities coincide, since at any time only one of the two options is in-the-money. A *chooser* option gives the holder the right to choose whether to enter into the call or the put contract. This is also known as an *as-you-like-it* option.

Special case If both the call and the put have the same strike K and maturity T, the chooser option can be constructed using call–put parity. At chooser date t, we have

$$\max\{C(S_t, K, t, T), P(S_t, K, t, T)\}$$
$$= \max\{C(S_t, K, t, T), C(S_t, K, t, T) + Ke^{-R_t T} - S_0 e^{-D_t T}\}$$
$$= C(S_t, K, t, T) + (Ke^{-R_t T} - S_t e^{-D_t T})^+$$
$$= C(S_t, K, t, T) + e^{-D_t T}(Ke^{-R_t T + D_t T} - S_t)^+,$$

where $C(S_s, K, s, T)$ and $P(S_s, K, s, T)$ denote the prices at time s of a call and a put with spot S_s, maturity T and strike K, respectively. Therefore today's price [64] of the chooser option is

$$C(S_0, K, 0, T) + e^{-D_t T} P(S_0, Ke^{-R_t T + D_t T}, 0, t).$$

General case Let $C(S_s, K_c, s, T_c)$ denote the price of a call with strike K_c, maturity T_c and asset price S_s at evaluation date s, and let $P(S_s, K_p, s, T_p)$ denote a corresponding put. At choosing date t, the value of the chooser option is

$$\max\{C(S_t, K_c, t, T_c), P(S_t, K_p, t, T_p)\},$$

where S_t is the value of the asset S at time t. Since calls are increasing and puts are decreasing functions of S_t, there is a unique value S^* that satisfies

$$C(S^*, K_c, t, T_c) = P(S^*, K_p, t, T_p).$$

The option holder will choose the call whenever $S_t > S^*$ and the put whenever $S_t < S^*$. Hence, the fair value of the chooser option today is

$$\int_0^{S^*} P(x, K_p, t, T_p)\mathbb{P}[S_t \in dx] + \int_{S^*}^{\infty} C(x, K_c, t, T_c)\mathbb{P}[S_t \in dx].$$

It is easy to determine S^* using bisection. Evaluation of the integrals gives

$$\int_0^{S^*} P(x, K_p, t, T_p)\mathbb{P}[S_t \in dx] = e^{-R_{T_p}} K_p N_2\left(-d_2(S^*, t), -d_2(K_p, T_p); -\frac{\Sigma_t}{\Sigma_{T_p}}\right)$$
$$- e^{-D_{T_p}} S_0 N_2\left(-d_1(S^*, t), -d_1(K_p, T_p); -\frac{\Sigma_t}{\Sigma_{T_p}}\right)$$

and

$$\int_{S^*}^{\infty} C(x, K_c, t, T_c)\mathbb{P}[S_t \in dx] = -e^{-R_{T_c}} K_c N_2\left(d_2(S^*, t), d_2(K_c, T_c); -\frac{\Sigma_t}{\Sigma_{T_c}}\right)$$
$$+ e^{-D_{T_c}} S_0 N_2\left(d_1(S^*, t), d_1(K_c, T_c); -\frac{\Sigma_t}{\Sigma_{T_c}}\right),$$

where N_2 is the bivariate cumulative normal function and where

$$d_1(K, t) = \frac{\log(S_0/K) + \tilde{\mathcal{N}}_t}{\Sigma_t} \quad \text{and} \quad d_2(K, t) = \frac{\log(S_0/K) + \mathcal{N}_t}{\Sigma_t}.$$

Here we have used the formula derived in Appendix A.2.

Volatility interpolation Since this product has two strikes and maturities, it is not clear which implied volatility should be used. The goal is to obtain a chooser option price close to the market price of the call (put) whenever the choice is likely to be the call (put).

Going beyond the Black–Scholes framework, one approach would be to retrieve the volatility σ_t up to the chooser date for a strike level $\frac{1}{2}(K_c + K_p)$ from all the market implied volatilities.

The value S^* is then determined using the forward volatilities calculated from that volatility and the implied volatilities of the call and put option. If one of the forward volatilities becomes negative, we replace the arithmetic average for the chooser date volatility interpolation by one of the two strikes.

3.5 Prolongation Options

A *prolongation* option consists of a plain vanilla option and a *prolongation feature*. The prolongation feature only comes into life if the plain vanilla option expires worthless, in which case the holder of the option will receive a new plain vanilla option.

Note that, in contrast to an *instalment* option, the prolongation feature will be paid for completely at the start of the plain vanilla option.

We distinguish two variants of the product.

- *Straight prolongation*: the strike of the new option is fixed at time 0.

- *Refixed prolongation*: the strike of the new option will be fixed at t at a certain percentage in- or out-of-the-money. This percentage (given by a constant factor α) has to be stipulated at time 0.

Refixed prolongation We first consider the simpler case of a *call* with *refixed prolongation*. The value of the call at t is

$$(S_t - K)^+ + X,$$

where
$$X = \mathbb{E}\big[e^{-R_{tT}}(S_T - \alpha S_t)^+ 1_{\{S_t < K\}} \mid \mathcal{F}_t\big] = S_t C(1, \alpha, t, T) 1_{\{S_t < K\}}$$

is the value at t of the prolongation feature and $C(S, K, s, T)$ denotes the price at s of a call with strike K, maturity T and spot S (see Section 2.1.2). The fair value of X today is

$$C(1, \alpha, t, T) \mathbb{E}[e^{-R_t} S_t 1_{\{S_t < K\}}] = S_0 e^{-D_t} C(1, \alpha, t, T) N(-d_1(K, t)),$$

with
$$d_1(K, t) = \frac{\log(S_0/K) + \tilde{\mathcal{N}}_t}{\Sigma_t} \quad \text{and} \quad d_2(K, t) = \frac{\log(S_0/K) + \mathcal{N}_t}{\Sigma_t},$$

and where we have used the results in Section 1.1.12. This is very much in the spirit of the asset-or-nothing call (see Section 2.3.2).

Hence the price of a call with refixed prolongation is

$$C(S_0, K, 0, t) + S_0 e^{-D_t} C(1, \alpha, t, T) N(-d_1(K, t)).$$

The corresponding put has the fair value

$$P(S_0, K, 0, t) + S_0 e^{-D_t} P(1, \alpha, t, T) N(d_1(K, t)).$$

Straight prolongation We now take a look at a *call* with *straight prolongation*. We assume the strike K to be refixed at a level $\bar{K}$. Its value at prolongation date t is

$$(S_t - K)^+ + X,$$

with

$$X = \mathbb{E}\big[e^{-R_{tT}}(S_T - \bar{K})^+ 1_{\{S_t < K\}} \mid \mathcal{F}_t\big] = 1_{\{S_t < K\}} C(S_t, \bar{K}, t, T)$$

denoting the value at t of the prolongation feature. Using the Black–Scholes formula (Section 2.1.2) for C as well as (A.2) with

$$\alpha = \Sigma_t / \Sigma_{tT}, \qquad \gamma = -d_2(K, t),$$
$$\beta_1 = \beta_2 + \tfrac{1}{2}\Sigma_{tT}, \qquad \beta_2 = d_2(\bar{K}, T),$$
$$f_1 = e^{-D_T - \Sigma_t^2/2} S_0, \qquad f_2 = -e^{-R_T} \bar{K},$$

we obtain the present value of the prolongation feature of the call

$$e^{-R_t}\mathbb{E}[X] = f_1 \int_{-\infty}^{\gamma} e^{\Sigma_t x} N(\alpha x + \beta_1) \varphi(x)\, dx + f_2 \int_{-\infty}^{\gamma} N(\alpha x + \beta_2) \varphi(x)\, dx$$
$$= e^{-D_T} S_0 N_2\big(-d_1(K, t), d_1(\bar{K}, T); -\Sigma_t / \Sigma_T\big)$$
$$- e^{-R_T} \bar{K} N_2\big(-d_2(K, t), d_2(\bar{K}, T); -\Sigma_t / \Sigma_T\big).$$

The corresponding put has the fair value

$$P(S_0, K, 0, t) + e^{-R_T} \bar{K} N_2\big(d_2(K, t), -d_2(\bar{K}, T); -\Sigma_t / \Sigma_T\big)$$
$$- e^{-D_T} S_0 N_2\big(d_1(K, t), -d_1(\bar{K}, T); -\Sigma_t / \Sigma_T\big).$$

3.6 Improving Options

An *improving* option has a maturity date T and a strike K like a plain vanilla option. In addition, a refixing date t prior to the maturity date as well as a factor α are stipulated at the start of the option. At the refixing date, the strike of the option is changed to α times the current spot if this is profitable for the holder of the option.

Derivation of formulas The payoff at maturity T of an improving call is given by

$$X = \big(S_T - \min(K, \alpha S_t)\big)^+.$$

Closed-Form Solutions for Non-Standard Products

At refixing date t, the value of the option is

$$e^{-R_{tT}} \mathbb{E}[X \mid \mathcal{F}_t] = C(S_t, \min(K, \alpha S_t), t, T)$$
$$= C(S_t, \alpha S_t, t, T) 1_{\{S_t \leq K/\alpha\}} + C(S_t, K, t, T) 1_{\{S_t > K/\alpha\}},$$

with $C(S, K, s, T)$ denoting the price at s of a call with strike K, maturity T and spot S. The value today of these terms can be obtained using the Black–Scholes formula (Section 2.1.2) and (A.2). Alternatively, we remark that

$$\left(S_T - \min(K, \alpha S_t)\right)^+ = (S_T - K)^+ + 1_{\{S_t \leq K/\alpha\}}[(S_T - \alpha S_t)^+ - (S_T - K)^+].$$

This means that an improving call can be synthesised by entering into a long position of a plain vanilla call and a refixing prolongation call and a short position in a straight prolongation call, as described in Section 3.5. A corresponding result holds for improving puts, since

$$\left(\max(K, \alpha S_t) - S_T\right)^+ = (K - S_T)^+ + 1_{\{S_t \geq K/\alpha\}}[(\alpha S_t - S_T)^+ - (K - S_T)^+].$$

The pricing formulas are then readily obtainable.

3.7 Power and Powered Options

A *power* option has the payoff of a plain vanilla option with the price of the underlying asset raised to the power α, where α is a positive exponent. In the case of a *powered* option, the payoff rather than the asset is raised to the power α. The parameter α is called the *gearing parameter*.

Power options The payoff of a power call at maturity T is

$$(S_T^\alpha - K)^+.$$

Since the asset at maturity is lognormally distributed, we have

$$S_T^\alpha = S_0^\alpha \exp\left(\alpha(R_T - D_T - \tfrac{1}{2}\Sigma_T^2 + \Sigma_T X)\right)$$
$$= S_0^\alpha \exp\left(R_T - [(1-\alpha)R_T + \alpha D_T - \tfrac{1}{2}\alpha(\alpha - 1)\Sigma_T^2] - \tfrac{1}{2}(\alpha\Sigma_T)^2 + \alpha\Sigma_T X\right),$$

where X is a standard normally distributed random variable. Therefore the option price is given by the Black–Scholes formula (Section 2.1.2) with modified dividend yield

$$D_T^{\text{pow}}(\alpha) = (1 - \alpha)R_T + \alpha D_T - \tfrac{1}{2}\alpha(\alpha - 1)\Sigma_T^2$$

and volatility $\alpha \Sigma_T$. Alternatively, Itô's lemma gives the same result via

$$d(\log S_t^\alpha) = \alpha\, d(\log S_t).$$

Powered options The payoff of a capped powered call at maturity T is

$$\left((S_T - K)^+\right)^\alpha 1_{\{S_T < H\}}.$$

We start by assuming the simple case of a zero strike, ie $K = 0$. Using a version of Girsanov's lemma, we obtain

$$\mathbb{E}[\, S_T^\alpha \, 1_{\{S_T < H\}} \,] = S_0^\alpha \exp\bigl(R_T - D_T^{\text{pow}}(\alpha)\bigr) \, \mathbb{P}[\, \tilde{S}_T < H \,],$$

where $\tilde{S}_T$ is related with S_T via

$$\log \frac{\tilde{S}_T}{S_T} = \alpha \Sigma_T.$$

The probability of $\{\tilde{S}_T < H\}$ is given by

$$\mathbb{P}[\, \tilde{S}_T < H \,] = N\!\left(\frac{1}{\Sigma_T} \left(\log \frac{H}{S_0} + \tfrac{1}{2}\Sigma_T^2 - R_T + D_T \right) - \alpha \Sigma_T \right).$$

In order to treat the case with strike, we use the Maclaurin expansion

$$(1 - x)^\alpha = \sum_{i=0}^{\infty} \binom{\alpha}{i} (-x)^i,$$

where the coefficients are given via the recursion

$$\binom{\alpha}{0} = 1, \qquad \binom{\alpha}{i} = \binom{\alpha}{i-1} \frac{\alpha - (i-1)}{i}, \quad i \geqslant 1.$$

This series converges for $|x| < 1$. Applying this identity yields

$$\mathbb{E}\bigl[\, ((S_T - K)^+)^\alpha \, 1_{\{S_T < H\}} \,\bigr]$$
$$= \mathbb{E}\bigl[\, 1_{\{K < S_T < H\}} (S_T - K)^\alpha \,\bigr]$$
$$= \sum_{i=0}^{\infty} \binom{\alpha}{i} (-1)^i K^i \bigl(\mathbb{E}[\, 1_{\{S_T < H\}} S_T^{\alpha - i} \,] - \mathbb{E}[\, 1_{\{S_T < K\}} S_T^{\alpha - i} \,] \bigr). \quad (3.2)$$

We have thus reduced the case with strike to the zero strike case treated above. For moderately sized α, eg $\alpha < 4$, the infinite series converges reasonably fast.

For the corresponding put

$$X = \bigl((K - S_T)^+\bigr)^\alpha \, 1_{\{S_T > L\}},$$

similar considerations lead to

$$\mathbb{E}[\, X \,] = \sum_{i=0}^{\infty} \binom{\alpha}{i} (-1)^i K^{\alpha - i} \bigl(\mathbb{E}[\, 1_{\{S_T < K\}} S_T^i \,] - \mathbb{E}[\, 1_{\{S_T < L\}} S_T^i \,] \bigr)$$

for $K > L$.

Usage of the options The interesting thing about power and powered options is their non-linear payoff and the increased leverage they consequently offer. For α significantly larger than 1, both power and powered options offer a larger maximal payoff per amount invested than plain vanilla options. As the greeks can also become large, power and powered options offer a comparatively cheap way to buy vega and gamma.

Closed-Form Solutions for Non-Standard Products

Practical pricing and hedging issues Since greeks can become large (especially for a large gearing parameter α), it is advisable to use only capped variants of the product. In the case of a power option, capping can be done by going long one power option and going short another power option with a different strike and the same gearing α. In the case of a powered option, capping can be accomplished by writing the payoff of a capped powered call option as

$$1_{\{M \geq S_T\}} \left((S_T - K)^+\right)^\alpha e^{-r(T-t)} + 1_{\{M < S_T\}} (M - K)^\alpha e^{-r(T-t)}, \quad M > K.$$

This means that a powered call option with a gearing of α and a strike of K capped at M can be priced by formula (3.2) and $(M - K)^\alpha$ digital call options with a strike of M. The powered put option is capped as

$$1_{\{S_T \geq M\}} \left((K - S_T)^+\right)^\alpha e^{-r(T-t)} + 1_{\{M > S_T\}} (K - M)^\alpha e^{-r(T-t)}, \quad K > M.$$

Further hedging issues are discussed in Section 10.3.5.

3.8 Compound Options

A *compound* option gives its holder the right to buy (or to sell) a plain vanilla option that expires at T and has strike K at some future date t for an agreed strike price k. In other words, it is an option to purchase an option.

We give a pricing formula for a European compound option on a European option [27]. American- or mixed-style compounds can be valued on a tree, using the closed-form European compound option prices as a control variate.

An iteration of this idea leads to *instalment* options: at stipulated dates t_i the holder of an instalment call has to pay instalments K_i ($i = 1, 2, \ldots, n - 1$) in order to receive the final option payoff $(S_{t_n} - K_n)^+$ at maturity $T = t_n$. These options can be priced on a tree (see Chapter 6).

Applications of compound options include:

- protection against a drop (rise) of volatility;

- deals involving several instalment payments rather than a single strike payment at maturity;

- structured trades using options with guaranteed repurchase price.

Pricing formula At time t, the holder of a call on a call will pay the instalment amount k if the value of the call with maturity T and strike K is greater than k. Since the call price increases with the stock price, there is a unique value S^* for the stock at time t such that the call price at time t equals k. This value can easily be determined by applying the Newton–Raphson method [61] to the plain vanilla formula. The fair value C^{call} of the compound option is given by

$$C^{\text{call}} = e^{-R_T} \mathbb{E}\left[(S_T - K)^+ 1_{\{S_t > S^*\}}\right] - e^{-R_t} \mathbb{E}[k 1_{\{S_t \leq S^*\}}].$$

The first term is given by

$$e^{-R_t} \int_{S^*}^{\infty} C(S_0 e^{\Sigma_t z + \mathcal{M}_t}, K, t, T) \varphi(z) \, dz$$
$$= S_0 e^{-D_T} N_2\big(-d_1(S^*, t), d_1(K, T); -\Sigma_t/\Sigma_T\big)$$
$$- K e^{-R_T} N_2\big(-d_2(S^*, t), d_2(K, T); -\Sigma_t/\Sigma_T\big),$$

where we have used the notation

$$d_1(K, t) = d_2(K, t) + \Sigma_t \quad \text{and} \quad d_2(K, t) = \frac{1}{\Sigma_t}\left(\log \frac{S_0}{K} + \mathcal{N}_t\right)$$

(compare the derivation in Section 3.2). The remaining term has previously been calculated. We can now summarise the pricing formula for the combinations of call or put options on calls or puts:

$$E_1^{E_2} = \chi_1 \chi_2 S_0 e^{-D_T} N_2\big(-\chi_1 \chi_2 d_1(S^*, t), \chi_2 d_1(K, T); -\chi_1 \Sigma_t/\Sigma_T\big)$$
$$- \chi_1 \chi_2 K e^{-R_T} N_2\big(-\chi_1 \chi_2 d_2(S^*, t), \chi_2 d_2(K, T); -\chi_1 \Sigma_t/\Sigma_T\big)$$
$$- \chi_1 k e^{-R_t} N\big(\chi_1 \chi_2 d_2(S^*, t)\big).$$

where χ_1 is 1 if the compound option is a call and -1 if it is a put. Similarly, χ_2 is 1 if the underlying option is call, and -1 if it is a put. Finally, $S^*_{\chi_2}$ is the stock price such that, at time t, the underlying option has a value of k.

Chapter 4
Closed-Form Solutions for Multi-Asset Products

Many exotic options traded today involve two or even more assets. The basic pricing methodology extends to the multi-asset case. The main feature that distinguishes these products from single-underlying products is that correlation has to be taken into account. In practice, it is very hard to hedge against movements in correlation; therefore traders often rely on a conservative estimation of the historical correlation.

In this chapter we shall present closed-form and near-closed-form solutions for various European multi-asset options [42, 51, 71]. Throughout we use the notation developed in Sections 1.2.1 and 1.3.1.

It has to be noted that, typically, we have to evaluate an n-variate cumulative normal distribution function, where n denotes the number of underlyings involved. This is only feasible if n is moderate. For higher dimensions, a Monte Carlo approach (see Chapter 7) is often preferable.

4.1 Exchange Options

One of the simplest multi-asset products is the option to exchange one asset for another. This means that at option maturity the option holder has the right to receive an asset S^1 in exchange for an asset S^2. Thus, the payoff at maturity T is given by

$$(S_T^1 - S_T^2)^+.$$

This option is also called a Margrabe option, after W. Margrabe, who gave its first pricing formula in [51].

In general, if X_1 and X_2 are random variables such that $\log X_1$ and $\log X_2$ are binormally distributed with means μ_1 and μ_2, variances σ_1 and σ_2, and correlation ρ, we have

$$\mathbb{E}[(X_1 - X_2)^+] = e^{\mu_1 + \sigma_1^2/2} N(d_1) - e^{\mu_2 + \sigma_2^2/2} N(d_2),$$

with

$$d_1 = \frac{\mu_1 - \mu_2 + \sigma_1^2 - \rho\sigma_1\sigma_2}{\tilde{\sigma}} \quad \text{and} \quad d_2 = \frac{\mu_1 - \mu_2 + \sigma_2^2 + \rho\sigma_1\sigma_2}{\tilde{\sigma}},$$

where $\tilde{\sigma}$ is given by

$$\tilde{\sigma}^2 = \sigma_1^2 + \sigma_2^2 - 2\rho\sigma_1\sigma_2.$$

So, in our case, this means that if we set

$$\tilde{\Sigma}_T^2 = (\Sigma_T^1)^2 - 2\Psi_T + (\Sigma_T^2)^2$$

then the value of an exchange option is given by

$$e^{-D_T^1} N(d_1) - e^{-D_T^2} N(d_2),$$

with

$$d_1 = \frac{\log(S_0^1/S_0^2) + D_T^2 - D_T^1 + (\Sigma_T^1)^2 - \Psi_T}{\tilde{\Sigma}_T}$$

and

$$d_2 = \frac{\log(S_0^1/S_0^2) + D_T^2 - D_T^1 + (\Sigma_T^2)^2 + \Psi_T}{\tilde{\Sigma}_T}.$$

4.2 Relative Digital Options

A *relative digital* option pays one unit if the price of stock 1 is higher than the price of stock 2. More specifically, a call has the payoff X with

$$X = 1_{\{S_T^1 > \lambda S_T^2\}}$$

and the corresponding put pays

$$X = 1_{\{S_T^1 < \lambda S_T^2\}}$$

at maturity T, where $\lambda > 0$ is a given constant. These options can be priced like standard digital options (see Section 2.3) once the parameters of the lognormal ratio process have been determined as in the following. More specifically, a payoff condition $S_T^1 > \lambda S_T^2$ translates to a condition on the ratio $\tilde{S} = S_T^1/S_T^2 > \lambda$.

Ratio and product processes If S_t^1 and S_t^2 are two lognormal processes, ie

$$S_t^i = S_0^i \exp(\mathcal{N}_t^i + \Sigma_t^i W_t^i), \quad i = 1, 2,$$

where W_t^1 and W_t^2 are two standard Brownian processes with correlation ρ_t, then the ratio process S^1/S^2, by the multi-dimensional version of Itô's lemma, satisfies the stochastic differential equation

$$d\log\left(\frac{S_t^1}{S_t^2}\right) = \left[d_t^2 + \tfrac{1}{2}(\sigma_t^2)^2 - d_t^1 - \tfrac{1}{2}(\sigma_t^1)^2\right] dt + \sigma_t^1 dW_t^1 - \sigma_t^2 dW_t^2.$$

However, the integral

$$\int_0^T (\sigma_u^1 dW_u^1 - \sigma_u^2 dW_u^2) du$$

is normally distributed with mean 0 and variance $\tilde{\Sigma}_T^2 = (\Sigma_T^1)^2 - 2\Psi_T + (\Sigma_T^2)^2$. Therefore the ratio can be written as

$$\frac{S_T^1}{S_T^2} = \frac{S_0^1}{S_0^2} \exp(\tilde{\mathcal{N}}_T + \tilde{\Sigma}_T X),$$

Closed-Form Solutions for Multi-Asset Products

where X denotes a standard normally distributed random variable. The volatility $\tilde{\Sigma}_T$ is

$$\tilde{\Sigma}_T^2 = (\Sigma_T^1)^2 - 2\Psi_T + (\Sigma_T^2)^2$$

and the drift $\tilde{\mathcal{N}}$ is given by

$$\tilde{\mathcal{N}}_T = \mathcal{N}_T^1 - \mathcal{N}_T^2.$$

For completeness, we also present results for a product process. The product $S^1 S^2$ is lognormal and can be written as

$$S_T^1 S_T^2 = S_0^1 S_0^2 \exp(\tilde{\mathcal{N}}_T + \tilde{\Sigma}_T X),$$

with drift

$$\tilde{\mathcal{N}}_T = \mathcal{N}_T^1 + \mathcal{N}_T^2$$

and volatility

$$\tilde{\Sigma}_T^2 = (\Sigma_T^1)^2 + 2\Psi_T + (\Sigma_T^2)^2.$$

4.3 Relative Outperformance Options

Outperformance options are options on the difference in performance of two potentially correlated assets. For example, if an investor believes that the French stock index CAC will have a better performance than the German index DAX, even if both decline, he or she might consider buying an outperformance option.

Outperformance options come in two flavours, either as *relative* or *absolute* variants. For the relative outperformance option, one considers the quotient of the performances of the two asset prices, while the absolute version looks at the difference in performance instead. Note that an absolute outperformance option has a strike, in contrast to a simple exchange option, which does not.

So a *relative outperformance* option is defined by the payoff

$$\left(\frac{S_T^1}{S_T^2} - K\right)^+$$

for a call and

$$\left(K - \frac{S_T^1}{S_T^2}\right)^+$$

for a put, with some strike K.

Example A German corporate client ABC issues its financial statements using euros. The client owns 5% of the US$-based stock of the American company XYZ. We want to solve the following problem:

- ABC wants to keep its stake in XYZ.

- For accounting reasons, ABC wants to protect the value of its stake in XYZ in euros.

- The stake of 5% in XYZ is too big to do a direct protection.

- ABC is bullish on XYZ's local market, ie on the North American market.

The following derivative product solves the problem.

ABC buys a quanto relative outperformance option of the S&P500 over XYZ, with strike and payoff in euros. So the payoff is

$$\max\left(\alpha \frac{S_T^1}{S_T^2} - K, 0\right),$$

where S^1 denotes the S&P500 index, S^2 denotes the XYZ stock price expressed in euros, α is a suitably chosen weighting and K is the strike in euros.

We now use the results from Section 4.2 on the ratio process. The value of the relative outperformance call and put are then the standard European call and put values with the underlying process being the lognormal ratio process. We reiterate the results below.

If X is normally distributed with mean μ and variance σ, then

$$\mathbb{E}[(e^X - K)^+] = e^{\mu + \sigma^2/2} N\left(\frac{-\log K + \mu + \sigma^2}{\sigma}\right) - KN\left(\frac{-\log K + \mu}{\sigma}\right).$$

Using the forward notation

$$F = \mathbb{E}[e^X] = e^{\mu + \sigma^2/2},$$

we can write alternatively

$$\mathbb{E}[(e^X - K)^+] = FN\left(\frac{\log(F/K) + \tfrac{1}{2}\sigma^2}{\sigma}\right) - KN\left(\frac{\log(F/K) - \tfrac{1}{2}\sigma^2}{\sigma}\right).$$

This gives us the call option value

$$C_{\text{rel-outp}} = e^{-R_T}[FN(d_1) - KN(d_2)],$$

with

$$F = \mathbb{E}\left[\frac{S_T^1}{S_T^2}\right] = \frac{S_0^1}{S_0^2} \exp(\mathcal{N}_T^1 - \mathcal{N}_T^2 + \tfrac{1}{2}\tilde{\Sigma}_T^2)$$

$$= \frac{S_0^1}{S_0^2} \exp(D_T^2 - D_T^1 + (\Sigma_T^2)^2 - \Psi_T^{12})$$

and

$$d_1 = \frac{\log(F/K) + \tfrac{1}{2}\tilde{\Sigma}_T^2}{\tilde{\Sigma}_T} \quad \text{and} \quad d_2 = \frac{\log(F/K) - \tfrac{1}{2}\tilde{\Sigma}_T^2}{\tilde{\Sigma}_T}.$$

Similarly, the corresponding put value is given by

$$P_{\text{rel-outp}} = e^{-R_T}[KN(-d_2) - FN(-d_1)].$$

4.4 Outperformance Options

As indicated earlier, an *(absolute) outperformance* option is an option on the weighted difference between the prices of two correlated assets. So, for example, an outperformance call pays at maturity T the difference between the weighted difference of the two assets and the strike. The strike is assumed to be positive.

If it is zero, we are back to the case of the exchange option, which we have already covered in Section 4.1.

In the present case, with a non-zero strike, a closed-form solution is not possible. However, our solution requires just the solution of an implicit one-dimensional equation and a one-dimensional numerical integration. Both can be performed very efficiently, the former using the Newton–Raphson method [61].

Outperformance call   The payoff of an outperformance call option is given by

$$X = (w_1 S_T^1 - w_2 S_T^2 - K)^+,$$

with strike $K > 0$ and weights $w_1 > 0$ and $w_2 > 0$. To simplify the calculations, we introduce the notation

$$M_1 = \log(w_1 S_0^1) + \mathcal{M}_T^1,$$
$$M_2 = \log(w_2 S_0^2) + \mathcal{M}_T^2.$$

Then the value C^{outp} of the option is given by

$$e^{R_T} C^{\text{outp}} = \int_{-\infty}^{\infty} \int_{-\infty}^{\infty} (e^{M_1 + \Sigma_T^1 x} - e^{M_2 + \Sigma_T^2 y} - K)^+ \varphi_2(x, y; \rho) \, dx \, dy$$

$$= e^{M_1} \int_{-\infty}^{\infty} \int_{A(y)}^{\infty} e^{\Sigma_T^1 x} \varphi_2(x, y; \rho) \, dx \, dy$$

$$- e^{M_2} \int_{-\infty}^{\infty} e^{\Sigma_T^2 y} \int_{A(y)}^{\infty} \varphi_2(x, y; \rho) \, dx \, dy$$

$$- K \int_{-\infty}^{\infty} \int_{A(y)}^{\infty} \varphi_2(x, y, \rho) \, dx \, dy$$

$$= e^{M_1 + (\Sigma_T^1)^2/2} \int_{-\infty}^{\infty} N\left(-\frac{A(y) - \rho y - \Sigma_T^1(1 - \rho^2)}{\sqrt{1 - \rho^2}}\right) \varphi(y - \rho \Sigma_T^1) \, dy$$

$$- e^{M_2} \int_{-\infty}^{\infty} e^{\Sigma_T^2 y} N\left(-\frac{A(y) - \rho y}{\sqrt{1 - \rho^2}}\right) \varphi(y) \, dy$$

$$- K \int_{-\infty}^{\infty} N\left(-\frac{A(y) - \rho y}{\sqrt{1 - \rho^2}}\right) \varphi(y) \, dy$$

$$= e^{M_1 + (\Sigma_T^1)^2/2} \int_{-\infty}^{\infty} N\left(-\frac{A(y + \rho \Sigma_T^1) - \rho y - \Sigma_T^1}{\sqrt{1 - \rho^2}}\right) \varphi(y) \, dy$$

$$- e^{M_2 + (\Sigma_T^2)^2/2} \int_{-\infty}^{\infty} N\left(-\frac{A(y + \Sigma_T^2) - \rho y - \rho \Sigma_T^2}{\sqrt{1 - \rho^2}}\right) \varphi(y) \, dy$$

$$- K \int_{-\infty}^{\infty} N\left(-\frac{A(y) - \rho y}{\sqrt{1 - \rho^2}}\right) \varphi(y) \, dy, \tag{4.1}$$

with $A(y)$ implicitly defined by

$$\exp(M_1 + \Sigma_T^1 A(y)) - \exp(M_2 + \Sigma_T^2 y) - K = 0.$$

We use (4.1) to compute the value of the outperformance call option by numerical integration. The functional form of the density $\varphi(y)$ allows this numerical integration to be performed using highly efficient specialised routines, such as the Gauss–Hermite numerical quadrature technique [61].

Note that if Σ_T^1, Σ_T^2 and K are all positive then, from the foregoing equation, $A(y)$ is well defined for all y and is given by

$$A(y) = \frac{1}{\Sigma_T^1}[\log(e^{M_2 + \Sigma_T^2 y} + K) - M_1] \qquad (4.2)$$

$$= \frac{1}{\Sigma_T^1}[\log(1 + Ke^{-M_2 - \Sigma_T^2 y}) + M_2 + \Sigma_T^2 y - M_1]. \qquad (4.3)$$

Even though expressions (4.2) and (4.3) are equivalent, they are very different with regard to their numerical stability. When $M_2 + \Sigma_T^2 y$ is negative, expression (4.2) is advantageous; but when it is positive, it is better to use expression (4.3).

Outperformance put The payoff of the corresponding outperformance put is defined as

$$X = e^{-R_T}(K - w_1 S_T^1 + w_2 S_T^2)^+.$$

By inspecting the computation above, one sees that its value can be obtained from formula (4.1) by changing the signs of the arguments of all the cumulative normal functions N, and then changing the sign of the resulting value.

4.5 European Digital Option on Best or Worst of Two Assets

Another useful product is the *digital option on the best of two assets* or *on the worst of two assets*. For example, a digital call on the best of two assets pays one unit of currency if the price of either of the two assets is above the strike K at maturity T, while the digital call on the worst pays one unit if both of the assets have a price above the strike. Similarly, a put on the worst of two assets pays one unit if the price of either of the two assets is below the strike K at maturity T. This product can also easily be generalised to the case of more than two assets. However, for simplicity, we present in the following just the case of two assets and leave the generalisation to the reader. See Section 4.6 for more details.

Derivation of formulas The payoff of a digital call on the best of two assets S^1 and S^2 with strike K and maturity T is given by

$$X = 1_{\{\max(S_T^1, S_T^2) \geq K\}}.$$

Hence the value $C^{\mathrm{dig}}_{\mathrm{best\text{-}of\text{-}2}}$ can be expressed as

$$e^{R_T} C^{\mathrm{dig}}_{\mathrm{best\text{-}of\text{-}2}} = \mathbb{P}[S_T^1 > S_T^2, S_T^1 \geq K] + \mathbb{P}[S_T^1 \leq S_T^2, S_T^2 \geq K]$$

$$= \mathbb{P}\left[\log\frac{S_T^1}{S_T^2} > 0, \log S_T^1 \geq \log K\right] + \mathbb{P}\left[\log\frac{S_T^1}{S_T^2} \leq 0, \log S_T^2 \geq \log K\right].$$

However, $\log(S_T^1/S_T^2)$ and $\log S_T^1$ are binormally distributed, with means

$$\tilde{M} = \mathbb{E}\left[\log\frac{S_T^1}{S_T^2}\right] = \log\frac{S_0^1}{S_0^2} + \mathcal{N}_T^1 - \mathcal{N}_T^2,$$

$$M^1 = \mathbb{E}[\log S_T^1] = \log S_0^1 + \mathcal{N}_T^1,$$

variances

$$\text{Var}\left[\log\frac{S_T^1}{S_T^2}\right] = \tilde{\Sigma}_T^2 = (\Sigma_T^1)^2 - 2\Psi_T + (\Sigma_T^2)^2,$$

$$\text{Var}[\log S_T^1] = (\Sigma_T^1)^2,$$

and covariance

$$\text{Cov}\left[\int_0^T \sigma_t^1 dW_t^1 - \int_0^T \sigma_t^2 dW_t^2, \int_0^T \sigma_t^1 dW_t^1\right] = \int_0^t \rho_s \sigma_s^1 \sigma_s^2 dt + (\Sigma_T^1)^2$$

$$= \Psi_T - (\Sigma_T^1)^2.$$

In addition, $\log(S_T^1/S_T^2)$ and $\log S_T^2$ are binormally distributed, with the mean M^2 and variances following directly from the above. Thus,

$$\mathbb{P}[S_T^1 > S_T^2, S_T^1 \geqslant K] = \mathbb{P}\left[\log\frac{S_T^1}{S_T^2} > 0, \log S_T^1 \geqslant \log K\right]$$

$$= N_2\left(\frac{\tilde{M}}{\tilde{\Sigma}_T}, \frac{M^1 - \log K}{\Sigma_T^1}; \frac{\Psi_T - (\Sigma_T^1)^2}{\tilde{\Sigma} \cdot \Sigma_T^1}\right).$$

Similarly,

$$\mathbb{P}[S_T^1 \leqslant S_T^2, S_T^2 \geqslant K] = N_2\left(\frac{-\tilde{M}}{\tilde{\Sigma}_T}, \frac{M^2 - \log K}{\Sigma_T^2}; \frac{\Psi_T - (\Sigma_T^2)^2}{\tilde{\Sigma}_T \cdot \Sigma_T^2}\right).$$

To summarise, we have

$$C_{\text{best-of-2}}^{\text{dig}} = e^{-R_T}\left[N_2\left(\frac{\tilde{M}}{\tilde{\Sigma}_T}, \frac{M^1 - \log K}{\Sigma_T^1}; \frac{\Psi_T - (\Sigma_T^1)^2}{\tilde{\Sigma}_T \cdot \Sigma_T^1}\right)\right.$$

$$\left.+ N_2\left(\frac{-\tilde{M}}{\tilde{\Sigma}_T}, \frac{M^2 - \log K}{\Sigma_T^2}; \frac{\Psi_T - (\Sigma_T^2)^2}{\tilde{\Sigma}_T \cdot \Sigma_T^2}\right)\right].$$

Summary of pricing formulas The price of a digital call on the worst of two, with payoff

$$1_{\{\min(S_T^1, S_T^2) \geqslant K\}},$$

and the corresponding put options are obtained analogously. We arrive at the following formulas:

$$C_{\text{best-of-2}}^{\text{dig}} = e^{-R_T}[N_2(a, b_1; \rho_1) + N_2(-a, b_2; \rho_2)],$$

$$C_{\text{worst-of-2}}^{\text{dig}} = e^{-R_T}[N_2(-a, b_1; -\rho_1) + N_2(a, b_2; -\rho_2)],$$

$$P_{\text{best-of-2}}^{\text{dig}} = e^{-R_T}[1 - N_2(a, b_1; \rho_1) - N_2(-a, b_2; \rho_2)],$$

$$P_{\text{worst-of-2}}^{\text{dig}} = e^{-R_T}[1 - N_2(-a, b_1; -\rho_1) - N_2(a, b_2; -\rho_2)],$$

with

$$a = \frac{\tilde{M}}{\tilde{\Sigma}_T},$$

$$b_1 = \frac{M^1 - \log K}{\Sigma_T^1}, \qquad b_2 = \frac{M^2 - \log K}{\Sigma_T^2},$$

$$\rho_1 = \frac{\Psi_T - (\Sigma_T^1)^2}{\tilde{\Sigma}_T \Sigma_T^1}, \qquad \rho_2 = \frac{\Psi_T - (\Sigma_T^2)^2}{\tilde{\Sigma}_T \Sigma_T^2}.$$

4.6 Best or Worst of Several Assets

Consider a customer who wishes to buy a call option on a stock and who has several stocks available that he might want as an underlying. Instead of buying the option on a particular stock, he could, for example, buy the option on the best of these underlyings. Of course, such an option will in general be relatively expensive. Another option he might buy is a call on the worst of these underlyings, if he expects *all* of them to perform very well. This option might be quite cheap.

So, a call on the best of several assets gives the holder the right to buy the most expensive of the assets at maturity T at a strike price K. Similarly, a call on the worst of several assets gives the right to buy the cheapest of the assets. Of course, there are corresponding puts on the best or on the worst of several assets.

Pricing We concentrate first on the case of a call on the best of n assets. The payoff is given by

$$X = \left(\max_{i=1,\ldots,n} S_T^i - K\right)^+.$$

Hence, the fair value C^{best} of this option is given by

$$e^{R_T} C^{\text{best}} = \mathbb{E}[X]$$

$$= \sum_{i=1}^n \mathbb{E}\left[S_T^i \mathbf{1}_{\{S_T^j < S_T^i \text{ for all } j \neq i\}} \mathbf{1}_{\{K < S_T^i\}}\right] - K\mathbb{P}\left[K < \max_{i=1,\ldots,n} S_T^i\right]$$

$$= \sum_{i=1}^n \left(e^{-D_T^i} S_0^i \tilde{\mathbb{P}}^i[S_T^j < S_T^i \text{ for all } j \neq i, K < S_T^i] \right.$$

$$\left. - e^{-R_T} K \mathbb{P}[\forall_{j:j\neq i} S_T^j < S_T^i \text{ for all } j \neq i, K < S_T^i]\right),$$

with $\tilde{\mathbb{P}}^i$ defined by

$$\frac{d\tilde{\mathbb{P}}^i}{d\mathbb{P}} = \exp(\sigma W_T^i - \tfrac{1}{2}\sigma_i^2 T).$$

Now we have

$$\mathbb{P}[S_T^j < S_T^i] = \mathbb{P}[\log S_0^j + \sigma_j W_T^j + \mu_j T < \log S_0^i + \sigma_i W_T^i + \mu_i T]$$

$$= N\left(\frac{\log(S_0^i/S_0^j) + (d_j - d_i + \frac{1}{2}\sigma_j^2 - \frac{1}{2}\sigma_i^2)T}{\sqrt{\sigma_j^2 - 2\sigma_j\sigma_i\rho_{ij} + \sigma_i^2}\sqrt{T}}\right)$$

$$= N\left(\frac{\log(S_0^i/S_0^j) + (d_j - d_i + \frac{1}{2}\sigma_j^2 - \frac{1}{2}\sigma_i^2)T}{\sigma_{ji}\sqrt{T}}\right)$$

$$= N(d_{ji}),$$

with

$$\sigma_{ji}^2 = \text{Cov}[\log S_T^j, \log S_T^i]/T = \sigma_j^2 - 2\sigma_j\sigma_i\rho_{ij} + \sigma_i^2 \quad (4.4)$$

if $i \neq j$, since

$$(W_T^j, W_T^i) \text{ with respect to } \mathbb{P} \stackrel{d}{=} \left(\sqrt{T}X, \sqrt{T}\left(\rho_{ij}X + \sqrt{1-\rho_{ij}^2}Y\right)\right),$$

with X and Y denoting two independent standard normally distributed random variables. Similarly, we obtain

$$\tilde{\mathbb{P}}^i[S_T^j < S_T^i] = \tilde{\mathbb{P}}^i[\log S_0^j + \sigma_j W_T^j + \mu_j T < \log S_0^i + \sigma_i W_T^i + \mu_i T]$$

$$= N\left(\frac{\log(S_0^i/S_0^j) + (d_j - d_i + \frac{1}{2}\sigma_j^2 + \frac{1}{2}\sigma_i^2 - \rho_{ij}\sigma_i\sigma_j)T}{\sqrt{\sigma_j^2 - 2\sigma_j\sigma_i\rho_{ij} + \sigma_i^2}\sqrt{T}}\right)$$

$$= N\left(\frac{\log(S_0^i/S_0^j) + (d_j - d_i + \frac{1}{2}\sigma_{ij}^2)T}{\sigma_{ji}\sqrt{T}}\right)$$

$$= N(\tilde{d}_{ji}^i)$$

since

(W_T^j, W_T^i) with respect to $\tilde{P}^i$

$$\stackrel{d}{=} (W_T^j + \rho_{ij}\sigma_i T, W_T^i + \sigma_i T) \text{ with respect to } \mathbb{P}$$

$$\stackrel{d}{=} \left(\sqrt{T}X + \rho_{ij}\sigma_i T, \sqrt{T}\left(\rho_{ij}X + \sqrt{1-\rho_{ij}^2}Y\right) + \sigma_i T\right).$$

We also have

$$\mathbb{P}[K < S_T^i] = \mathbb{P}[K < \log S_0^i + \sigma_i W_T^i + \mu_i T]$$

$$= N\left(\frac{\log(S_0^i/K) + (r - d_i - \frac{1}{2}\sigma_i^2)T}{\sigma_i\sqrt{T}}\right)$$

$$=: N(d_{*i})$$

and

$$\tilde{\mathbb{P}}^i[K < S_T^i] = \tilde{\mathbb{P}}^i[K < \log S_0^i + \sigma_i W_T^i + \mu_i T]$$

$$= N\left(\frac{\log(S_0^i/K) + (r - d_i + \frac{1}{2}\sigma_i^2)T}{\sigma_i\sqrt{T}}\right)$$

$$=: N(\tilde{d}_{*i}^i).$$

For the correlations, we have (see (4.4))

$$\rho_{ijk} := \mathrm{Corr}\left[\log\frac{S_T^i}{S_T^j}, \log\frac{S_T^i}{S_T^k}\right]$$

$$= \frac{\sigma_i^2 - \rho_{ij}\sigma_i\sigma_j - \rho_{ik}\sigma_i\sigma_k + \rho_{jk}\sigma_j\sigma_k}{\sigma_{ij}\sigma_{ik}}$$

if the indices i, j and k are all distinct, as well as

$$\rho_{i*j} := \mathrm{Corr}\left[\log S_T^i, \log\frac{S_T^i}{S_T^j}\right]$$

$$= \frac{\sigma_i - \rho_{ij}\sigma_j}{\sigma_{ij}}$$

if $i \neq j$, both with respect to $\mathbb{P}$ and with respect to $\tilde{\mathbb{P}}^i$. Now we conclude

$$C^{\mathrm{best}} = \sum_{i=1}^{n}\left[e^{-D_T^i}S_0^i N_n(\tilde{D}_i^{cb}; R_i^{cb}) - e^{-R_T}K N_n(D_i^{cb}; R_i^{cb})\right],$$

with

$$\tilde{D}_i^{cb}(j) = \tilde{d}_{ij}^i, \quad i \neq j,$$

$$\tilde{D}_i^{cb}(i) = \tilde{d}_{*i}^i,$$

$$D_i^{cb}(j) = d_{ij}, \quad i \neq j,$$

$$D_i^{cb}(i) = d_{*i},$$

$$R_i^{cb}(j,j) = 1,$$

$$R_i^{cb}(j,k) = \rho_{ijk}, \quad i \neq j \neq k \neq i,$$

$$R_i^{cb}(i,j) = R_i^{cb}(j,i) = \rho_{i*j}, \quad i \neq j.$$

Here $N_n(D, R)$ stands for the n-variate normal distribution evaluated at the vector D and with correlation matrix R.

For the price C^{worst} of the call on the worst of n assets, we have

$$C^{\text{worst}} = e^{-R_T} \mathbb{E}\left[\left(\min_{i=1,\ldots,n} S_T^i - K\right)^+\right]$$

$$= e^{-R_T} \sum_{i=1}^{n} \Big(\mathbb{E}\big[S_T^i \mathbf{1}_{\{S_T^j > S_T^i, j \neq i\}} \mathbf{1}_{\{K < S_T^i\}} \big]$$

$$- e^{-R_T} K \mathbb{P}\big[K < \max_{i=1,\ldots,n} S_T^i \big]\Big)$$

$$= \sum_{i=1}^{n} \Big(e^{-D_T^i} S_0^i \tilde{\mathbb{P}}\big[\forall_{j\,:\,j\neq i}\, S_T^j > S_T^i, K < S_T^i \big]$$

$$- e^{-R_T} K \mathbb{P}\big[\forall_{j\,:\,j\neq i}\, S_T^j > S_T^i, K < S_T^i \big]\Big).$$

Hence, we have

$$C^{\text{worst}} = \sum_{i=1}^{n}\left[e^{-d_i T} S_0^i N_n(\tilde{D}_i^{cw}; R_i^{cw}) - e^{-rT} K N_n(D_i^{cw}; R_i^{cw}) \right],$$

with

$$\tilde{D}_i^{cw}(j) = -\tilde{d}_{ij}^i, \quad i \neq j,$$

$$\tilde{D}_i^{cw}(i) = \tilde{d}_{*i}^i,$$

$$D_i^{cw}(j) = -d_{ij}, \quad i \neq j,$$

$$D_i^{cw}(i) = d_{*i},$$

$$R_i^{cw}(j,j) = 1,$$

$$R_i^{cw}(j,k) = \rho_{ijk}, \quad i \neq j \neq k \neq i,$$

$$R_i^{cw}(i,j) = R_i^{cw}(j,i) = -\rho_{i*j}, \quad i \neq j.$$

The price P^{best} for a put on the best of n assets is

$$P^{\text{best}} = \sum_{i=1}^{n}\left[-e^{-d_i T} S_0^i N_n(\tilde{D}_i^{pb}; R_i^{pb}) + e^{-rT} K N_n(D_i^{pb}; R_i^{pb}) \right],$$

with

$$\tilde{D}_i^{pb}(j) = \tilde{d}_{ij}^i, \quad i \neq j,$$

$$\tilde{D}_i^{pb}(i) = -\tilde{d}_{*i}^i,$$

$$D_i^{pb}(j) = d_{ij}, \quad i \neq j,$$

$$D_i^{pb}(i) = -d_{*i},$$

$$R_i^{pb}(j,j) = 1,$$

$$R_i^{pb}(j,k) = \rho_{ijk}, \quad i \neq j \neq k \neq i,$$

$$R_i^{pb}(i,j) = R_i^{pb}(j,i) = -\rho_{i*j}, \quad i \neq j,$$

while the corresponding put on the worst $\mathrm{P}^{\mathrm{worst}}$ is given by

$$\mathrm{P}^{\mathrm{worst}} = \sum_{i=1}^{n}[-e^{-d_iT}S_0^i N_n(\tilde{D}_i^{pw}; R_i^{pw}) + e^{-rT}KN_n(D_i^{pw}; R_i^{pw})],$$

with

$$\tilde{D}_i^{pw}(j) = -\tilde{d}_{ij}^i, \quad i \neq j,$$

$$\tilde{D}_i^{pw}(i) = -\tilde{d}_{*i}^i,$$

$$D_i^{pw}(j) = -d_{ij}, \quad i \neq j,$$

$$D_i^{pw}(i) = -d_{*i},$$

$$R_i^{pw}(j,j) = 1,$$

$$R_i^{pw}(j,k) = \rho_{ijk}, \quad i \neq j \neq k \neq i,$$

$$R_i^{pw}(i,j) = R_i^{pw}(j,i) = \rho_{i*j}, \quad i \neq j.$$

4.7 Basket Options

The underlying of an option can be a basket of assets rather than a single asset. A *basket* is effectively a weighted arithmetic average of several assets. So a call on a basket gives the holder the right to buy the basket at maturity T for a previously agreed price K.

If the assets are assumed to have a lognormal price process, then the basket itself does *not* necessarily follow a lognormal process. However, if the basket consists of many assets not too unevenly weighted, it is reasonable to assume that it follows a lognormal process. In other cases, however, we need a more sophisticated approximation technique. We now present one possible approach, which has similarities with a possible approach to Asian option valuation (see Section 2.5).

Note that options on stock indices are effectively options on baskets. For example, an option on the DAX is an option on a weighted basket of 30 stocks. It is very common to assume that such an index behaves lognormally.

Pricing The value of the basket at maturity T is given by the arithmetic mean

$$A = \sum_{i=1}^{n} w_i S_T^i$$

with w_i denoting the weight of asset i. We assume the weights $w_1, \ldots, w_n$ to be positive and to satisfy $w_1 + \cdots + w_n = 1$. The corresponding geometric mean is

$$G = \prod_{i=1}^{n}(S_T^i)^{w_i}.$$

A call on the basket has the payoff

$$(A - K)^+$$

at maturity T. The price C^{basket} of this call is then the discounted expectation of this payoff. However, because we expect to have more control over the geometric mean, we will write this price as an integral of the expectation conditional on the geometric mean:

$$\begin{aligned} C^{\text{basket}} &= e^{-R_T} \mathbb{E}[(A-K)^+] \\ &= e^{-R_T} \int_0^\infty \mathbb{E}[(A-K)^+ \mid G=x] g(x)\, dx \\ &= e^{-R_T} \left(\int_0^K \mathbb{E}[(A-K)^+ \mid G=x] g(x)\, dx \right. \\ &\qquad \left. + \int_K^\infty \mathbb{E}[A-K \mid G=x] g(x)\, dx \right) \\ &=: e^{-R_T}(C_1 + C_2), \end{aligned} \qquad (4.5)$$

with $g(x)$ denoting the density of G at $x \in \mathbb{R}$. Here we have used the inequality between the geometric and the arithmetic means,

$$A \geqslant G,$$

which is a consequence of Jensen's inequality. Now we define

$$X_i = \log S_T^i, \quad i = 1, \ldots, n,$$

$$X = \log G = \sum_{i=1}^n w_i X_i.$$

Then the mean μ_X and variance σ_X^2 of X are given by

$$\mu_X = \sum_{i=1}^n w_i (\mathcal{N}_T^i + \log S_0^i),$$

$$\sigma_X^2 = \sum_{i=1}^n \sum_{j=1}^n w_i w_j \Psi_T^{ij},$$

while the covariance ρ_X^i between X and X_i is given by

$$\rho_X^i = \sum_{j=1}^n w_j \Psi_T^{ij}.$$

Since $(X, X_i, i = 1, \ldots, n)$ is Gaussian, the conditional distribution of X_i given $X = x$ is normal with mean

$$\mathcal{N}_T^i + \frac{\rho_X^i}{\sigma_X^2}(x - \mu_X)$$

and variance

$$(\Sigma_T^i)^2 - \frac{(\rho_X^i)^2}{\sigma_X^2}.$$

Hence, the second integral C_2 is

$$C_2 = \int_{\log K}^{\infty} \left(\mathbb{E}[A \mid G = e^x] - K\right)\varphi_X(x)\,dx$$

$$= \sum_{i=1}^{n} w_i \int_{\log K}^{\infty} \mathbb{E}[X_i \mid X = x]\varphi_X(x)\,dx - K\int_{\log K}^{\infty} \varphi_X(x)\,dx$$

$$= \sum_{i=1}^{n} w_i \int_{\log K}^{\infty} \exp\left[\mathcal{N}_T^i + \frac{\rho_X^i}{\sigma_X^2}(x - \mu_X) + \tfrac{1}{2}\left((\Sigma_T^i)^2 - \frac{(\rho_X^i)^2}{\sigma_X^2}\right)\right]\varphi_X(x)\,dx$$

$$- KN\left(\frac{\mu_X - \log K}{\sigma_X}\right)$$

$$= \sum_{i=1}^{n} w_i \exp[\mathcal{N}_T^i + \tfrac{1}{2}(\Sigma_T^i)^2]\,N\left(\frac{\mu_X - \log K}{\sigma_X} + \frac{(\rho_X^i)^2}{\sigma_X^2}\right) - KN\left(\frac{\mu_X - \log K}{\sigma_X}\right),$$

with $\varphi_X(x)$ denoting the density of X at x.

We now want to evaluate the first integral C_1 in (4.5). For this, we define the vector $w^{\mathsf{T}} = [w_1, \ldots, w_n]$ and the $(n+1) \times n$ matrix

$$W = \begin{bmatrix} I^{n \times n} \\ w^{\mathsf{T}} \end{bmatrix} = \begin{bmatrix} 1 & & 0 \\ & \ddots & \\ 0 & & 1 \\ w_1 & \cdots & w_n \end{bmatrix},$$

where $I^{n \times n}$ denotes the $n \times n$ identity matrix, and the superscript T indicates transposition. We also define the random vectors $Y^{\mathsf{T}} = [X_1, \ldots, X_n]$ and $Y_+^{\mathsf{T}} = [X_1, \ldots, X_n, X]$. Then $Y_+ = W \cdot Y$. Let $C = (\Psi_T^{ij})_{ij}$ be the covariance matrix of Y. Then Y_+ has the covariance matrix

$$WCW^{\mathsf{T}} = \begin{bmatrix} C & Cw \\ w^{\mathsf{T}}C & w^{\mathsf{T}}Cw \end{bmatrix}.$$

The ith element of the vector Cw is ρ_X^i. We also have $w^{\mathsf{T}}Cw = \sigma_X^2$. It is known (see [43]) that the covariance matrix of Y given X is

$$\hat{C} = C - \frac{Cww^{\mathsf{T}}C}{w^{\mathsf{T}}Cw}.$$

Hence, the mean of A conditional on $X = \log K$ is given by

$$\hat{\mu}_A = \sum_{i=1}^{n} w_i \exp(\hat{\mu}_i + \tfrac{1}{2}\hat{C}_{ii}),$$

with

$$\hat{\mu}_i = \mathcal{N}_T^i + \frac{\rho_X^i}{\sigma_X^2}(\log K - \mu_X).$$

Closed-Form Solutions for Multi-Asset Products

The conditional variance of A given $X = \log K$ is

$$\hat{\sigma}_A^2 = \text{Var}\left[\sum_{i=1}^n w_i e^{X_i} \,\bigg|\, X = \log K\right]$$

$$= \sum_{i=1}^n \sum_{j=1}^n w_i w_j \,\text{Cov}[e^{X_i}, e^{X_j} \mid X = \log K]$$

$$= \sum_{i=1}^n \sum_{j=1}^n w_i w_j \big(\mathbb{E}[e^{X_i + X_j} \mid X = \log K]$$

$$- \mathbb{E}[e^{X_i} \mid X = \log K]\,\mathbb{E}[e^{X_j} \mid X = \log K]\big)$$

$$= \sum_{i=1}^n \sum_{j=1}^n w_i w_j \big(\exp(\hat{\mu}_i + \hat{\mu}_j + \tfrac{1}{2}\hat{C}_{ii} + \tfrac{1}{2}\hat{C}_{jj} + \hat{C}_{ij})$$

$$- \exp(\hat{\mu}_i + \tfrac{1}{2}\hat{C}_{ii}) \cdot \exp(\hat{\mu}_j + \tfrac{1}{2}\hat{C}_{jj})\big).$$

Now we use an approximation in assuming that the difference between the two means conditional on the geometric mean being equal to the strike K is lognormally distributed. More succinctly, the variable

$$\mathcal{E} = (A - G \mid G = K)$$

is assumed to be lognormally distributed. Since the mean and variance of a lognormally distributed random variable with parameters β and γ^2 are given by $\exp(\beta + \tfrac{1}{2}\gamma^2)$ and $(\exp \gamma^2 - 1)\exp(2\beta + \gamma^2)$ respectively, the values $\hat{\mu}_\mathcal{E} = \hat{\mu}_A - K$ and $\hat{\sigma}_\mathcal{E}^2 = \hat{\sigma}_A^2$ for the mean and variance of $\mathcal{E}$ can be used to obtain the parameters $\beta_\mathcal{E}$ and $\gamma_\mathcal{E}$:

$$\gamma_\mathcal{E}^2 = \log\left(\frac{\hat{\sigma}_\mathcal{E}^2}{\hat{\mu}_\mathcal{E}^2} + 1\right),$$

$$\beta_\mathcal{E} = \log \hat{\mu}_\mathcal{E} - \tfrac{1}{2}\gamma_\mathcal{E}^2.$$

Note that β and γ depend on K, ie $\beta = \beta(K)$ and $\gamma = \gamma(K)$.

We get an approximated value C_1^e as follows:

$$C_1 = e^{-R_T} \int_0^K \mathbb{E}[(A - G - (K - x))^+ \mid G = x]\, g(x)\, dx$$

$$= e^{-R_T} \int_0^K \mathbb{E}[(A - G - x)^+ \mid G = K - x]\, g(K - x)\, dx$$

$$\approx e^{-R_T} h \sum_{i=0}^m C^e\big(ih, \beta_\mathcal{E}(K - ih), \gamma_\mathcal{E}(K - ih)\big) g(K - ih)$$

$$= C_1^e,$$

with h denoting the interval width, g denoting the density of G, and assuming option values for strikes greater than mh to be negligible. Here we use the plain vanilla call value $C^e(K, \beta, \gamma)$ of a call with strike K on a lognormal asset with mean β and variance γ^2 and without discounting, which is given by

$$C^e(K, \beta, \gamma) = \exp(\beta + \tfrac{1}{2}\gamma^2) N\left(\frac{\beta - \log K}{\gamma} + \gamma\right) - KN\left(\frac{\beta - \log K}{\gamma}\right).$$

The value of a put on the basket can be obtained by the call–put parity from the value of the call.

4.8 Hindsight Options

Assume that a customer wants to buy some options on the performance of various underlyings. When the option matures, it will be known which of the equities (or indexes) performed best, and with hindsight the customer would have chosen more of the corresponding options and less of the others. A *hindsight* option allows him to do exactly that. So this product is a generalisation of an option on the best or the worst of several assets. A special case of a hindsight option would, for example, be an option on the performance of the third-best performing asset out of seven assets.

So, a *hindsight call* option pays out a (linear) combination of the *ordered* relative positive performances of n assets. More precisely, the payoff is given by

$$\alpha_1 \cdot \left(\frac{S_T^{\pi_1}}{S_0^{\pi_1}} - 1\right)^+ + \cdots + \alpha_n \cdot \left(\frac{S_T^{\pi_n}}{S_0^{\pi_n}} - 1\right)^+,$$

where π is a permutation (rearrangement) of $\{1, \ldots, n\}$ such that

$$\frac{S_T^{\pi_1}}{S_0^{\pi_1}} \geq \cdots \geq \frac{S_T^{\pi_n}}{S_0^{\pi_n}}.$$

Analogously, a *hindsight put* option pays

$$\alpha_1 \cdot \left(1 - \frac{S_T^{\pi_n}}{S_0^{\pi_n}}\right)^+ + \cdots + \alpha_m \cdot \left(1 - \frac{S_T^{\pi_1}}{S_0^{\pi_1}}\right)^+.$$

Since we are considering *relative* performance, we can simplify the notation by assuming that $S_0^1 = \cdots = S_0^n = 1$. Thus, $S_T^{\pi_1} \geq \cdots \geq S_T^{\pi_n}$.

Breaking up the problem We concentrate on the case of a call first. The value C^{hs} of the option is given by

$$e^{R_T} \cdot C^{hs} = \mathbb{E}[\alpha_1 \cdot (S_T^{\pi_1} - 1)^+ + \cdots + \alpha_n \cdot (S_T^{\pi_n} - 1)^+]$$
$$= \alpha_1 \cdot \mathbb{E}[(S_T^{\pi_1} - 1)^+] + \cdots + \alpha_n \cdot \mathbb{E}[(S_T^{\pi_n} - 1)^+].$$

Thus, we have to evaluate $\mathbb{E}[(S_T^{\pi_i} - 1)^+]$ for $i = 1, \ldots, n$. To do this, note that

$$\mathbb{E}[(S_T^{\pi_i} - 1)^+] = \sum_{j=1}^{n} \mathbb{E}[(S_T^j - 1) \cdot 1_{A_{ij}}],$$

where $A_{ij} = \{\omega : \pi_i(\omega) = j \text{ and } S_T^j(\omega) \geq 1\}$. In the following, we write $\mu_j = r_t^j - d_t^j$. Choosing the measure $\mathbb{P}_j$ with Radon–Nikodym derivative

$$\frac{d\mathbb{P}_j}{d\mathbb{P}} = \exp(\sigma_j W_T^j - \tfrac{1}{2}\sigma_j^2 T) = S_T^j e^{-\mu_j T},$$

we get

$$\mathbb{E}[(S_T^j - 1) \cdot 1_{A_{ij}}] = \mathbb{E}[S_T^j \cdot 1_{A_{ij}}] - \mathbb{E}[1_{A_{ij}}] = e^{\mu_j T}\mathbb{P}_j(A_{ij}) - \mathbb{P}(A_{ij}).$$

So it remains to evaluate $\mathbb{P}(A_{ij})$ and $\mathbb{P}_j(A_{ij})$.

Computing $\mathbb{P}(A_{ij})$ We have

$$\mathbb{P}(A_{ij}) = \sum_{B \not\ni j,\, |B|=i-1} \mathbb{P}(S_T^j \leq S_T^k \text{ for } k \in B,\, S_T^j \geq S_T^k \text{ for } k \notin B,\, S_T^j \geq 1),$$

where we sum over subsets B of $\{1,\ldots,n\}$.

Now, in the following, we fix j and some $B \subset \{1,\ldots,n\}$ with $j \notin B$. Let $\varepsilon_k = +1$ for $k \in B$, and $\varepsilon_k = -1$ otherwise. We consider the n-dimensional normally distributed random vector $\tilde{W}^1, \ldots, \tilde{W}^n$ defined by

$$\tilde{W}^k = \frac{\varepsilon_k}{\sqrt{T}} \cdot \begin{cases} \dfrac{\sigma_j W_T^j - \sigma_k W_T^k}{\sigma_{jk}} & \text{for } k \neq j, \\ W_T^j & \text{for } k = j, \end{cases}$$

with $\sigma_{jk}^2 = \sigma_j^2 - 2\rho_{jk}\sigma_j\sigma_k + \sigma_k^2$.

Each $\tilde{W}^k$ has a mean of zero. The correlation between $\tilde{W}^k$ and $\tilde{W}^l$ is given by

$$\varepsilon_k \varepsilon_l \cdot \begin{cases} \dfrac{\rho_{kl}\sigma_k\sigma_l - \rho_{jk}\sigma_j\sigma_k - \rho_{jl}\sigma_j\sigma_l + \sigma_j^2}{\sigma_{jk}\sigma_{jl}} & \text{if } k \neq j \neq l, \\ \dfrac{\sigma_j - \rho_{jk}\sigma_k}{\sigma_{jk}} & \text{if } k \neq j = l, \\ \dfrac{\sigma_j - \rho_{jl}\sigma_l}{\sigma_{jl}} & \text{if } k = j \neq l, \\ 1 & \text{if } k = j = l. \end{cases}$$

We denote the correlation matrix by R_B^j.

So we can compute

$$\mathbb{P}(S_T^j \leq S_T^k \text{ for } k \in B,\, S_T^j \geq S_T^k \text{ for } k \notin B,\, S_T^j \geq 1)$$
$$= \mathbb{P}\left(\tilde{W}^k \leq \frac{\varepsilon_k[\mu_k - \mu_j + \tfrac{1}{2}(\sigma_j^2 - \sigma_k^2)]\sqrt{T}}{\sigma_{jk}} \text{ for } k \neq j,\, \tilde{W}^j \leq \frac{(\mu_j - \tfrac{1}{2}\sigma_j^2)\sqrt{T}}{\sigma_j}\right)$$
$$= N_n(D_B^j, R_B^j),$$

with the n-dimensional vector $D^j = D^j_B$ given by

$$D^j_k = \varepsilon_k \sqrt{T} \cdot \begin{cases} \dfrac{\mu_k - \mu_j + \frac{1}{2}(\sigma_j^2 - \sigma_k^2)}{\sigma_{jk}} & \text{if } k \neq j, \\ \dfrac{-\mu_j + \frac{1}{2}\sigma_j^2}{\sigma_j} & \text{if } k = j. \end{cases}$$

Here $N_n(D, R)$ stands for the n-variate normal distribution evaluated at the vector D and with correlation matrix R.

So, we have

$$\mathbb{P}(A_{ij}) = \sum_{B \not\ni j,\, |B|=i-1} N_n(D^j_B, R^j_B).$$

Computing $\mathbb{P}_j(A_{ij})$ The computation of $\mathbb{P}_j(A_{ij})$ proceeds analogously. Under $\mathbb{P}_j$, the geometric Brownian motion S^k_t has drift $\mu_k + \rho_{kj}\sigma_k\sigma_j$. So, we can repeat the calculation of the preceding sector, but with μ_k replaced by $\mu_k + \rho_{kj}\sigma_k\sigma_j$. This gives rise to a new vector $\tilde{D}^j$ and a correlation matrix that is unchanged.

Explicitly, we have

$$\mathbb{P}_j(A_{ij}) = \sum_{B \not\ni k,\, |B|=i-1} N_n(\tilde{D}^j_B, R^j_B),$$

with

$$\tilde{D}^j_k = \varepsilon_k \sqrt{T} \cdot \begin{cases} \dfrac{\mu_k - \mu_j - \frac{1}{2}\sigma_{jk}^2}{\sigma_{jk}} & \text{for } k \neq j, \\ \dfrac{-\mu_j - \frac{1}{2}\sigma_j^2}{\sigma_j} & \text{for } k = j. \end{cases}$$

Putting the pieces together To summarise, the price of a hindsight call is given by

$$C^{hs} = e^{-R_T} \sum_{i=1}^{n} \alpha_i \sum_{j=1}^{n} \left(\sum_{B \not\ni j,\, |B|=i-1} [e^{\mu_j T} \cdot N_n(\tilde{D}^j_B, R^j_B) - N_n(D^j_B, R^j_B)] \right).$$

The valuation of a hindsight put proceeds analogous. The formula is

$$P^{hs} = e^{-R_T} \sum_{i=1}^{n} \alpha_i \sum_{j=1}^{n} \left(\sum_{B \ni j,\, |B|=n-i+1} [N_n(D^j_B, R^j_B) - e^{\mu_j T} \cdot N_n(\tilde{D}^j_B, R^j_B)] \right).$$

Computational complexity When pricing a hindsight option with the formula above, one needs to evaluate $n \cdot 2^n$ multivariate normal distributions. Whether this can be done sufficiently quickly depends on the quality of the numerical routine for computing multivariate normal distributions and on the dimension, ie the number of underlyings of the option.

Closed-Form Solutions for Multi-Asset Products

4.9 Outside Barrier Options

For standard barrier options, the same underlying asset determines the payoff as well as whether the barrier event happens. It is both reasonable and possible to relax this restriction. The resulting option is called an *outside barrier* option.

We denote the asset triggering the barrier event S^1, and the asset determining the payoff by S^2. So, for example, a down-and-out outside barrier call pays at maturity T the amount

$$(S_T^2 - K)^+ \cdot 1_{\{S_t^1 > H \text{ for all } 0 \leqslant t \leqslant T\}}.$$

One can also define *outside digital* options. A down-and-out outside digital call pays

$$1_{\{S_T^2 > K\}} \cdot 1_{\{S_t^1 > H \text{ for all } 0 \leqslant t \leqslant T\}}.$$

The closed-form solutions The solutions can easily be derived using formula (1.9) in Section 1.1.19. We only state the results.

Out options For an up-and-out call, we have

$$e^{R_T} C^{\text{out}} =$$

$$K \exp\left(\frac{2\mathcal{N}_T^1 \log(H/S_0^1)}{(\Sigma_T^1)^2}\right)$$

$$\cdot N_2\left(\frac{\log(S_0^2/K) + \mathcal{N}_T^2 + 2\rho \frac{\Sigma_T^2}{\Sigma_T^1}\log(H/S_0^1)}{\Sigma_T^2}, \frac{\log(S_0^1/H) - \mathcal{N}_T^1}{\Sigma_T^1}; -\rho\right)$$

$$- K N_2\left(\frac{\log(S_0^2/K) + \mathcal{N}_T^2}{\Sigma_T^2}, \frac{-\log(S_0^1/H) - \mathcal{N}_T^1}{\Sigma_T^1}; -\rho\right)$$

$$+ S_0^2 \exp(\mathcal{M}_T^2)$$

$$\cdot N_2\left(\frac{\log(S_0^2/K) + \tilde{\mathcal{N}}_T^2}{\Sigma_T^2}, \frac{-\log(S_0^1/H) - \mathcal{N}_T^1 - \Psi_T}{\Sigma_T^1}; -\rho\right)$$

$$- S_0^2 \exp\left(\mathcal{M}_T^2 + \frac{2(\mathcal{N}_T^1 + \Psi_T)\log(H/S_0^1)}{(\Sigma_T^1)^2}\right)$$

$$\cdot N_2\left(\frac{\log(S_0^2/K) + \tilde{\mathcal{N}}_T^2 + 2\rho \frac{\Sigma_T^2}{\Sigma_T^1}\log(H/S_0^1)}{\Sigma_T^2}, \frac{\log(S_0^1/H) - \mathcal{N}_T^1 - \Psi_T}{\Sigma_T^1}; -\rho\right).$$

For an up-and-out put, we have

$$e^{R_T}\mathbf{P}^{\text{out}} =$$

$$KN_2\left(\frac{-\log(S_0^2/K) - \mathcal{N}_T^2}{\Sigma_T^2}, \frac{-\log(S_0^1/H) - \mathcal{N}_T^1}{\Sigma_T^1}; \rho\right)$$

$$- K\exp\left(\frac{2\mathcal{N}_T^1\log(H/S_0^1)}{(\Sigma_T^1)^2}\right)$$

$$\cdot N_2\left(\frac{-\log(S_0^2/K) - \mathcal{N}_T^2 - 2\rho\frac{\Sigma_T^2}{\Sigma_T^1}\log(H/S_0^1)}{\Sigma_T^2}, \frac{\log(S_0^1/H) - \mathcal{N}_T^1}{\Sigma_T^1}; \rho\right)$$

$$+ S_0^2 \exp\left(\mathcal{M}_T^2 + \frac{2(\mathcal{N}_T^1 + \Psi_T)\log(H/S_0^1)}{(\Sigma_T^1)^2}\right)$$

$$\cdot N_2\left(\frac{-\log(S_0^2/K) - \tilde{\mathcal{N}}_T^2 - 2\rho\frac{\Sigma_T^2}{\Sigma_T^1}\log(H/S_0^1)}{\Sigma_T^2}, \frac{\log(S_0^1/H) - \mathcal{N}_T^1 T - \Psi_T}{\Sigma_T^1}; \rho\right)$$

$$- S_0^2 \exp(\mathcal{M}_T^2)$$

$$\cdot N_2\left(\frac{-\log(S_0^2/K) - \tilde{\mathcal{N}}_T^2}{\Sigma_T^2}, \frac{-\log(S_0^1/H) - \mathcal{N}_T^1 T - \Psi_T}{\Sigma_T^1}; \rho\right).$$

For a down-and-out call, we have

$$e^{R_T}\mathbf{C}_{\text{out}} =$$

$$K\exp\left(\frac{2\mathcal{N}_T^1\log(H/S_0^1)}{(\Sigma_T^1)^2}\right)$$

$$\cdot N_2\left(\frac{\log(S_0^2/K) + \mathcal{N}_T^2 + 2\rho\frac{\Sigma_T^2}{\Sigma_T^1}\log(H/S_0^1)}{\Sigma_T^2}, \frac{-\log(S_0^1/H) + \mathcal{N}_T^1}{\Sigma_T^1}; \rho\right)$$

$$- KN_2\left(\frac{\log(S_0^2/K) + \mathcal{N}_T^2}{\Sigma_T^2}, \frac{\log(S_0^1/H) + \mathcal{N}_T^1}{\Sigma_T^1}; \rho\right)$$

$$+ S_0^2 \exp(\mathcal{M}_T^2)$$

$$\cdot N_2\left(\frac{\log(S_0^2/K) + \tilde{\mathcal{N}}_T^2}{\Sigma_T^2}, \frac{\log(S_0^1/H) + \mathcal{N}_T^1 + \Psi_T}{\Sigma_T^1}; \rho\right)$$

$$- S_0^2 \exp\left(\mathcal{M}_T^2 + \frac{2(\mathcal{N}_T^1 + \Psi_T)\log(H/S_0^1)}{(\Sigma_T^1)^2}\right)$$

$$\cdot N_2\left(\frac{\log(S_0^2/K) + \tilde{\mathcal{N}}_T^2 + 2\rho\frac{\Sigma_T^2}{\Sigma_T^1}\log(H/S_0^1)}{\Sigma_T^2}, \frac{-\log(S_0^1/H) + \mathcal{N}_T^1 + \Psi_T}{\Sigma_T^1}; \rho\right).$$

Closed-Form Solutions for Multi-Asset Products

Finally, for a down-and-out put, we have

$$e^{R_T} P_{\text{out}} =$$

$$KN_2\left(\frac{-\log(S_0^2/K) - \mathcal{N}_T^2}{\Sigma_T^2}, \frac{\log(S_0^1/H) + \mathcal{N}_T^1}{\Sigma_T^1}; -\rho\right)$$

$$- K\exp\left(\frac{2\mathcal{N}_T^1 \log(H/S_0^1)}{(\Sigma_T^1)^2}\right)$$

$$\cdot N_2\left(\frac{-\log(S_0^2/K) - \mathcal{N}_T^2 - 2\rho\frac{\Sigma_T^2}{\Sigma_T^1}\log(H/S_0^1)}{\Sigma_T^2}, \frac{-\log(S_0^1/H) + \mathcal{N}_T^1}{\Sigma_T^1}; -\rho\right)$$

$$+ S_0^2 \exp\left(\mathcal{M}_T^2 + \frac{2(\mathcal{N}_T^1 + \Psi_T)\log(H/S_0^1)}{(\Sigma_T^1)^2}\right)$$

$$\cdot N_2\left(\frac{-\log(S_0^2/K) - \tilde{\mathcal{N}}_T^2 - 2\rho\frac{\Sigma_T^2}{\Sigma_T^1}\log(H/S_0^1)}{\Sigma_T^2}, \frac{-\log(S_0^1/H) + \mathcal{N}_T^1 + \Psi_T}{\Sigma_T^1}; -\rho\right)$$

$$- S_0^2 \exp(\mathcal{M}_T^2)$$

$$\cdot N_2\left(\frac{-\log(S_0^2/K) - \tilde{\mathcal{N}}_T^2}{\Sigma_T^2}, \frac{\log(S_0^1/H) + \mathcal{N}_T^1 + \Psi_T}{\Sigma_T^1}; -\rho\right).$$

In options In options can be calculated as the difference between an otherwise identical Black–Scholes plain vanilla option and the corresponding out option. Explicitly, we have

$$C^{\text{in}} = C - C^{\text{out}},$$

$$P^{\text{in}} = P - P^{\text{out}},$$

$$C_{\text{in}} = C - C_{\text{out}},$$

$$P_{\text{in}} = P - P_{\text{out}}.$$

Hedge ratios This type of option has also a correlation exposure. For $\rho = 1$, inside options and outside options are the same. For $\rho < 1$, an outside option is worth less than an inside option and the price difference increases as ρ decreases.

4.10 Outside Digital Options

In the spirit of outside barrier options (Section 4.9), one can define *outside digital* options. For example, a down-and-out outside digital call pays

$$1_{\{S_T^2 > K\}} \cdot 1_{\{S_t^1 > H \text{ for all } 0 \leqslant t \leqslant T\}}.$$

Summary of formulas   We arrive at the following formulas. For an up-and-out outside digital call $C^{\text{out}}_{\text{os-dig}}$, we have

$$e^{R_T} C^{\text{out}}_{\text{os-dig}} = -K N_2 \left(\frac{\log(S_0^2/K) + \mathcal{N}_T^2}{\Sigma_T^2}, \frac{\log(H/S_0^1) - \mathcal{N}_T^1}{\Sigma_T^1}; -\rho \right)$$
$$+ S_0^2 e^{\mathcal{M}_T^2} \cdot N_2 \left(\frac{\log(S_0^2/K) + \mathcal{M}_T^2}{\Sigma_T^2}, \frac{\log(H/S_0^1) - \mathcal{N}_T^1 - \Psi_T}{\Sigma_T^1}; -\rho \right).$$

For an up-and-out outside digital put $P^{\text{out}}_{\text{os-dig}}$, we have

$$e^{R_T} P^{\text{out}}_{\text{os-dig}} = K N_2 \left(-\frac{\log(S_0^2/K) + \mathcal{N}_T^2}{\Sigma_T^2}, \frac{\log(H/S_0^1) - \mathcal{N}_T^1}{\Sigma_T^1}; \rho \right)$$
$$- S_0^2 e^{\mathcal{M}_T^2} \cdot N_2 \left(\frac{-\log(S_0^2/K) + \mathcal{M}_T^2}{\Sigma_T^2}, \frac{\log(H/S_0^1) - \mathcal{N}_T^1 - \Psi_T}{\Sigma_T^1}; \rho \right).$$

For a down-and-out outside digital call $C^{\text{os-dig}}_{\text{out}}$, we have

$$e^{R_T} C^{\text{os-dig}}_{\text{out}} = -K N_2 \left(\frac{\log(S_0^2/K) + \mathcal{N}_T^2}{\Sigma_T^2}, \frac{-\log(H/S_0^1) - \mathcal{N}_T^1}{\Sigma_T^1}; \rho \right)$$
$$+ S_0^2 e^{\mathcal{M}_T^2} \cdot N_2 \left(\frac{\log(S_0^2/K) + \mathcal{M}_T^2}{\Sigma_T^2}, \frac{-\log(H/S_0^1) - \mathcal{N}_T^1 - \Psi_T}{\Sigma_T^1}; \rho \right).$$

For a down-and-out outside digital put $P^{\text{os-dig}}_{\text{out}}$, we have

$$e^{R_T} P^{\text{os-dig}}_{\text{out}} = K N_2 \left(-\frac{\log(S_0^2/K) + \mathcal{N}_T^2}{\Sigma_T^2}, \frac{-\log(H/S_0^1) - \mathcal{N}_T^1}{\Sigma_T^1}; -\rho \right)$$
$$- S_0^2 e^{\mathcal{M}_T^2} \cdot N_2 \left(\frac{-\log(S_0^2/K) + \mathcal{M}_T^2}{\Sigma_T^2}, \frac{-\log(H/S_0^1) - \mathcal{N}_T^1 - \Psi_T}{\Sigma_T^1}; -\rho \right).$$

As before, in options can be calculated as the difference between an otherwise identical digital option and the corresponding out option. Explicitly, we have

$$C^{\text{in}}_{\text{os-dig}} = C_{\text{digital}} - C^{\text{out}}_{\text{os-dig}},$$
$$P^{\text{in}}_{\text{os-dig}} = P_{\text{digital}} - P^{\text{out}}_{\text{os-dig}},$$
$$C^{\text{os-dig}}_{\text{in}} = C_{\text{digital}} - C^{\text{os-dig}}_{\text{out}},$$
$$P^{\text{os-dig}}_{\text{in}} = P_{\text{digital}} - P^{\text{os-dig}}_{\text{out}}.$$

Chapter 5
Closed-Form Fixed Income and Hybrid Products

In this chapter we derive a few closed-form solutions for pure fixed income as well as hybrid products. Throughout we adopt the assumption that the short rate process follows the generalised Vasicek model, as introduced in Section 1.5. In particular, the short rate process follows a normal distribution and the bond price process a lognormal distribution.

5.1 Bond Options and Swaptions

Bond options play a similar role in the fixed income world as plain vanilla equity options play in the equity world. A *bond* option is the option to buy (call) or sell (put) a certain bond at some maturity T for a fixed strike price K.

We first concentrate on the case of a zero bond. The price of such an instrument is lognormally distributed under the extended Vasicek model we are using. Therefore, we can derive a formula analogous to the plain vanilla Black–Scholes formula.

Options on zero bonds Recall that we denote the price at time t of a zero bond paying the amount 1 at time S by $P(t, S)$. For a call on such a zero bond maturing at time S, the payoff at maturity T is given by

$$\left(P(T, S) - K\right)^+.$$

Therefore, the option value is given by

$$\mathbb{E}\left[B_T^{-1} \cdot \left(P(T, S) - K\right)\right],$$

where B_T denotes the cash bond. Since we have stochastic discounting (by a discount factor $1/B_T$) that depends on the short rate process, the integral we have to evaluate does not look very straightforward. However, we can simplify our task and avoid stochastic discounting by using the forward measure approach. We recall that the forward measure $\mathbb{P}_T$ is given by

$$\frac{d\mathbb{P}_T}{d\mathbb{P}} = \frac{1}{P(0, T)B_T}.$$

Using this measure, we can simplify

$$\mathbb{E}\left[B_T^{-1} \cdot (P(T,S) - K)^+ \right] = P(0,T) \cdot \mathbb{E}_{P_T}\left[(P(T,S) - K)^+ \right].$$

To evaluate this integral, we need the mean μ_P and variance σ_P^2 of the process $\log P(T, S)$. For this, we recall that

$$P(T, S) = \frac{P(0, S)}{P(0, T)} \exp\left(-\int_0^T \sigma_s [B_1(s, S) - B_1(s, T)] dW_s \right.$$
$$\left. - \frac{1}{2} \int_0^T \sigma_s^2 [B_1^2(s, S) - B_1^2(s, T)] ds \right).$$

Thus $\log P(T, S)$ has mean

$$\mu_P = \log P(0, S) - \log P(0, T) - \frac{1}{2} \int_0^T \sigma_s^2 [B_1(s, S) - B_1(s, T)]^2 ds$$

and variance

$$\sigma_P^2 = \int_0^T \sigma_s^2 [B_1(s, S) - B_1(s, T)]^2 ds$$

under the forward measure $\mathbb{P}_T$.

Now, if X is a normally distributed random variable with mean μ and variance σ, we have

$$\mathbb{E}\left[(X - K)^+ \right] = e^{\mu + \sigma^2/2} \cdot N\left(\frac{-\log K + \mu + \sigma^2}{\sigma}\right) - K \cdot N\left(\frac{-\log K + \mu}{\sigma}\right).$$

From this, it follows that the value of the call on a zero bond is given by

$$P(0, T)\left[e^{\mu_P + \sigma_P^2/2} \cdot N\left(\frac{-\log K + \mu_P + \sigma_P^2}{\sigma_P}\right) - K \cdot N\left(\frac{-\log K + \mu_P}{\sigma_P}\right) \right].$$

Similarly, a put on a zero bond has the value

$$P(0, T)\left[K \cdot N\left(-\frac{-\log K + \mu_P}{\sigma_P}\right) - e^{\mu_P + \sigma_P^2/2} \cdot N\left(-\frac{-\log K + \mu_P + \sigma_P^2}{\sigma_P}\right) \right].$$

Options on coupon bonds Now let us consider the case where the bond that underlies the option has coupons. Assume that at time S_i the bond pays an amount α_i, where the index i runs from 1 to n. So, in practice, we will often have $\alpha_i = cd$ for $i < n$, and $\alpha_n = 1 + cd$, where c is the coupon amount and d is the number of years between two coupon dates, using a certain day count convention. We can and will regard the coupon bond as a portfolio of zero bonds.

If a one-factor model for the short rate is used, a decomposition due to Jamshidian can be used to price a coupon bond option. Note that the value $P(T, S) = P(T, r_T, S)$ at time T of a zero bond maturing at time S is a deterministic and monotone decreasing function of the short rate r_T. In particular, it can be written as $P(T, S) = A \exp(-Br_T)$, where A and B are positive constants depending only on the model parameter and on the times T

and S. So, the value of the coupon bond depends monotonically on the short rate R_T. This implies that, for each strike $K > 0$, there is a unique short rate r_* at time T that makes the value of the coupon bond equal to K, ie

$$\sum_{i=1}^{n} \alpha_i P(T, S_i, r_*) = K.$$

We set $K_i = \alpha_i P(T, S_i, r_*)$, so that $K_1 + \cdots + K_n = K$. Hence, if $r_T > r_*$ we have $\alpha_i P(T, S_i, r_T) < K_i$, and if $r_T < r_*$ we have $\alpha_i P(T, S_i, r_T) > K_i$. This implies that

$$\left(\sum_{i=1}^{n} \alpha_i P(T, S_i, r_T) - K\right)^+ = \sum_{i=1}^{n} (\alpha_i P(T, S_i, r_T) - K_i)^+.$$

This in turn means that we can value the option on the coupon bond as a portfolio of options on zero bonds if we know r_*. Moreover, this implicitly given value can be efficiently computed using the Newton–Raphson method [61].

Swaptions A *swaption* is an option to enter a swap contract at a fixed swap rate on the option maturity date. A *swap* is a contract to exchange fixed interest and floating interest payments on a fixed notional for specified periods. We will indicate how the pricing of a swaption can be reduced to the pricing of an option on a coupon bond.

Let us consider a call that allows the holder to enter a long swap position, ie to receive fixed payments and make floating payments. Without affecting pricing, we can assume that the notional is exchanged at the swap maturity.

The floating interest side initially has a value equal to the notional amount, as demonstrated in Section 2.1.3 for equity swaps. The fixed interest side has the same cash flows as a coupon bond. So, effectively, exercising the swaption amounts to giving up the notional amount in exchange for a coupon bond. In other words, the swaption is a call on a coupon bond, with a strike equal to the notional.

Market prices of swaptions are generally quoted as implied volatilities. These volatilities are calculated by assuming that the swap rate is lognormally distributed. This assumption is called the *Black swaption model*. Section 5.2 discusses this market parametrisation for caps and floors.

5.2 Caps and Floors

Caplets A caplet with strike K on an interest rate R for the period $[T, T + t_R]$ has the payoff

$$(R - K)^+ f_a$$

at maturity T and pays at $T + t_R$. Here f_a is the accrual period in years calculated using applicable calendar and day count conventions. It is generally assumed that R is a simply compounded rate based on a day count fraction f_R calculated from t_R, again using some calendar and day count conventions. In

other words, we have
$$\frac{1}{1+Rf_R} = P(T, T+t_R),$$

where $P(t, T)$ denotes the price at t of a zero bond expiring at T. Using this relation, we obtain

$$(R-K)^+ \frac{f_a}{1+Rf_R} = \frac{f_a}{f_R}(1+Kf_R)\left(\frac{1}{1+Kf_R} - P(T, T+t_R)\right)^+.$$

This means that a caplet is worth $(1 + Kf_R)f_a/f_R$ times the price of a European put on a zero bond with bond expiry $t + t_R$, strike $1/(1 + Kf_R)$ and maturity T. For the pricing of options on zero bonds, we refer the reader to Section 5.1.

Caps and floors A *cap* is a series of n caplets based on a sequence $0 < T_0 < T_1 < \cdots < T_n$. Here caplet i caps the rate on $[T_{i-1}, T_i]$, matures at T_{i-1}, and pays at T_i. A *floor* is a series of floorlets, each having a payoff of the form
$$(K-R)^+ f_a.$$

Market conventions Dates in the definition of caps, floors and swaptions are often specified using tenors of the form mM and nY, meaning m months or n years. In order to calculate actual cashflow dates from these tenors, calendars are commonly used which may exclude weekends and local bank holidays.

Day count conventions are used to calculate the year fraction $t(D_1, D_2)$ between two dates D_1 and D_2. Examples for day count conventions are

- **30/360:**
  One year is assumed to have 360 days and each month 30 days. Hence the difference is given by the formula
  $$t(D_1, D_2) = 360(y_2 - y_1) + 30 * (m_2 - m_1) + [\min(d_2, 30) - \min(d_1, 30)],$$
  where (d_i, m_i, y_i) denotes day, month and year of date D_i.

- **Act/360:**
  $$t(D_1, D_2) = (D_2 - D_1)/360.$$

- **Act/365:**
  $$t(D_1, D_2) = (D_2 - D_1)/365.$$

Black's model Market prices of caps and floors are usually quoted as implied volatilities. These volatilities are calculated assuming that the interest rate R is lognormally distributed. The forward is given by the forward rate R_f obtained from the yield curve, applying appropriate calendar and day count conventions. The volatility σ is then obtained, assuming that the volatility for caplet i is $\sigma\sqrt{T_i}$, by inverting Black's formula [6] for caplets:
$$C^{\text{rate}} = f_a e^{-R_{T+t_R}}[R_f N(d_1) - K(d_2)],$$
with
$$d_1 = \frac{1}{\sigma\sqrt{T}}\left(\log\frac{R_f}{K} + \frac{\sigma^2 T}{2}\right) \quad \text{and} \quad d_2 = \frac{1}{\sigma\sqrt{T}}\left(\log\frac{R_f}{K} - \frac{\sigma^2 T}{2}\right).$$

The strike is usually set at-the-money, ie at R_f, thus equating caplet and floorlet prices.

5.3 European Options (Merton Formula)

In our hybrid set-up, the asset distribution is lognormal. The usual Black–Scholes formula is not valid any more, since the mean reversion λ and the short rate volatility σ influence the forward and volatility of the asset. As shown in Section 1.5.2, the mean of $\log(S_T/S_0)$ changes from $R_T - D_T - \frac{1}{2}\Sigma_T^2$ in the standard model to

$$\mathbb{E}[\log B_T] - D_T - \tfrac{1}{2}\Sigma_T^2$$

$$= R_T - D_T - \tfrac{1}{2}B_2(0,T) - \int_0^T \rho_s \sigma_s \sigma_s^S B_1(s,T)\,ds - \tfrac{1}{2}\Sigma_T^2$$

in the hybrid model. The variance increases from Σ_T^2 in the standard model to

$$v(0,T)^2 = \Sigma_T^2 + B_2(0,T) + 2\int_0^T \rho_s \sigma_s \sigma_s^S B_1(s,T)\,ds.$$

Therefore the fair price C of the call is

$$C = S_0 e^{-D_T} N(d_1) - e^{-R_T} K N(d_2),$$

with

$$d_1 = d_2 + v(0,T) \quad \text{and} \quad d_2 = \frac{1}{v(0,T)}\left(\log\frac{S_0}{K} + R_T - D_T - \frac{v(0,T)^2}{2}\right),$$

The put price P is

$$P = e^{-R_T} K N(-d_2) - S_0 e^{-D_T} N(-d_1).$$

5.4 Equity/Bond Outperformance Options

An interesting product is the option on the outperformance between an equity and a bond. Such a product allows the buyers to profit from their opinion on relative differences between fixed income and equity markets.

The pricing of such a product, in both its absolute and relative versions, can be performed in our hybrid framework.

The payoffs In the case of a call, an *equity-bond (absolute) outperformance* option with strike K pays

$$\left(w_1 \cdot \frac{S_T}{S_0} - w_2 \cdot \frac{P(T,S)}{P(0,S)} - K\right)^+$$

at option maturity $T \leqslant S$. The corresponding put has a payoff of

$$\left(K - w_1 \cdot \frac{S_T}{S_0} + w_2 \cdot \frac{P(T,S)}{P(0,S)}\right)^+.$$

The product can be generalised to coupon bonds, paying a coupon α_i at time S_i. The payoff of a call is then

$$\left(w_1 \cdot \frac{S_T}{S_0} - w_2 \cdot \frac{\sum_{i=1}^n \alpha_i P(T, S_i)}{\sum_{i=1}^n \alpha_i P(0, S_i)} - K \right)^+.$$

The *equity–bond relative outperformance* option pays

$$\left(\frac{S_T/S_0}{P(T, S)/P(0, S)} - K \right)^+$$

as a call, and

$$\left(K - \frac{S_T/S_0}{P(T, S)/P(0, S)} \right)^+$$

as a put.

For notational simplicity, we will restrict our attention to calls in this section. Valuations for the corresponding puts can be derived in a similar fashion. Alternatively, one can also use a version of the call–put parity.

Absolute outperformance with zero strike We shall again use the forward measure approach. So, we get, for example,

$$\mathbb{E}_P \left[B_T^{-1} \cdot \left(w_1 \cdot \frac{S_T}{S_0} - w_2 \cdot \frac{P(T, S)}{P(0, S)} - K \right)^+ \right]$$

$$= P(0, T) \cdot \mathbb{E}_{P_T} \left[\left(w_1 \cdot \frac{S_T}{S_0} - w_2 \cdot \frac{P(T, S)}{P(0, S)} - K \right)^+ \right],$$

where $\mathbb{P}_T$ is the forward measure given by

$$\frac{d\mathbb{P}_T}{d\mathbb{P}} = \frac{1}{P(0, T) B_T}.$$

To evaluate our integral, we use the fact that the two random variables

$$\log \frac{S_T}{S_0} \quad \text{and} \quad \log P(T, S)$$

form a two-dimensional Gaussian random vector. We will need the first and second moments of these variables under the forward measure $\mathbb{P}_T$.

The mean and variance for $\log(S_t/S_0)$ can be found in Section 1.2.1. The same moments for $\log P(T, S)$ under the forward measure are given in Section 5.1. The covariance of $\log(S_T/S_0)$ and $\log P(T, S)$ is given by

$$\text{Cov}\left[\int_0^T \sigma_s^S dW_s^S + \int_0^T \sigma_s B_1(s, T) dW_s, \int_0^T \sigma_s [B_1(s, T) - B_1(s, S)] dW_s \right]$$

$$= \int_0^T \rho_s \sigma_s^S \sigma_s [B_1(s, T) - B_1(s, S)] ds + \int_0^t \sigma_s^2 [B_1(s, T)^2 - B_1(s, T) B_1(s, S)] ds.$$

Closed-Form Fixed Income and Hybrid Products 145

In general, if X_1 and X_2 are random variables such that $\log X_1$ and $\log X_2$ are binormally distributed with means μ_1 and μ_2, variances σ_1 and σ_2, and correlation ρ, we have

$$\mathbb{E}\left[(w_1 X_1 - w_2 X_2)^+\right] = w_1 e^{\mu_1 + \sigma_1^2/2} N(d_1) - w_2 e^{\mu_2 + \sigma_2^2/2} N(d_2),$$

with

$$d_1 = \frac{\log(w_1/w_2) + \mu_1 - \mu_2 + \sigma_1^2 - \rho \sigma_1 \sigma_2}{\tilde{\sigma}}$$

and

$$d_2 = \frac{\log(w_1/w_2) + \mu_1 - \mu_2 + \sigma_2^2 + \rho \sigma_1 \sigma_2}{\tilde{\sigma}},$$

where $\tilde{\sigma}$ is given by

$$\tilde{\sigma}^2 = \sigma_1^2 - 2\rho \sigma_1 \sigma_2 + \sigma_2^2.$$

Using the moments for $\log(S_T/S_0)$ and $\log P(T, S)$, we can use this formula to price an absolute outperformance option if the strike K is zero.

Relative outperformance Using the notation X_1 and X_2 from the previous section, $\log(X_1/X_2)$ is normal with mean $\mu_1 - \mu_2$ and variance $\tilde{\sigma}^2$. Thus

$$\mathbb{E}\left[\left(\frac{X_1}{X_2} - K\right)^+\right] = e^{\mu_1 - \mu_2 + \tilde{\sigma}^2/2} N(d_1) - K N(d_2),$$

with

$$d_2 = \frac{-\log K + \mu_1 - \mu_2}{\tilde{\sigma}} \quad \text{and} \quad d_1 = d_2 + \tilde{\sigma}.$$

Using the moments for $\log(S_T/S_0)$ and $\log P(T, S)$ again, we can now price a relative outperformance option.

Absolute outperformance with coupons and strike So far, we have made two assumptions for the absolute outperformance option:

- The bond is not allowed to pay any coupons.

- The strike is supposed to be zero.

In this section we will remove these restrictions. Note, however, that a non-zero strike can be treated in the same way as a negative coupon, so that we can restrict ourselves to the case of a coupon bond with zero strike. Assume the bond pays at time S_i the amount α_i.

We let $X = \log(S_T/S_0)$ (with moments μ_1 and σ_1 under the forward measure) and $Y = r_T$ (the short rate, with moments μ_2 and σ_2). Their correlation is denoted by ρ. We also let $P(T, y, S_i) = A_i \exp(B_i y)$ denote the price of a zero bond at time T paying 1 at time T, given a short rate $y = r_T$ at time T. The price of the option (without discounting) is given by

$$\int_{-\infty}^{\infty} \int_{-\infty}^{\infty} \left(w_1 e^x - w_2 \sum_i \alpha_i P(T, y, S_i)\right)^+ \mathbb{P}[X \in dx, Y \in dy]$$

$$= \int_{-\infty}^{\infty} \int_{-\infty}^{\infty} \left(w_1 e^x - w_2 \sum_i \alpha_i P(T, y, S_i)\right)^+ \mathbb{P}[Y \in dy \mid X \in dx] \mathbb{P}[X \in dx].$$

Note that the conditional distribution of Y given X is normal, with moments

$$\mathbb{E}[\,Y \mid X\,] = \mu_2 + \frac{\rho \sigma_2}{\sigma_1}(X - \mu_1)$$

and

$$\mathrm{Var}[\,Y \mid X\,] = \sigma_2^2(1 - \rho^2).$$

This can be verified immediately or the details can be found in [70]. We denote such a distribution by Y_x.

The inner integral can be evaluated in a similar way to a coupon bond option, using Jamshidian's decomposition.

If r is normally distributed, then

$$\mathbb{E}\left[\left(K - \sum_{i=1}^{n} A_i e^{B_i r}\right)^+\right] = \sum_{i=1}^{n} \mathbb{E}\left[(K_i - A_i e^{B_i r})^+\right],$$

where $K_i = A_i e^{B_i r_*}$ with the (unique) r_* that solves

$$K = \sum_{i=1}^{n} A_i e^{B_i r_*}.$$

The number r_* can be determined efficiently with the Newton–Raphson method [61].

Hence the value of our option is

$$\int_{-\infty}^{\infty} \mathbb{E}\left[\left(w_1 e^x - w_2 \sum_i \alpha_i P(T, Y_x, T_i)\right)^+\right] \mathbb{P}[\,X \in dx\,]$$

$$= \int_{-\infty}^{\infty} \sum_{i=1}^{n} \mathbb{E}\left[\left(K_i(x) - A_i e^{B_i Y_x}\right)^+\right] \mathbb{P}[\,X \in dx\,].$$

The outer integral can be solved by numerical integration.

Chapter 6
The Tree Approach

A *tree* is a numerical scheme that approximates a process by a random walk with a finite number of states, as defined in Section 1.1.6. If one process has to be modelled, there may be two states corresponding to up and down movements or three states incorporating an additional stationary state. Trees of this nature are denoted *one-factor binomial* and *trinomial* trees respectively. In the case of two processes, the corresponding tree is called a *two-factor* tree. The convergence of the processes modelled through tree schemes to Gaussian processes is stated in Donsker's theorem (see Section 1.1.13). Trees may be viewed as a special case of explicit finite difference schemes (see Chapter 8).

There are several reasons for the use of tree schemes for option pricing.

- Trees are the obvious tool for obtaining American- or Bermudan-style option prices.

- They allow the incorporation of discrete dividends.

- It is possible to model a variety of products on the same tree.

- They allow time-dependent drift and volatility.

- Convergence and numerical stability are good, provided that the boundary conditions are "nice".

- Greeks can be obtained within the scheme.

There are some inevitable downsides:

- Trees, in particular trees with more than one factor, are relatively slow compared with analytical approaches.

- The greeks are not very stable.

- Discontinuities in the boundary conditions, such as barriers or digital payoff functions, typically lead to a "sawtooth" pattern in the solution as a function of spot. We present a method to smooth the solution in Section 6.3.

- It is not straightforward to incorporate a volatility smile into the scheme. For literature on implied trees, we refer the reader to [66, 17, 22]. An alternative is the finite difference scheme described in Chapter 8.

This chapter expands and generalises the example of the discrete one-step binomial arbitrage-free pricing framework discussed in Section 1.2.

6.1 Setting Up the Tree

The first stage in the tree approach to option pricing is the initialisation of the tree structure corresponding to the underlying process or processes. We now provide the mechanism for achieving this for one- and two-factor trees based on either a Gaussian or lognormal process.

6.1.1 Binomial Tree for a Gaussian Process

In this section we show how to model a process X of the form

$$X_t = \int_0^t \sigma_s \, dW_s, \quad t \in [0, T],$$

on a binomial tree, with W denoting a Brownian motion and $\sigma > 0$ a deterministic volatility. From each node emanate two nodes that correspond to an up (u) and a down (d) movement, as illustrated in Figure 6.1. We denote the corresponding probabilities by p_u and p_d. Let the grid spacing in the time direction be denoted by Δt and the step size in the space direction by Δx.

Thus, assuming a location x of the process at time t, at time $t + \Delta t$ the

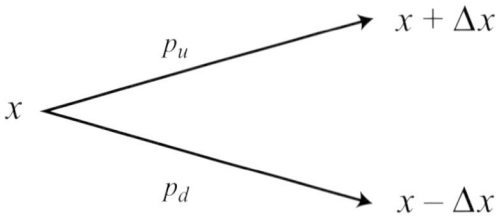

6.1 Evolution in a binomial tree.

process moves to $x + \Delta x$ with probability p_u and to $x - \Delta x$ with probability p_d. Since the sum of the probabilities is 1 and the mean of $X_{t+\Delta t} - X_t$ is 0, we have $p_u + p_d = 1$ and $p_u \Delta x + p_d(-\Delta x) = 0$. Hence

$$p_u = p_d = \tfrac{1}{2}.$$

In order to obtain a *recombining tree*, Δx has to be constant in time. Hence the time dependence of the variance of the process X has to be reflected in the choice of the time steps.

We let N denote the number of time steps to be used to cover the horizon $[0, T]$. We claim that the right choice for Δx is given by

$$(\Delta x)^2 = \int_0^T \sigma_s^2 \, ds / N.$$

Let a sequence of time steps

$$(\Delta t)_n = t_n - t_{n-1}, \quad n = 1, 2, \ldots, N,$$

The Tree Approach

be defined by

$$t_n = \inf_t \left\{ \int_0^t \sigma_s^2 \, ds \geq n(\Delta x)^2 \right\}, \quad n = 0, 1, 2, \ldots, N.$$

We then have $t_0 = 0$ and $t_n = T$. Moreover, the variance equation is satisfied, since

$$p_u(\Delta x)^2 + p_d(-\Delta x)^2 = (\Delta x)^2 = \int_{t_{n-1}}^{t_n} \sigma_s^2 \, ds, \quad n = 1, 2, \ldots, N.$$

In the special case of constant σ, the calculation reduces to

$$\Delta t = T/N, \qquad \Delta x = \sigma\sqrt{\Delta t}.$$

6.1.2 Binomial Tree for a Lognormal Process

In this section we show how to model the lognormal process X defined by

$$X_t = \exp\left(\mathcal{N}_t + \int_0^t \sigma_s \, dW_s\right), \quad t \in [0, T],$$

on a binomial tree, with W denoting a Brownian motion and $\sigma > 0$ a deterministic volatility. Again, from each node emanate two nodes that correspond to an up (u) and a down (d) movement. We denote the corresponding probabilities by p_u and p_d. We let the grid spacing in the time direction be denoted by Δt and the step size in the space direction by Δx. An up movement starting in e^x therefore leads to $e^{x+\Delta x}$, while a down movement ends in $e^{x-\Delta x}$. Since the sum of the probabilities is 1 and the mean of $X_{t+\Delta t}/X_t$ is $e^{\mathcal{M}_{t,t+\Delta t}}$, we have $p_u + p_d = 1$ and $p_u e^{\Delta x} + p_d e^{-\Delta x} = e^{\mathcal{M}_{t,t+\Delta t}}$. Hence

$$p_u = \frac{e^{\mathcal{M}_{t,t+\Delta t}} - e^{-\Delta x}}{e^{\Delta x} - e^{-\Delta x}} \quad \text{and} \quad p_d = \frac{e^{\Delta x} - e^{\mathcal{M}_{t,t+\Delta t}}}{e^{\Delta x} - e^{-\Delta x}}.$$

The variance equation reads

$$p_u(\Delta x)^2 + p_d(-\Delta x)^2 = \Sigma_{t,t+\Delta t}^2.$$

This yields

$$\Delta x = \Sigma_{t,t+\Delta t}.$$

In order to obtain a recombining tree, Δx has to be constant in time. Therefore the time steps have to be adjusted as in the previous section.

6.1.3 Trinomial Tree

In this section we show how to model the Gaussian process X given by

$$X_t = \int_0^t \sigma_s \, dW_s, \quad t \in [0, T],$$

with W denoting a Brownian motion and $\sigma > 0$ a deterministic volatility on a trinomial tree. Now, from each node emanate *three* other nodes corresponding to an increasing (u), a constant (m) and a decreasing (d) movement (see Figure 6.2). We denote the corresponding probabilities by p_a with $a \in \{u, m, d\}$.

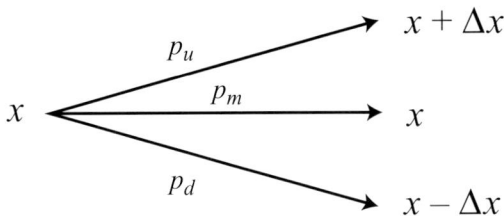

6.2 Gaussian evolution in a trinomial tree.

Additionally, we denote the grid spacing in the time direction by Δt and the step size in the space direction by Δx.

The (time-dependent) tree probabilities are chosen in such a way as to match the first and second moments of the process X, ie

$$1 = p_u + p_m + p_d \quad \text{and} \quad 0 = p_u \Delta x + p_d(-\Delta x)$$

for normalisation and mean, and

$$p_u(\Delta x)^2 + p_d(-\Delta x)^2 = \sigma_t^2 \Delta t$$

for the second moment. Solving for p_u, p_m and p_d yields

$$p_u = \frac{\sigma_t^2 \Delta t}{2(\Delta x)^2}, \quad p_m = 1 - 2p_u, \quad p_d = p_u.$$

In order to obtain non-negative probabilities, we have to ensure that

$$(\Delta x)^2 \geqslant \sigma_t^2 \Delta t.$$

We thus set the (time-independent) grid spacings to

$$\Delta t = T/N, \quad \Delta x = \frac{c}{\sqrt{\Delta t}} \max_{t \in [0,T]} \sigma_t,$$

with N denoting the number of time steps of the tree and $c \geqslant 1$ a constant. Choosing a larger constant c results in a wider spacing Δx and a higher probability p_m of the middle node. In the case of $c = 1$ and constant volatility σ, the trinomial tree reduces to a binomial tree.

Application: one asset Let S denote an asset process with time-dependent volatility σ, dividend yield d and interest rate r. Then, we can write

$$S_t = S_0 \exp(\mathcal{M}_t + X_t), \quad t \in [0, T].$$

Hence the process S is a deterministic function of the process X.

6.1.4 Two-Factor Tree

In this section we show how to model a two-dimensional Gaussian process, ie a pair of processes (X^1, X^2) of the form

$$X_t^1 = \int_0^t \sigma_s^1 \, dW_s^1, \quad X_t^2 = \int_0^t \sigma_s^2 \, dW_s^2, \quad t \in [0, T],$$

with W^1, W^2 denoting two Brownian motions with correlation

$$\langle dW_t^1, dW_t^2 \rangle = \rho_t \, dt, \quad t \in [0, T],$$

and $\sigma^1, \sigma^2 > 0$ two deterministic volatilities. We construct a tree that is trinomial in both dimensions. From each node emanate nine other nodes corresponding to an increasing (u), a constant (m) and a decreasing (d) primary/secondary process. We denote the corresponding probabilities by p_{ab}, with $a \in \{u, m, d\}$ referring to the primary process X^1 movement and $b \in \{u, m, d\}$ referring to the movement of X^2. We denote the (time-independent) grid spacing in the time direction by Δt and the (time-independent) step sizes of the primary and secondary processes by Δx^1 and Δx^2. The (time-dependent) tree probabilities are chosen in such a way as to match the first and second moments of the processes X^1 and X^2.

Tree probabilities We have the equations

$$p_{uu} + p_{mu} + p_{du} + p_{um} + p_{mm} + p_{dm} + p_{ud} + p_{md} + p_{dd} = 1,$$
$$(p_{uu} + p_{um} + p_{ud})\Delta x^1 + (p_{du} + p_{dm} + p_{dd})(-\Delta x^1) = 0,$$
$$(p_{uu} + p_{mu} + p_{du})\Delta x^2 + (p_{ud} + p_{md} + p_{dd})(-\Delta x^2) = 0,$$

for normalisation and mean, and

$$(p_{uu} + p_{um} + p_{ud})(\Delta x^1)^2 + (p_{du} + p_{dm} + p_{dd})(-\Delta x^1)^2 = (\sigma_t^1)^2 \Delta t,$$
$$(p_{uu} + p_{mu} + p_{du})(\Delta x^2)^2 + (p_{ud} + p_{md} + p_{dd})(-\Delta x^2)^2 = (\sigma_t^2)^2 \Delta t,$$
$$(p_{uu} + p_{dd})\Delta x^1 \Delta x^2 - (p_{ud} + p_{du})\Delta x^1 \Delta x^2 = \sigma_t^1 \sigma_t^2 \rho_t \Delta t,$$

for the second moments. Since the processes are driftless, we add the symmetry requirements

$$p_{dm} = p_{um} \quad \text{and} \quad p_{mu} = p_{md}.$$

From this, we conclude from

$$p_{uu} + p_{um} + p_{ud} = \mathbb{P}[\text{asset 1 up}] = \mathbb{P}[\text{asset 1 down}]$$
$$= p_{du} + p_{dm} + p_{dd}$$

that $p_{uu} = p_{dd}$ must also hold. A similar argument for asset 2 yields $p_{ud} = p_{du}$.

We thus arrive at a system of eight independent equations for the nine unknown probabilities, namely

$$p_{uu} + p_{mu} + p_{du} + p_{um} + p_{mm} + p_{dm} + p_{ud} + p_{md} + p_{dd} = 1,$$
$$(p_{uu} + p_{um} + p_{ud}) - (p_{du} + p_{dm} + p_{dd}) = 0,$$
$$(p_{uu} + p_{mu} + p_{du}) - (p_{ud} + p_{md} + p_{dd}) = 0,$$
$$(p_{uu} + p_{um} + p_{ud}) + (p_{du} + p_{dm} + p_{dd}) = s_1^2,$$
$$(p_{uu} + p_{mu} + p_{du}) + (p_{ud} + p_{md} + p_{dd}) = s_2^2,$$
$$(p_{uu} + p_{dd}) - (p_{ud} + p_{du}) = \rho_t s_1 s_2,$$
$$p_{dm} = p_{um},$$
$$p_{mu} = p_{md}.$$

Here s_1 and s_2 are given by

$$s_1 = \frac{\sigma_t^1 \sqrt{\Delta t}}{\Delta x^1}, \qquad s_2 = \frac{\sigma_t^2 \sqrt{\Delta t}}{\Delta x^2}.$$

Eliminating eight of the nine variables (all except p_{mm}) yields

$$p_{ud} = \tfrac{1}{4}(p_{mm} + s_1^2 + s_2^2 - \rho_t s_1 s_2 - 1).$$

An equivalent independent equation to the system of eight equations is

$$p_{uu} = \tfrac{1}{4}(p_{mm} + s_1^2 + s_2^2 + \rho_t s_1 s_2 - 1).$$

A good value for the middle node is the non-correlated solution

$$(1 - s_1^2)(1 - s_2^2).$$

Solving the system then yields

$$p_{uu} = p_{dd} = \tfrac{1}{4}(s_1^2 s_2^2 + \rho_t s_1 s_2),$$
$$p_{ud} = p_{du} = \tfrac{1}{4}(s_1^2 s_2^2 - \rho_t s_1 s_2),$$
$$p_{um} = p_{dm} = \tfrac{1}{2}s_1^2(1 - s_2^2),$$
$$p_{mu} = p_{md} = \tfrac{1}{2}s_2^2(1 - s_1^2),$$
$$p_{mm} = (1 - s_1^2)(1 - s_2^2).$$

We choose

$$\Delta x^1 = c_1 \sqrt{\Delta t} \max_{0 \leqslant u \leqslant T} \sigma_u^1, \qquad \Delta x^2 = c_2 \sqrt{\Delta t} \max_{0 \leqslant u \leqslant T} \sigma_u^2,$$

with $c_1 \geqslant 1$ and $c_2 \geqslant 1$. This ensures that $0 \leqslant s_1 \leqslant 1$ and $0 \leqslant s_2 \leqslant 1$. In the special case of constant volatilities σ^1 and σ^2 and $c_1 = c_2 = 1$, the tree reduces to a two-factor binomial tree. In order to put more weight on the middle nodes, we may increase c_1 and c_2. However, since the numbers p_{ab} must be probabilities, we have the constraint

$$|\rho_t| \leqslant s_1 s_2 = \frac{\sigma_t^1 \sigma_t^2}{c_1 c_2 (\max_{0 \leqslant u \leqslant T} \sigma_u^1)(\max_{0 \leqslant u \leqslant T} \sigma_u^2)}, \qquad t \in [0, T].$$

In the special case of constant variances σ^1 and σ^2, this can always be achieved via a suitable choice of c_1 and c_2. If this constraint is violated, the formulas for the probabilities need to be modified. We present the following modifications:

- $\rho_t > s_1 s_2$ and $|\rho_t| < \min(s_1/s_2, s_2/s_1)$.

  If $\rho_t > s_1 s_2$, the off-diagonal probabilities p_{ud} and p_{du} become negative. We therefore abandon our value for p_{mm} and choose $p_{ud} = 0$ instead. The solution to the above problem now reads

$$p_{uu} = p_{dd} = \tfrac{1}{2}\rho_t s_1 s_2,$$
$$p_{du} = p_{ud} = 0,$$
$$p_{dm} = p_{um} = \tfrac{1}{2}s_1(s_1 - \rho_t s_2),$$
$$p_{md} = p_{mu} = \tfrac{1}{2}s_2(s_2 - \rho_t s_1),$$
$$p_{mm} = (1 - s_1^2)(1 - s_2^2) + (\rho - s_1 s_2)s_1 s_2,$$

and is non-negative if $|\rho_t| < \min(s_1/s_2, s_2/s_1)$.

- $\rho_t < -s_1 s_2$ and $|\rho_t| < \min(s_1/s_2, s_2/s_1)$.

  If $\rho_t < -s_1 s_2$, the diagonal probabilities p_{uu} and p_{dd} become negative. We therefore abandon our value for p_{mm} and choose $p_{uu} = 0$ instead. The solution to the above problem now reads

$$p_{dd} = p_{uu} = 0,$$
$$p_{du} = p_{ud} = -\tfrac{1}{2}\rho_t s_1 s_2,$$
$$p_{dm} = p_{um} = \tfrac{1}{2}s_1(s_1 + \rho_t s_2),$$
$$p_{md} = p_{mu} = \tfrac{1}{2}s_2(s_2 + \rho_t s_1),$$
$$p_{mm} = (1 - s_1^2)(1 - s_2^2) - (\rho + s_1 s_2)s_1 s_2,$$

and is non-negative if $|\rho_t| < \min(s_1/s_2, s_2/s_1)$.

- If $|\rho_t| > \min(s_1/s_2, s_2/s_1)$, we obtain negative probabilities. As this only happens in extreme cases, a reduction of ρ_t may solve this problem.

Application: two assets Let S^1 and S^2 denote two asset processes with time-dependent volatilities σ^i, correlation ρ, dividend yields d^i and interest rate r. Then, we can write

$$S_t^i = S_0^i \exp(\mathcal{M}_t^i + X_t^i), \qquad t \in [0, T], \quad i = 1, 2.$$

Hence the processes S^1 and S^2 are deterministic functions of the processes X^1 and X^2.

Application: asset and short rate Let S denote an asset process and r the corresponding short rate process, as described in Section 1.5. The processes $\tilde{r}$

and $\tilde{S}$, with

$$\tilde{r}_t = \Lambda_t r_t + r_0 - \mathbb{E}[\Lambda_t r_t]$$
$$= \Lambda_t r_t - \int_0^t \theta_s \Lambda_s \, ds$$
$$= \Lambda_t(r_t - f(0, t)) + r_0 - \int_0^t \sigma_s^2 \Lambda_s B_1(s, t) \, ds$$

and

$$\tilde{S}_t = \log\left(\frac{S_t}{S_0}\right) - \Lambda_t B_1(0, t) r_t + A_t,$$

where

$$A_t = -\mathbb{E}[\log B_t - r_0 B_1(0, t)] + B_1(0, t)\big(\mathbb{E}[\Lambda_t r_t] - r_0\big) + \int_0^t [d_s + \tfrac{1}{2}(\sigma_s^S)^2] \, ds$$
$$= -\int_0^t \theta_s B_1(s, t) \, ds + B_1(0, t) \int_0^t \theta_s \Lambda_s \, ds + \int_0^t [d_s + \tfrac{1}{2}(\sigma_s^S)^2] \, ds,$$

are deterministic functions of the short rate and asset processes r and S. Using Section 1.5, it is easy to verify that these processes satisfy

$$d\tilde{r}_t = \sigma_t \Lambda_t \, dW_t$$

and

$$d\tilde{S}_t = \sigma_t^S \, dW_t^S - \sigma_t \Lambda_t B_1(0, t) \, dW_t.$$

The process for $\tilde{S}_t$ can be rewritten as

$$d\tilde{S}_t = \hat{\sigma}_t \, d\hat{W}_t,$$

with

$$\hat{\sigma}_t^2 = \sigma_t^2 \Lambda_t^2 B_1^2(0, t) - 2\rho_t \sigma_t \sigma_t^S \Lambda_t B_1(0, t) + (\sigma_t^S)^2,$$

where $\hat{W}_t$ is a standard Brownian motion with

$$\langle dW_t, d\hat{W}_t \rangle = \frac{\rho_t \sigma_t^S - \sigma_t \Lambda_t B_1(0, t)}{\hat{\sigma}_t} \, dt.$$

6.2 Option Pricing Using Trees

Generally in option pricing, the final payoff of the instrument is known, and it is the value today of this payoff that we require. This is achieved in the tree approach by "running backwards" through the tree.

Here we provide a general procedure for running backwards through the tree that allows for possible early exercise features that are present with American and Bermudan options.

6.2.1 Running Through the Tree

To compute the fair value of an option, the computation runs backwards through the tree, starting at maturity T and proceeding backwards in time steps of size Δt. At every node (S_t, t), with t denoting the time slice and S_t being the process value (if two processes are modelled, this reads $S_t = (S_t^1, S_t^2)$), the option has an inner value I, an option value V and a holding value H.

- The inner value $I(S_t, t)$ is the amount that the holder receives at time t on exercising the option. It can be expressed as a function of time t, the modelled process S_t at time t and other deterministic parameters. The function I is part of the option contract. Typical examples are the inner values for an American call or an American digital call:

$$\max(S_t - K_t, 0), \qquad 1_{\{S_t > H_t\}}.$$

In the case of two assets, inner values may be of the best of two, outperformance or relative outperformance types:

$$\max\bigl(\max(S_t^1, S_t^2) - K_t, 0\bigr), \quad \max\left(\frac{S_t^1}{S_0^1} - \frac{S_t^2}{S_0^2}, 0\right), \quad \max\left(\frac{S_t^1}{S_t^2} - K_t, 0\right).$$

- There is a holding value H assigned to node (S_t, t) that can be calculated from option values $V(S_{t+\Delta t}, t + \Delta t)$ of the next time slice. If the possible states to move to from S_t are $(S_{t+\Delta t}^a, a \in A)$ with probabilities $(p_a, a \in A)$, then the holding value is

$$H(S_t, t) = e^{-R_{t,t+\Delta t}} \sum_{a \in A} p_a V(S_{t+\Delta t}^a, t + \Delta t), \tag{6.1}$$

where $e^{-R_{t,t+\Delta t}}$ is the discount factor applicable on the time interval $[t, t + \Delta t]$.

- The option values are obtained from the inner values and the holding values. At maturity T, the option has to be exercised, so

$$V(S_T, T) = I(S_T, T).$$

At any time slice t prior to T, the option value is

$$V(S_t, t) = \max\bigl(I(S_t, t), H(S_t, t)\bigr)$$

if there is an exercise right at t. If no early exercise is allowed at t, the option value is set to the holding value, ie

$$V(S_t, t) = H(S_t, t).$$

The fair price of the option at time $t = 0$ is $V(S_0, 0)$.

6.2.2 Tree Greeks

A standard procedure for calculating greeks is perturbation. While this is reliable in the case of closed-form solutions, it is not advisable for tree numerical schemes since the pricing function is in general not smooth. Another downside of perturbation is calculation time.

Instead, the tree approach can be used to calculate delta, gamma and theta. We show how to obtain greeks in the case of a one-factor binomial or trinomial lognormal tree.

First, the number of time steps has to be increased by two, such that the node representing the option value is the middle node at time index 2. More formally, we denote the option price of the extended model at time node i_t and space node i_s by $V(i_t, i_s)$, with

$$i_t \in \{0, 1, 2, \ldots, N, N+1, N+2\},$$
$$i_s(i_t) \in \{0, 1, 2, \ldots, I_s(i_t)\}.$$

The number of spatial nodes $I_s + 1$ is dependent on the time node and given by $I_s(i_t) = i_t$ in the binomial and $I_s(i_t) = 2i_t$ in the trinomial one-factor case. We then obtain the price Π, delta Δ, gamma Γ and theta Θ from the tree as

$$\Pi = V(2, 1),$$
$$\Delta = \frac{1}{S_0(u^2 - u^{-2})} \left\{ V(2, 2)u^{-2} + V(2, 1)(u^2 - u^{-2}) - V(2, 0)u^2 \right\},$$
$$\Gamma = \frac{2}{S_0^2(u^2 - u^{-2})} \left\{ \frac{V(2, 2) - V(2, 1)}{u^2 - 1} - \frac{V(2, 1) - V(2, 0)}{1 - u^{-2}} \right\},$$
$$\Theta = \tfrac{1}{2}\{V(4, 2) + V(2, 1)\}.$$

6.3 Barrier Options

We now apply these ideas to the pricing of barrier options. As already stated, pricing barriers can pose some difficulties because of the discontinuous behaviour near the barrier level, a problem we also address here.

6.3.1 Knock-Out Options

We let the option contract to be priced on the tree contain a barrier $H = (H_t, t \in [0, T])$. Here H_t is understood to be a region of the state space of the process S where the option becomes worthless, eg $H_t = [h_t, \infty)$ for an up-and-out option or $H_t = [0, h_t^1] \times [0, h_t^2]$ for an option with down-an-out barriers on two assets in the case of a two-factor tree. Assume that, if the option knocks out at time t, a rebate G_t is paid to the investor.

In these cases the option values V described in Section 6.2.1 have to be changed. The option value at maturity is

$$V(S_T, T) = I(S_T, T) 1_{\{S_T \notin H_T\}} + G_T 1_{\{S_T \in H_T\}}.$$

At any time slice t prior to T, the option value is

$$V(S_t, t) = \max(I(S_t, t), H(S_t, t)) 1_{\{S_t \notin H_t\}} + G_t 1_{\{S_t \in H_t\}}$$

if there is an exercise right at t. If no early exercise is allowed at t, the option value is set to

$$V(S_t, t) = H(S_t, t) 1_{\{S_t \notin H_t\}} + G_t 1_{\{S_t \in H_t\}}.$$

If the payment of the rebate is deferred to the maturity of the option, then G_t has to be replaced by $G_t e^{-R_{t,T}}$.

The calculation of inner values I and holding values H remains unchanged.

6.3.2 Knock-In Options

Knock-in options are worthless (or pay a no-hit rebate G) until the process S enters the region described by the barrier H. They require a so-called *thick tree* with two layers, ie the main pricing layer and an auxiliary layer. The *auxiliary layer* contains the holding, inner and option values V^a, I^a and H^a described in Section 6.2.1 for the corresponding product without the barrier H. The main layer (V^m, H^m, I^m) is set up via

$$H^m(S_t, t) = e^{-R_{t+\Delta t}} \sum_{a \in A} \{ p_a V^a(S^a_{t+\Delta t}, t + \Delta t) 1_{\{S^a_{t+\Delta t} \in H_{t+\Delta t}\}}$$
$$+ p_a V^m(S^a_{t+\Delta t}, t + \Delta t) 1_{\{S^a_{t+\Delta t} \notin H_{t+\Delta t}\}} \}$$

for the holding value,

$$I^m(S_t, t) = I^a(S_t, t) 1_{\{S_t \in H_t\}}$$

for the inner value, and

$$V^m(S_t, t) = \max(I^m(S_t, t), H^m(S_t, t))$$

if there is an exercise right at t. If no early exercise is allowed at t, the option value is set to

$$V^m(S_t, t) = H^m(S_t, t).$$

We remark that the no-hit rebate G cannot be collected if the holder of the option exercises early.

6.3.3 Probability Fitting Near the Barrier

In order to avoid the well-known "sawtooth" shape of the solution as a function of the spot of the secondary underlying, we alter the probabilities p_{ab} derived above to $\tilde{p}_{ab}$ ($a, b \in \{u, m, d\}$).

We let the barrier level B be below S_t^2 and

$$S_*^2 = \min_{s \in [t, t+\Delta t]} S_s^2.$$

In the case of a node being close to the barrier, we alter the secondary stock process S^2 to $\tilde{S}^2$, with

$$\tilde{S}^2_{t+\Delta t} = S^2_{t+\Delta t} 1_{\{S_*^2 > B\}} + B 1_{\{S_*^2 <= B\}}.$$

We remark that $(S^1, \tilde{S}^2)$ is still a Markov process and we calculate the first two moments

$$\mathbb{E}_t[\tilde{S}^2_{t+\Delta t}] = S_t^2 F_t^2 \mathbb{E}[e^{\sigma_t^2 \sqrt{\Delta t} X - (\sigma_t^2)^2 \Delta t / 2} 1_{\{S_*^2 > B\}}] + B \mathbb{P}[S_*^2 \leq B]$$
$$= S_t^2 F_t^2 (1 - p_B^+) + B p_B$$
$$\mathbb{E}_t[(\tilde{S}^2_{t+\Delta t})^2] = (S_t^2 F_t^2)^2 e^{(\sigma_t^2)^2 \Delta t} (1 - p_B^{++}) + B p_B,$$

where F_t^i denotes the expectation of $S_{t+\Delta t}^i/S_t^i$, X is a standard normally distributed random variable, and

$$p_B = N\left(\pm\frac{-\alpha_t\,dx + \tilde{\mu}_t\Delta t}{\sigma_t^2\sqrt{\Delta t}}\right) + e^{2\alpha_t\tilde{\mu}_t\,dx/(\sigma_t^2)^2}N\left(\pm\frac{-\alpha_t\,dx - \tilde{\mu}_t\Delta t}{\sigma_t^2\sqrt{\Delta t}}\right),$$

$$p_B^+ = N\left(\pm\frac{-\alpha_t\,dx + \tilde{\mu}_t^+\Delta t}{\sigma_t^2\sqrt{\Delta t}}\right) + e^{2\alpha_t\tilde{\mu}_t^+\,dx/(\sigma_t^2)^2}N\left(\pm\frac{-\alpha_t\,dx - \tilde{\mu}_t^+\Delta t}{\sigma_t^2\sqrt{\Delta t}}\right),$$

$$p_B^{++} = N\left(\pm\frac{-\alpha_t\,dx + \tilde{\mu}_t^{++}\Delta t}{\sigma_t^2\sqrt{\Delta t}}\right) + e^{2\alpha_t\tilde{\mu}_t^{++}\,dx/(\sigma_t^2)^2}N\left(\pm\frac{-\alpha_t\,dx - \tilde{\mu}_t^{++}\Delta t}{\sigma_t^2\sqrt{\Delta t}}\right),$$

with

$$\tilde{\mu}_t = r_t - d_t^2 - \tfrac{1}{2}(\sigma_t^2)^2,$$
$$\tilde{\mu}_t^+ = r_t - d_t^2 + \tfrac{1}{2}(\sigma_t^2)^2,$$
$$\tilde{\mu}_t^{++} = r_t - d_t^2 + \tfrac{3}{2}(\sigma_t^2)^2.$$

The distance to the barrier B at time t is measured by

$$\alpha_t = \frac{\log(B/\tilde{S}_t^2)}{\Delta x^2}, \qquad \alpha_{t+\Delta t} = \frac{\log(B/\tilde{S}_{t+\Delta t}^2)}{\Delta x^2}.$$

Note that p_B is the absorption probability of S^2 from above ($\pm = +$) or below ($\pm = -$) during $[t, t+\Delta t]$. As mentioned above, we only detail the case $S_t^2 > B$.

Using a lognormal approximation of the distribution of

$$(\tilde{S}_{t+\Delta t}^2 \mid S_t^1, \tilde{S}_t^2) = (\tilde{S}_{t+\Delta t}^2 \mid \tilde{S}_t^2),$$

we obtain the modified moments

$$e_B\Delta t = \log\left(F_t^2(1 - p_B^+) + \frac{B}{S_t^2}p_B\right) - \tfrac{1}{2}\sigma_B^2\Delta t - [r_t - d_t^2 - \tfrac{1}{2}(\sigma_t^2)^2]\Delta t,$$

$$\sigma_B^2\Delta t = \log\left(\frac{\mathbb{E}_t[(\tilde{S}_{t+\Delta t}^2)^2]}{\mathbb{E}_t[\tilde{S}_{t+\Delta t}^2]^2}\right)$$

of

$$(\tilde{X}_{t+\Delta t}^2 \mid \tilde{S}_t^2) = \log\left(\frac{\tilde{S}_{t+\Delta t}^2}{\tilde{S}_t^2}\right) - [r_t - d_t^2 - \tfrac{1}{2}(\sigma_t^2)^2]\Delta t$$

on $\{\tilde{S}_t^2 > B\}$. The modified correlation ρ_B of $(\tilde{X}_{t+\Delta t}^2 \mid S_t^1, \tilde{S}_t^2)$ and $(X_{t+\Delta t}^1 \mid S_t^1, \tilde{S}_t^2)$ is calculated in Section 6.3.4.

For every node, we calculate 10 probabilities, namely the nine probabilities $\tilde{p}_{ab}$ ($a, b \in \{u, m, d\}$) for up, middle and down movements of the two assets, and the absorption probability p_B. As for the case without a barrier, we aim at a match of the first two moments. The equations are

$$1 = \tilde{p}_{uu} + \tilde{p}_{mu} + \tilde{p}_{du} + \tilde{p}_{um} + \tilde{p}_{mm} + \tilde{p}_{dm} + \tilde{p}_{ud} + \tilde{p}_{md} + \tilde{p}_{dd} + p_B,$$

$$0 = (\tilde{p}_{uu} + \tilde{p}_{um} + \tilde{p}_{ud})\Delta x^1 + (\tilde{p}_{du} + \tilde{p}_{dm} + \tilde{p}_{dd})(-\Delta x^1),$$

$$e_B\Delta t = (\tilde{p}_{uu} + \tilde{p}_{mu} + \tilde{p}_{du} + p_B\alpha_{t+\Delta t})\Delta x^2 + (\tilde{p}_{ud} + \tilde{p}_{md} + \tilde{p}_{dd})(-\Delta x^2),$$

for normalisation and mean, and

$$(\sigma_t^1)^2 \Delta t = (\tilde{p}_{uu} + \tilde{p}_{um} + \tilde{p}_{ud})(\Delta x^1)^2 + (\tilde{p}_{du} + \tilde{p}_{dm} + \tilde{p}_{dd})(-\Delta x^1)^2,$$

$$(\sigma_B^2)^2 \Delta t + (e_B \Delta t)^2 = (\tilde{p}_{uu} + \tilde{p}_{mu} + \tilde{p}_{du})(\Delta x^2)^2$$
$$+ (\tilde{p}_{ud} + \tilde{p}_{md} + \tilde{p}_{dd} + p_B \alpha_{t+\Delta t}^2)(-\Delta x^2)^2,$$

$$\sigma_t^1 \sigma_t^2 \rho_B \Delta t = (\tilde{p}_{uu} + \tilde{p}_{dd})\Delta x^1 \Delta x^2 - (\tilde{p}_{ud} + \tilde{p}_{du})\Delta x^1 \Delta x^2,$$

for the second moments. We introduce the notation

$$e_2 = e_B \frac{\Delta t}{\Delta x^2} - p_B \alpha_{t+\Delta t},$$

$$s_1^2 = (\sigma_t^1)^2 \frac{\Delta t}{(\Delta x^1)^2},$$

$$s_2^2 = [(\sigma_B^2)^2 + e_B^2 \Delta t] \frac{\Delta t}{(\Delta x^2)^2},$$

$$\rho_{\text{loc}} = \sigma_t^1 \sigma_t^2 \rho_B \frac{\Delta t}{\Delta x^1 \Delta x^2}.$$

Since we assume that $S_t^2 > B$, we have $\alpha_t < 0$ as well as $e_2 > 0$. There are the additional constraints

$$\tilde{p}_{uu} = \tilde{p}_{mu} = \tilde{p}_{du} = 0, \quad 1 < \alpha_{t+\Delta t},$$
$$\tilde{p}_{um} = \tilde{p}_{mm} = \tilde{p}_{dm} = 0, \quad 0 < \alpha_{t+\Delta t},$$
$$\tilde{p}_{ud} = \tilde{p}_{md} = \tilde{p}_{dd} = 0, \quad -1 < \alpha_{t+\Delta t},$$

as well as the condition of positive probabilities. We obtain the following solution:

- If $-1 > \alpha_{t+\Delta t}$, ie no movement passes the barrier, we treat the knock-out event as independent of the asset movements, so

$$\tilde{p}_{ab} = p_B p_{ab}, \quad a, b \in \{u, m, d\}.$$

- If $-1 \leq \alpha_{t+\Delta t} < 0$, ie only the down movement passes the barrier, we have

$$\tilde{p}_{uu} = \tfrac{1}{2}(m_{\text{pos}} + \rho_{\text{pos}}),$$
$$\tilde{p}_{mu} = 0,$$
$$\tilde{p}_{du} = \tfrac{1}{2}(m_{\text{pos}} - \rho_{\text{pos}}),$$
$$\tilde{p}_{um} = \tfrac{1}{2}(w_{\text{pos}} - \rho_{\text{pos}}),$$
$$\tilde{p}_{mm} = 0,$$
$$\tilde{p}_{dm} = \tfrac{1}{2}(w_{\text{pos}} + \rho_{\text{pos}}),$$
$$\tilde{p}_{ud} = \tilde{p}_{md} = \tilde{p}_{dd} = 0,$$

with

$$m_{\text{pos}} = \min(e_2, 1 - p_B),$$

$$w_{\text{pos}} = 1 - p_B - m_{\text{pos}},$$

$$\rho_{\text{pos}} = \text{sign}(\rho) \min(m_{\text{pos}}, w_{\text{pos}}, \rho_{\text{loc}}).$$

The first moments are correct if $m_{\text{pos}} = e_2 \Delta t \geqslant 0$. The correlation is matched if $\rho_{\text{loc}} = \rho_{\text{pos}}$.

- If $0 \leqslant \alpha_{t+\Delta t} < 1$, ie the middle and down movements pass the barrier, we have

$$\tilde{p}_{uu} = \tilde{p}_{du} = \tfrac{1}{2}(1 - p_B - \tilde{p}_{mu}),$$

$$\tilde{p}_{mu} = \max(1 - p_B - s_1^2, 0),$$

$$\tilde{p}_{um} = \tilde{p}_{mm} = \tilde{p}_{dm} = \tilde{p}_{ud} = \tilde{p}_{md} = \tilde{p}_{dd} = 0.$$

The first moment of S^1 is matched, and the second moment of S^1 is correct if $\tilde{p}_{md} = 1 - p_B - s_1^2$.

- If $1 < \alpha_{t+\Delta t}$, ie every movement passes the barrier, we set $p_B = 1$ and $\tilde{p}_{ab} = 0$ ($a, b \in \{u, m, d\}$).

In order to reduce computation time, we define a threshold $\alpha_* \geqslant 1$ for α: if $\alpha_t < -\alpha_*$ and $\alpha_{t+\Delta t} < -\alpha_*$, we use the probabilities p_{ab} ($a, b \in \{u, m, d\}$) from Section 6.1.4.

It is straightforward to derive the corresponding probabilities in the case of an up-and-out barrier $B > S_t^2$.

Sloping barriers In case of a time-dependent barrier ($B_t, t \in [0, T]$) rather than a constant level B, we assume a locally exponentially sloping barrier

$$B_s = B_t e^{g_t(s-t)}, \quad s \in [t, t + \Delta t].$$

The correct moments are now obtained if p_B, p_B^+ and p_B^{++} are calculated from

$$\tilde{\mu}_t = r_t - d_t^2 - \tfrac{1}{2}(\sigma_t^2)^2 - g_t,$$

$$\tilde{\mu}_t^+ = r_t - d_t^2 + \tfrac{1}{2}(\sigma_t^2)^2 - g_t,$$

$$\tilde{\mu}_t^{++} = r_t - d_t^2 + \tfrac{3}{2}(\sigma_t^2)^2 - g_t.$$

See Section 2.4.1.

6.3.4 Tree Correlation for Barriers

Here we calculate the correlation ρ_B of $(\tilde{X}^2_{t+\Delta t} \mid S^1_t, \tilde{S}^2_t)$ and $(X^1_{t+\Delta t} \mid S^1_t, \tilde{S}^2_t)$. We have

$$E_t[S^1_{t+\Delta t}\tilde{S}^2_{t+\Delta t}]$$
$$= F^1_t F^2_t E_t\left[\exp\left\{\sigma^1_t\left(\rho_t X + \sqrt{1-\rho^2_t}\, Y\right) - (\sigma^1_t)^2 + \sigma^2_t X - (\sigma^2_t)^2\right\} 1_{\{S^2_* > B\}}\right]$$
$$+ B p_B E_t[S^1_{t+\Delta t} 1_{\{S^2_* \leq B\}}]$$
$$= F^1_t F^2_t e^{\sigma^1_t \sigma^2_t \rho_t \Delta t}(1 - p^\rho_B)$$
$$+ F^1_t B p_B \left\{ N\left(\frac{\log(F^2_t/B) + [-\frac{1}{2}(\sigma^2_t)^2 + \rho_t \sigma^1_t \sigma^2_t]\Delta t}{\sigma^2_t \sqrt{\Delta t}}\right)\right.$$
$$\left. + \left(\frac{B}{S^2_t}\right)^{(2\tilde{\mu}^2_t + \rho_t \sigma^1_t)/\sigma^2_t} N\left(\frac{-\log(F^2_t/B) + [-\frac{1}{2}(\sigma^2_t)^2 + \rho_t \sigma^1_t \sigma^2_t]\Delta t}{\sigma^2_t \sqrt{\Delta t}}\right)\right\},$$

with

$$p^\rho_B = N\left(\pm\frac{-\alpha\, dx + \tilde{\mu}^\rho_t \Delta t}{\sigma^2_t \sqrt{\Delta t}}\right) + e^{2\alpha\tilde{\mu}^\rho_t dx/(\sigma^2_t)^2} N\left(\pm\frac{-\alpha\, dx - \tilde{\mu}^\rho_t \Delta t}{\sigma^2_t \sqrt{\Delta t}}\right),$$
$$\tilde{\mu}^\rho_t = r_t - d^2_t + \rho_t \sigma^1_t \sigma^2_t + \tfrac{1}{2}(\sigma^2_t)^2.$$

See the closed-form solution for a down-and-out outside barrier call (with zero strike) in Section 4.9. Now, using a lognormal approximation of $\tilde{S}^2$, we obtain the correlation ρ_B of the returns from

$$\sigma^1_t \sigma^2_t \rho_B \Delta t$$
$$= \log\left\{ e^{\sigma^1_t \sigma^2_t \rho_t \Delta t}(1 - p^\rho_B) \right.$$
$$+ \frac{B}{F^2_t} p_B \left[N\left(\frac{\log(S^2_t/B) + (\tilde{\mu}^2_t + \rho_t \sigma^1_t \sigma^2_t)\Delta t}{\sigma^2_t \sqrt{\Delta t}}\right)\right.$$
$$\left.\left. + \left(\frac{B}{S^2_t}\right)^{(2\tilde{\mu}^2_t + \rho_t \sigma^1_t)/\sigma^2_t} N\left(\frac{-\log(S^2_t/B) + (\tilde{\mu}^2_t + \rho_t \sigma^1_t \sigma^2_t)\Delta t}{\sigma^2_t \sqrt{\Delta t}}\right)\right]\right\}.$$

6.4 Bermudan Asian Options

A Bermudan Asian call (put) gives, like a plain vanilla call (put), the right to buy (sell) a risky asset S^A for a previously fixed strike price K. In contrast to a plain vanilla call, $(S^A_t, t \in [0, T])$ is the arithmetic average process of the stock S based on previously agreed fixing dates $(t_i, i = 1, 2, \ldots)$ with $t_i < t_{i+1}$, ie

$$S^A_t = \frac{1}{n+1}\sum_{k=0}^{n} S_{t_k}, \qquad t \in [0, T], \quad t_n \leq t < t_{n+1}. \tag{6.2}$$

The holder has the right to exercise the option at previously agreed dates or time intervals, eg

- only at maturity (European case);
- at any time between evaluation date and maturity (American case);
- at previously agreed dates or time intervals (Bermudan case).

The product allows a global cap, ie the replacement of S_t^A by $\min(S_t^A, C)$, with C denoting the cap.

Finally, the product allows one or several resets of the average. If, for example, the fixing date t_i were a reset date, the average S_t^A for $t = t_k \geq t_i$ would be the running average including only the fixing dates $t_i, t_{i+1}, \ldots t_k$ (if t_i is the only reset date in $\{t_i, t_{i+1}, \ldots, t_k\}$).

Concept and notation Since the option has a path-dependent payoff, the usual tree with nodes indexed by spot and time is not sufficient. The basic idea now is to introduce a third index for the average. It turns out that using a geometric average rather than an arithmetic average leads to a recombining tree. In order to improve the approximation further, we will finally adjust the first moment of the geometric average as described in Section 2.5.1, ie replace the process (S_t^A) (see (6.2)) by

$$S_t^G - \mathbb{E}[S_t^G] + \mathbb{E}[S_t^A], \quad t \in [0, T],$$

with S_t^G denoting the geometric average

$$S_t^G = \left(\prod_{k=0}^n S_{t_k}\right)^{\frac{1}{n+1}}, \quad t \in [0, T], \quad t_n \leq t < t_{n+1}. \tag{6.3}$$

We introduce the following notation:

$$i_t = \text{time index},$$
$$I_t = \text{number of time steps},$$
$$i_a = \text{average index},$$
$$I_a(i_t) = \text{number of averages},$$
$$i_s = \text{stock index},$$
$$I_s(i_t) = \text{number of stock prices},$$
$$\Delta = T/I_t,$$
$$u = e^{\sigma\sqrt{\Delta}},$$
$$p = \frac{e^{(r-d)\Delta} - 1/u}{u - 1/u},$$
$$df = e^{-r\Delta}.$$

Basic case In order to describe the tree, we first focus on the case of a geometric Asian call with averaging at each time step. We remark that, at $i_t \Delta$, the price of the asset assumes one of the values $S(i_t, i_s)$, $i_s \in \{0, 1, \ldots, I_s(i_t)\}$,

The Tree Approach

with
$$I_s(i_t) = i_t$$
and
$$S(i_t, i_s) = S_0 u^{-i_t + 2i_s}. \tag{6.4}$$

Now, the path of a stock is entirely described by a sequence of stock indexes
$$(i_s(k), k = 0, 1, \ldots, I_t).$$

The geometric average $S^G_{i_t}$ at $i_t \Delta$ for that path is (see (6.4))

$$\log S^G_{i_t} = \frac{1}{i_t + 1} \sum_{k=0}^{i_t} \log S(k, i_s(k))$$
$$= \log S_0 + \log u \left(-\tfrac{1}{2} i_t + 2 \frac{i_s(0) + i_s(1) + i_s(2) + \cdots + i_s(i_t)}{i_t + 1} \right).$$

Since $i_s(k)$ can assume the values $0, 1, \ldots, k$, the sum $i_s(0) + i_s(1) + \cdots + i_s(i_t)$ is some $i_a \in \{0, 1, \ldots, I_a(i_t)\}$, with

$$I_a(i_t) = \sum_{k=0}^{i_t} k = \tfrac{1}{2} i_t (i_t + 1).$$

We define the geometric average $G(i_t, i_a)$ associated with time index i_t and average index i_a by

$$G(i_t, i_a) = S_0 u^{-i_t/2 + 2i_a/(i_t+1)}. \tag{6.5}$$

We now detail the calculation in pseudo-code. Let $V(i_t, i_s, i_a)$ denote the value of the option at time index i_t, stock index i_s and average index i_a. At maturity $i_t = I_t$, we do

for $(i_a = 0, i_a \leqslant I_a(I_t), i_a = i_a + 1)$
  for $(i_s = 0, i_s \leqslant I_s(I_t), i_s = i_s + 1)$
    $V(I_t, i_s, i_a) = (G(I_t, i_a) - K)^+.$

We remark that the option price does not depend on the spot index. Now, we proceed backwards $i_t = I_t - 1, \ldots, 0$ as follows:

for $(i_t = I_t - 1, i_t \geqslant 0, i_t = i_t - 1)$
  for $(i_a = 0, i_a \leqslant I_a(i_t), i_a = i_a + 1)$
    for $(i_s = 0, i_s \leqslant I_s(i_t), i_s = i_s + 1)$
      $V(i_t, i_s, i_a) =$
        $df \left(p V(i_t + 1, i_s + 1, i_a + i_s + 1) + (1 - p) V(i_t + 1, i_s, i_a + i_s) \right)$
      if (exercise right at $i_t \Delta$)
        $V(i_t, i_s, i_a) = \max(V(i_t, i_s, i_a), G(i_t, i_a) - K).$

The fair price Π of the Asian call with fixings at every time grid point between evaluation date 0 and maturity is

$$\Pi = V(0,0,0).$$

General case We now treat the more general case of fixings at arbitrary dates (including past dates) and of the arithmetic rather than the geometric average. In addition, we allow a Bermudan exercise schedule and resets at fixing dates.

We let $s_{-m} < s_{1-m} < \cdots < s_0 \leqslant 0 < s_1 < s_2 < \cdots < s_n = T$ denote the fixings and assume that $s_k = f_k \Delta$, $k = 1, \ldots, I_f(I_t)$, with $I_f(i_t)$ denoting the number of fixings in $(0, i_t \Delta]$ (hence $I_f(I_t) = n$). We fix the path of a stock by a sequence of stock indices $\big(i_s(k), k = 0, 1, \ldots, I_t\big)$. The geometric average $S_{i_t}^G$ of the future spot prices at $t_i \Delta$ for that path is (see (6.4))

$$\log S_{i_t}^G = \frac{1}{I_f(i_t)} \sum_{k=1}^{I_f(i_t)} \log S\big(f_k, i_s(f_k)\big)$$

$$= \log S_0 + \log u \bigg(-\frac{I_a(i_t)}{I_f(i_t)} + 2 \frac{i_s(f_1) + i_s(f_2) + \cdots + i_s(f_{I_f(i_t)})}{I_f(i_t)} \bigg).$$

The average index can assume values between 0 and

$$I_a(i_t) = f_1 + f_2 + \cdots + f_{I_f(i_t)}.$$

Moving the first moment of the geometric average to the first moment of the arithmetic average leads to the definition of the approximate arithmetic average (see Section 2.5.1) at time index i_t and average index i_a:

$$A(i_t, i_a) = \frac{1}{m + 1 + I_f(i_t)} \bigg(\sum_{k=0}^{m} S_{t_{-k}} \bigg) + \frac{I_f(i_t)}{m + 1 + I_f(i_t)} \bigg(S_0 u^{-\frac{I_a(i_t) + 2i_a}{I_f(i_t)}} + M(i_t) \bigg),$$

with

$$M(i_t) = \bigg(\frac{S_0}{I_f(i_t)} \sum_{k=1}^{I_f(i_t)} e^{(r-d)f_k \Delta} \bigg) - e^{M^G(i_t) + V^G(i_t)/2},$$

$$M^G(i_t) = \log S_0 + \frac{(r - d - \tfrac{1}{2}\sigma^2)\Delta}{I_f(i_t)} \sum_{k=1}^{I_f(i_t)} f_k,$$

$$V^G(i_t) = \frac{\sigma^2 \Delta}{I_f(i_t)^2} \sum_{k,l=1}^{I_f(i_t)} \min(f_k, f_l).$$

To calculate the option price, we perform

for $(i_a = 0, i_a \leqslant I_a(I_t), i_a = i_a + 1)$

  for $(i_s = 0, i_s \leqslant I_s(I_t), i_s = i_s + 1)$

    $V(I_t, i_s, i_a) = \big(\max(A(I_t, i_a), C) - K\big)^+,$

thus initialising the option payoff at maturity. At that stage, I_f and I_a, as well as the derived quantities average A and mean correction M, are calculated based

The Tree Approach

on the set of fixing indexes f_k $(k = r, \ldots, I_f(I_t))$, with $t_r = f_r \Delta$ being the last reset date before (or at) maturity T.

We now proceed backwards in time according to the scheme

for $(i_t = I_t - 1, i_t \geq 0, i_t = i_t - 1)$

    if $(i_t + 1$ is a reset date$)$

        update I_f, I_a, A and M so that all the fixings from the last reset date

        prior to $i_t + 1$ are included.

    for $(i_a = 0, i_a \leq I_a(i_t), i_a = i_a + 1)$

        for $(i_s = 0, i_s \leq I_s(i_t), i_s = i_s + 1)$

            if $((i_t + 1)\Delta$ is a fixing date and no reset date$)$

$$V(i_t, i_s, i_a) =$$
$$df \left(p\, V(i_t + 1, i_s + 1, i_a + i_s + 1) + (1 - p)\, V(i_t + 1, i_s, i_a + i_s) \right)$$

            else if $((i_t + 1)\Delta$ is a reset date$)$

$$V(i_t, i_s, i_a) = df \left(p\, V(i_t + 1, i_s + 1, i_s + 1) + (1 - p)\, V(i_t + 1, i_s, i_s) \right)$$

            else

$$V(i_t, i_s, i_a) = df \left(p\, V(i_t + 1, i_s + 1, i_a) + (1 - p)\, V(i_t + 1, i_s, i_a) \right)$$

            if (exercise right at $i_t \Delta$)

$$V(i_t, i_s, i_a) = \max\left(V(i_t, i_s, i_a), \max(A(i_t, i_a), C) - K\right).$$

The fair price Π of the general Asian call is

$$\Pi = V(0, 0, 0).$$

Discrete dividends Assume discrete dividend amounts $(d_j, j = 1, 2, \ldots, m)$ to be paid at dates $(s_j, j = 1, 2, \ldots, m)$. In order to incorporate discrete dividend payments into the model, we change the asset process S described in Section 1.3.2 to

$$S_t = \tilde{S}_t + D_t,$$

$$\frac{d\tilde{S}_t}{\tilde{S}_t} = \sigma\, dW_t + (r - d)\, dt, \quad t \in [0, T],$$

$$\tilde{S}_0 = S_0 - D_0,$$

where D_t is the value at t of all dividends to be paid in $[t, T]$, ie

$$D_t = \sum_{j:\, s_j \in [t, T]} d_j e^{-r(s_j - t)}, \quad t \in [0, T].$$

As before, σ and r denote volatility and (continuously compounded) interest rate. Here d denotes the (continuously compounded) repo rate. The process S can also be written as

$$S_t = \tilde{S}_0 \exp[\sigma W_t + (r - d - \sigma)t] + D_t, \quad t \in [0, T].$$

We obtain an arithmetic average based on fixing dates $(t_i, i = 1, 2, \ldots, n)$ as

$$\frac{1}{n}\sum_{i=1}^{n} S_{t_i} = \left(\frac{1}{n}\sum_{i=1}^{n} \tilde{S}_{t_i}\right) + \left(\frac{1}{n}\sum_{i=1}^{n} D_{t_i}\right).$$

Hence, in the tree, if an average $A(i_t, i_a)$ at a time and average index i_t and i_a is calculated, we use $\tilde{S}_0$ instead of S_0. We then add $(1/n)\sum_{i=1}^{n} D_{t_i}$ to $A(i_t, i_a)$, with $t_1, t_2, \ldots, t_n$ denoting the fixing dates of the average $A(i_t, i_a)$.

Greeks The above tree can be used to obtain greeks. We increase the number of time steps I_t by two. Accordingly, we shift all the fixings, reset dates and exercise periods by adding 2Δ. Fixings in the past, however, have to remain in the past. Thus, at time index $i_t = 2$, we have

$$I_a(i_t) = 0, \qquad I_s(i_t) = 2.$$

In order not to alter the price obtained, as described in the previous section, we do not use the additional nodes for the mean correction. Hence, if we have a sequence $\big(M(1), M(2), \ldots, M(I_t)\big)$ of mean corrections in the model with I_t time steps, we use the sequence $\big(0, 0, M(1), M(2), \ldots, M(I_t)\big)$ in the model with $I_t + 2$ time steps.

Now, we obtain price Π_B, delta Δ_B, gamma Γ_B and theta Θ_B from the tree (see [36]):

$$\Pi_B = V(2, 1, 0),$$

$$\Delta_B = \frac{1}{S_0(u^2 - u^{-2})}\left\{V(2, 2, 0)u^{-2} + V(2, 1, 0)(u^2 - u^{-2}) - V(2, 0, 0)u^2\right\},$$

$$\Gamma_B = \frac{2}{S_0^2(u^2 - u^{-2})}\left\{\frac{V(2, 2, 0) - V(2, 1, 0)}{u^2 - 1} - \frac{V(2, 1, 0) - V(2, 0, 0)}{1 - u^{-2}}\right\},$$

$$\Theta_B = \tfrac{1}{2}\big\{V(4, 2, 0) + V(2, 1, 0)\big\}.$$

6.5 Convertible Bonds

A *convertible bond* is a security that combines the features of a bond and a stock option. In addition to yielding a fixed stream of coupons and a final redemption payment, a convertible bond gives to investors the right to exchange the bond for stock during predetermined periods in the lifetime of the bond. Frequently issued by corporates to raise finance, and convertible into the stock of the issuer, the convertible bond is an important example of a product valued within a tree framework.

Convertibles can be attractive to investors because they provide the opportunity to participate in the positive performance of the issuer's stock, through the call option embedded in the bond. Should the stock underperform, however, the value of the convertible does not necessarily fall proportionately, as would the value of a direct investment in the stock, but declines more slowly since the coupons will, in any event, still be received. As the stock price falls, the value of optionality in the convertible bond tends to zero and the convertible

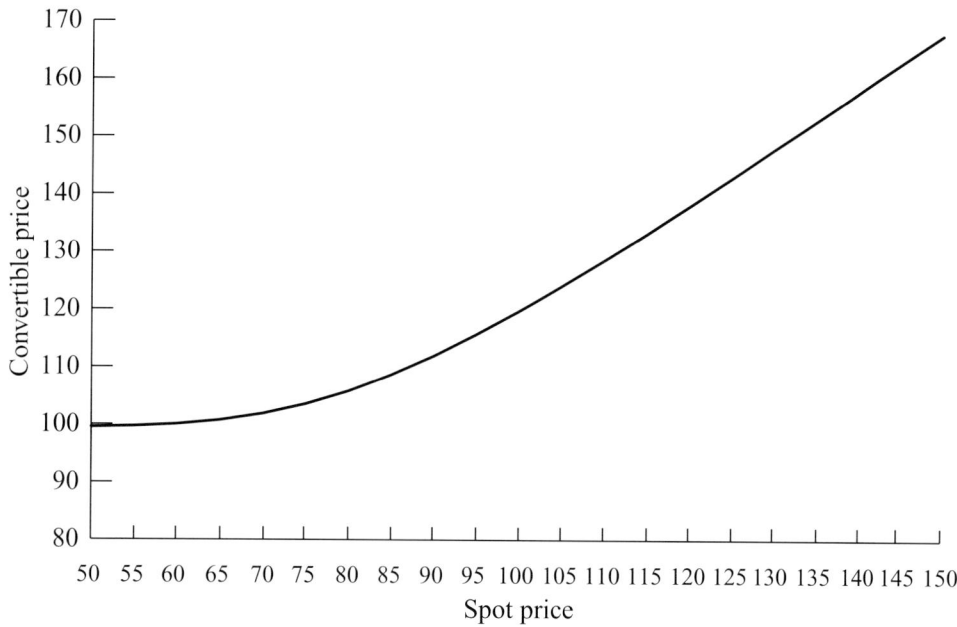

6.3 Bond price against spot without call feature or credit risk.

value approaches the value of the corresponding pure bond, known as its *bond floor*. The coupons can be seen as providing a cushion for the investor in the event of underperformance by the stock (see Figure 6.3 for an example).

Furthermore, while an investment in the equivalent notional of the stock would yield dividends not known with certainty long in advance, the convertible offers stability of income from the coupons. Investors may thus regard a convertible as a combination of the underlying stock and a swap which exchanges the dividends for fixed coupons, combined with a put on the stock.

Convertibles can be attractive to issuers as means of financing, because they allow a lower coupon to be paid than would be the case if they issued a simple bond. This is because, in addition to the coupons, the issuer sells a call on its own stock, the value of which lowers the debt payments that the issuer must otherwise make. The value of the call, of course, depends on the volatility of the stock, which will therefore influence the terms of a convertible issue.

6.5.1 Features of Convertible Bonds

While the principal features of the convertible bond are the coupon payments and the right to convert, in practice most issues have other features, only some of which are listed here.

- *Conversion feature.* The holder of the convertible may exchange the bond during stipulated time intervals for a specified number of the underlying equities. The number of shares exchanged for one bond is known as the conversion ratio, and the equivalent conversion price is defined as the Notional/Conversion ratio, assuming that the bond's notional is paid for stock. This conversion price can vary over time.

- *Issuer call feature.* Many corporate bonds include a provision that allows the firm to repurchase (ie call) the bond during specified periods for a specified price (the call price), with or without the accrued coupon. The effect of this is to limit the investor's participation in the positive performance of the stock, and to decrease the value of the callable bond in comparison with the corresponding non-callable bond. In some cases, the call feature may only be exercised conditional on the stock price exceeding some predetermined threshold level, and subject to a notice period. Such an additional feature limits the call rights of the issuer and so increases the bond price relative to a callable bond with no call protection.

- *Put feature.* The holder of the bond may, during stipulated time intervals or on specified dates, return it to the issuer in exchange for a known payment. The holder may thus force early redemption. The put is exercisable at the discretion of the investor; since it is a right sold by the issuer to the investor, it acts to increase the value of the bond. An investor might exercise the put option in the event that the stock had underperformed since issue and the option feature was therefore essentially worthless (far out-of-the-money). Whether exercise of the put is optimal in such a circumstance will also depend on the prevailing interest rate environment, since the investor will compare the put value, received immediately, with the discounted value of future coupons. Adequate modelling of this feature might therefore require that the stochasticity of interest rates be taken into account.

- *Redemption.* The convertible bond holder receives periodic coupon payments from the issuer and, if not converted before maturity, receives the whole principal at that time. Some issues offer a premium redemption, ie the final redemption payment is not 100% of notional but some greater amount.

Figure 6.4 shows the effects on bond price of call features, both with and without an equity trigger threshold. The following features are evident:

- At low values of spot, the bond is never called and the call provisions are irrelevant.

- The hard-callable bond is called early at high asset values, resulting in early conversion and loss of the associated option time value.

- The soft-callable bond is an intermediate case. At extreme asset values it behaves exactly as the hard-callable bond, in that at low values it is a pure bond and at high values the trigger level is certain to be exceeded, and therefore ineffective. At intermediate asset values, the call protection is valuable and has a high probability of inhibiting the early call.

Some convertible bonds have strike reset features, which allow the conversion price to be reset, on specified dates, to lower values. This acts to recover the optionality of the bond if the stock has seriously underperformed such that the option component is worth virtually nothing: the convertible bond in such a situation has lost the investor advantages associated with its equity content. A strike reset feature acts to insure the investor against this.

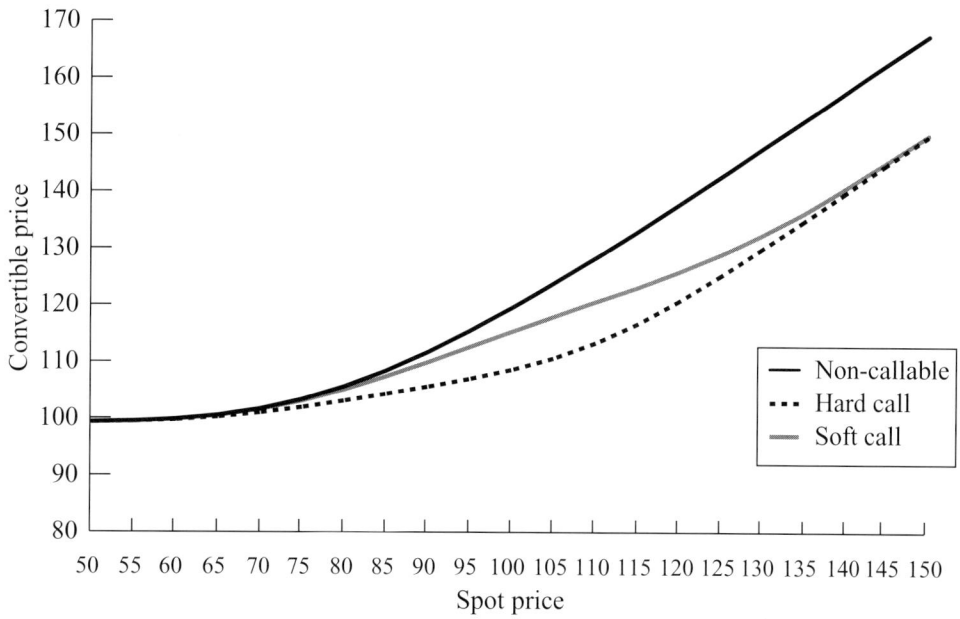

6.4 The effects of call provision on bond price.

6.5.2 Valuing Convertible Bonds on a Tree

The approach to valuation of convertible bonds on a tree may be developed by comparison with the American or Bermudan option. Since the optionality has American or Bermudan character (the conversion right can either extend throughout the lifetime of the bond or be limited to subintervals), a tree approach using backward induction is appropriate.

Furthermore, we may consider valuing the bond on either a single-factor or two-factor tree. The single-factor tree, of course, models the underlying stock as the single stochastic factor (see Section 6.1.3). One approach using a two-factor tree models the corresponding short rate as a second stochastic factor (see Section 6.1.4), although the short rate is not the only possible choice for this second stochastic factor.

In either case, the calculation of the bond value at each tree node proceeds as indicated below. The only impact of a stochastic interest rate would be to change the calculation of the hold value $H(S_t, t)$ at each node, which calculation is in any event not particularly related to the convertible bond *per se*, inasmuch as it is common to all option types (see equation (6.6)).

At maturity, the bond is worth either the redemption payment or the conversion value, if greater. Thus the initialisation of the terminal nodes uses the function

$$\max(\alpha_T S_t, R), \qquad (6.6)$$

in which α_T is the conversion ratio at maturity (zero if conversion is forbidden at that date) and R is the redemption amount.

During the lifetime of the bond, at any given value of S_t, t, the various possibilities are as follows.

- The holder can convert and receive an amount $\alpha_t S_t$.

- The holder can exercise a put feature and receive P_t.

- The holder can take no action, retaining the bond for a hold value $H(S_t, t)$.
- The issuer can call the bond for an amount C_t.

Note that C_t may include a component for the accruing coupon and that $H(S_t, t)$ will include a component for the coupon, if any, paid in the interval $(t, t + \Delta t)$. At times t such that any of the above features is not active, we may assume

- no conversion available: $\alpha_t = 0$;
- no put available: $P_t = 0$;
- no call available: C_t arbitrarily large;

At each tree node, we execute the logic

$$O(S_t, t) = \max\left(\alpha_t S_t, \min(H(S_t, t), C_t), P_t\right).$$

The above assumes that conversion has priority over call (which accords with the fact that issuer calls are generally subject to a notice period during which the investor has the option to convert or accept the call), and makes the simplification that the call is not subject to any trigger feature.

6.5.3 Incorporating Credit Spread into the Valuation

The impact of the issuer's creditworthiness on the value of a convertible bond can hardly be overstated, and a convertible bond pricing model must take account of it. We define a *credit spread* to be the additional discounting rate applied to coupon payments to reflect the issuer's probability of default and consequent non-payment of the coupons.

We may calibrate this parameter from market prices of non-convertible bonds issued by the issuer of the convertible bond. If the coupon payments C_i were not risky, the pure bond price would be

$$P = \sum_i C_i \exp(R_{t_i}),$$

and, in the case of risky payments, we define the credit spread s above risk-free rates by inversion of the equation

$$P_{\text{Mkt}} = \sum_i C_i \exp(R_{t_i} + s),$$

and it is a small generalisation to regard this as time-dependent. If pure bonds issued by the convertible issuer are not available, then equally rated companies' bonds can be used to get an estimate for s.

As applied to convertible valuation, the commonest method of taking into account credit spread is to assume it to be deterministic and to incorporate it into the discounting applied to find $H(S_t, t)$. However, unconditionally applying the full s to $H(S_t, t)$ implies that future cashflows resulting from conversion are regarded as just as risky as those from coupon payments, whereas the model assumes that it fully accounts for the uncertainty in conversion value through the stochasticity of the stock price.

An approach to this issue is to incorporate only a fraction of s into the hold value $H(S_t, t)$. This fraction is determined by the probability of conversion as seen at a given node: given this probability p_t^j at node j and time t, we apply a discounting rate

$$r_{t,t+\Delta t} + (1 - p_t^j)s$$

at that node.

The evaluation of that probability is a separate problem from the primary valuation problem, and may be approached by carrying at each node, in addition to the bond value $O(S_t, t)$, an additional value p_t^j, and evolving it backwards through the tree in a similar manner to the bond price. It is initially set to zero or one at the terminal nodes according to the result of the max function in (6.6). Thereafter, at each node, its value is calculated from the values at the previous nodes

$$p_t^j = \sum_{a \in A} p_a p_{t+\Delta t}^a$$

(compare equation (6.6)) in the same way as the hold value of an option is calculated from values at connected nodes. Since this additional value is associated with each tree node, it follows that if the bond valuation tree is seen as an array of nodes lying in one plane, the associated array of conversion probabilities can be seen as lying in a parallel plane, leading to the name *thick tree* for this situation where more than one value is associated with a given node. The same approach also appears in other valuation problems. The effect of credit spread, modelled thus, on the bond price, is shown in Figure 6.5. It can be seen that the spread affects the bond floor, which is risky, but not the conversion.

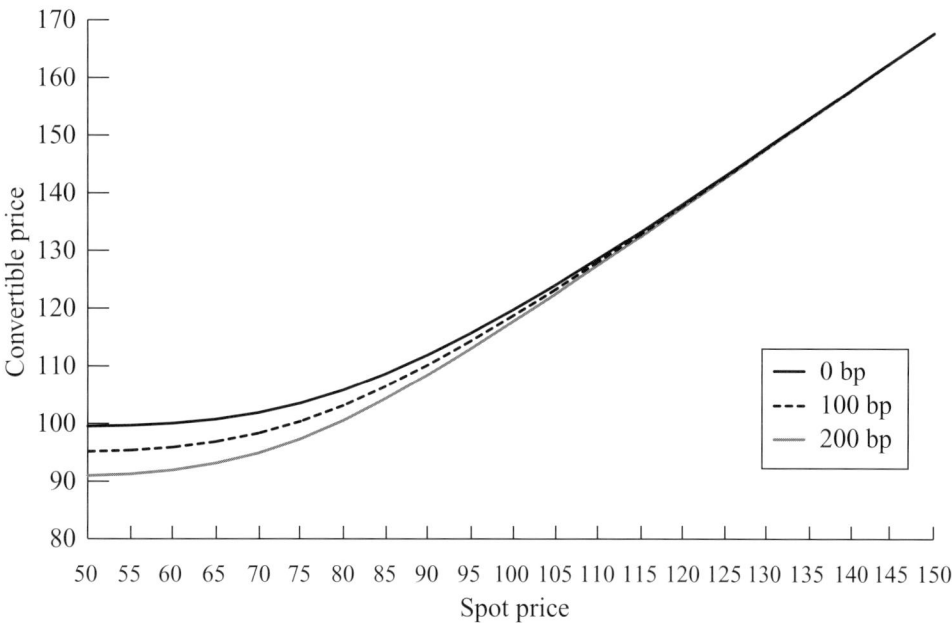

6.5 Bond price against spot and credit spread.

The assumption that issuer creditworthiness can be accounted for by a deterministic credit spread is simplistic. In practice, it is common for a company's credit rating to change over time in accordance with rating agencies' perceptions of it, which suggests that it is most naturally treated as a stochastic process. Furthermore, it is reasonable to assume that this process will be correlated with the stock price process, since the same factors that cause a company's stock to underperform might also adversely affect its credit rating. This effect could dramatically affect the bond price as the stock falls, since the bond floor would fall at the same time and the cushioning effect, noted above to be the effect of the coupons flooring the bond price, would fall as the option price fell.

Proceeding with an approach to convertible valuation using stochastic credit rating, such as the discrete Markov process approach given in [40], gives a considerable improvement in the quality of the model. In practice it is found that, even for moderate default probabilities, the effect of the stochasticity of credit spread can be very pronounced. Comparing such models with two-factor models in which the second stochastic factor is the interest rate rather than the default probability, it is found in general that the credit spread is the more important factor in the sense that, for realistic scenarios, it produces a greater deviation from the price due to the single-factor model.

Chapter 7
Monte Carlo Methods

One of the most flexible and easily implemented approaches to valuing complex derivatives is the Monte Carlo method. It allows the pricing of nearly any imaginable path-dependent European option with one or several underlyings, requiring only very modest programming effort. These features often make it the method of choice for pricing derivatives that are new on the market and for which it is not feasible to develop a more specialised model.

This flexibility comes at a price, of course. One of the drawbacks of the Monte Carlo procedure is the computation time necessary to arrive at a sufficiently precise result. There exist various methods for increasing the precision that can be achieved with a given number of Monte Carlo simulations, or, in other words, to reduce the number of steps needed to arrive at a certain precision.

While Monte Carlo methods are well suited to European exercise styles, it is much harder to use them for American or Bermudan options. The reason is that to make an exercise decision one needs the price of the derivative at all exercise times, which is not directly provided by the method. However, there do exist enhancements that make the method work with early exercise features.

Mathematically, the Monte Carlo method amounts to an numerical evaluation of an expected value, ie an integral. This is of use for us, since the risk-neutral valuation of derivatives amounts to the computation of an expectation under a risk-neutral measure.

7.1 The Basic Method

The purpose of the *Monte Carlo method* is to approximately evaluate an integral of the form

$$\mathbb{E}[f] = \int_\Omega f(\omega)\,\mathbb{P}[d\omega],$$

where $\mathbb{P}$ is a probability measure on Ω and f is a random variable on Ω. In our context, usually $\Omega = \mathbb{R}^n$ and $\mathbb{P}$ is some Borel measure that has a density. The random variable f should be well behaved (eg in terms of the smoothness of its derivatives) for the method to work successfully.

We approximate this integral by the finite sum

$$\frac{1}{N}\sum_{i=1}^{N} f(x_i),$$

where $x_1, \ldots, x_N$ are independent pseudo-random numbers drawn according to the probability measure $\mathbb{P}$. These pseudo-random numbers can be generated using one of the routines from [61].

Let us illustrate the method in the case of a simple plain vanilla call with payoff $(S_T - K)^+$. Recall that

$$S_T = S_0 \cdot \exp\left(\mathcal{N}_T + \int_0^T \sigma_u \, dW_u\right).$$

Now $X = \int_0^T \sigma_u \, dW_u$ is normally distributed with mean zero and variance Σ_T^2. So let us draw independent random numbers $x_1, \ldots, x_N$ from the standard normal distribution. Then

$$S_0 \cdot \exp(\mathcal{N}_T + \Sigma_T x_i), \quad i = 1, \ldots, n,$$

are sample values for S_T. Thus, we can get an approximation for the option value by taking the average of the values

$$\left(S_0 \cdot \exp(\mathcal{N}_T + \Sigma_T x_i) - K\right)^+, \quad i = 1, \ldots, n,$$

and then discounting by e^{-R_T}. To summarise, our Monte Carlo value is computed as

$$e^{-R_T} \cdot \frac{1}{N}\sum_{i=1}^{N} \left(S_0 \cdot \exp(\mathcal{N}_T + \Sigma_T x_i) - K\right)^+.$$

This generalises readily to payoffs f depending on the asset value at a finite number of fixing times $0 \leq t_1 < \cdots < t_m = T$. Let us, for example, consider a discretely sampled up-and-out call

$$f(S_{t_0}, \ldots, S_{t_m}) = (S_T - K)^+ \cdot 1_{\{\max\{S_{t_i} : i=1,\ldots,m\} < H\}}$$

with barrier H. We need to simulate the joint distribution of the asset prices $S_{t_1}, \ldots, S_{t_m}$. For this, we note that

$$\log \frac{S_{t_{i+1}}}{S_{t_i}} = \mathcal{N}_{t_i, t_{i+1}} + \int_{t_i}^{t_{i+1}} \sigma_u \, dW_u.$$

Therefore, to get a single price path, we draw m independent standard normal random numbers $x_1, \ldots, x_m$. We then define $S_{t_1}, \ldots, S_{t_m}$ recursively by

$$S_{t_{i+1}} = S_{t_i} \cdot \exp(\mathcal{N}_{t_i, t_{i+1}} + \Sigma_{t_i, t_{i+1}} x_{i+1}).$$

We can also extend this method to multiple assets. For this, one needs to generate random numbers from an n-dimensional correlated normal distribution. Suppose that the correlation matrix is denoted by C. We can use a Cholesky decomposition $C = LL^T$, where L is a lower triangular matrix, which is easy to compute and which can be stored before the paths are generated. If X is a standard normal random vector, then $n \to X$ gives a correlated normal random vector, where the entries have correlation matrix C.

7.2 Speeding Up Monte Carlo

To apply the Monte Carlo method successfully, one usually needs to employ techniques to improve convergence. There exists a variety of such methods, of which we present a sample.

7.2.1 Antithetic Paths

The simplest approach is the antithetic path method. Assume that we have drawn the numbers $x_1, \ldots, x_m$ from standard normal distributions, and we have used these numbers to generate a price path $S_{t_1}, \ldots, S_{t_m}$. We then use the negatives $-x_1, \ldots, -x_m$ of these numbers to generate another price path $\tilde{S}_{t_1}, \ldots, \tilde{S}_{t_m}$. Note that the negative of a standard normally distributed random variable is again standard normally distributed. We evaluate our derivative contract on both of these price paths. This reduces the variance of the result, and at a low additional cost since no new random numbers have to be generated. The antithetic method makes sure that the sample mean of the normal random numbers drawn is exactly zero.

7.2.2 Control Variate Technique

Assume that we want to evaluate a difficult integral $\mathbb{E}[f]$ for which there is no closed form available. However, suppose that there is a function g such that $\mathbb{E}[g]$ can be evaluated in closed or near-closed form and such that g is close to f in the sense that $\mathbb{E}[f - g]$ is small. Then we can run a Monte Carlo simulation for $\mathbb{E}[f - g]$ and add $\mathbb{E}[g]$ to the result, to get an approximation for $\mathbb{E}[f]$. This method is very powerful and can lead to much improved results.

As an indication of why this method works, consider

$$\mathrm{Var}[f - g] = \mathrm{Var}[f] + \mathrm{Var}[g] - 2\,\mathrm{Cov}[f, g],$$

which is less than $\mathrm{Var}[f]$ if $\mathrm{Var}[g] < 2\,\mathrm{Cov}[f, g]$.

A typical example is the valuation of arithmetic Asian options, which cannot easily be valued in closed form. However, Asian options using the geometric instead of the arithmetic average *can* be valued in closed form, since the geometric average of lognormal random variables is itself lognormal. For a mildly volatile sequence, the arithmetic and the geometric average are quite close. Therefore, geometric Asian options provide an ideal control variate for arithmetic Asian options.

7.2.3 Low-Discrepancy Sequences

So far we have used pseudo-random numbers, which are constructed to behave like proper random numbers under as many different statistical tests as possible.

However, for the purpose of Monte Carlo simulations, pseudo-random numbers are not always the optimal choice. Often, convergence can be improved by the use of low-discrepancy numbers (also known as quasi-random numbers), for which Sobol numbers and Faure numbers are well-known examples.

In the following, we will provide precise definitions to get a feeling for the subject. A good reference on the subject is the book by Tezuka [72]. For a finite subset X of the k-dimensional unit interval $[0, 1]^k$, we define the *discrepancy* as

$$D^{(k)}(X) = \sup_{x_1,\ldots,x_k \in [0,1]} \left| \frac{|X \cap [0, x_1) \times \cdots \times [0, x_k)|}{|X|} - x_1 \cdots x_k \right|.$$

Further, if $x = (x_1, x_2, \ldots)$ is a sequence in $[0, 1]^k$, we write

$$D_N^{(k)}(x) = D^{(k)}(\{x_1, \ldots, x_N\}).$$

So, the discrepancy measures in some sense how far points in a sequence are from being equally distributed on $[0, 1]^k$. Indeed, one says that a sequence $x = (x_1, x_2, \ldots)$ in $[0, 1]^k$ is *uniformly distributed* if $D_N^{(k)}(x)$ tends to zero when $n \to \infty$.

Now, a sequence $x = (x_1, x_2, \ldots)$ in $[0, 1]^k$ is called a *low-discrepancy sequence* if, for all integers $N > 1$, we have

$$D_N^{(k)}(x) \leqslant c \cdot \frac{(\log N)^{k-1}}{N}$$

for some constant c.

Indeed, it is an open conjecture whether there exists a constant c_k, depending only on the dimension k, such that

$$D^{(k)}(X) > c_k \cdot \frac{(\log |X|)^{k-1}}{|X|}$$

holds for any finite subset X of $[0, 1]^k$. This is known to be true for $k = 1$ and $k = 2$.

There are various classical families of low-discrepancy sequences. One of the simplest is *Halton sequences* $H_1, H_2, \ldots$, defined as follows. Choose any number base b (so for binary numbers $b = 2$). Write the integer n in b-ary expansion as $a_m a_{m-1} \cdots a_0$. Then set

$$H_n = \frac{a_0}{b} + \frac{a_1}{b^2} + \cdots + \frac{a_m}{b^{m+1}}.$$

For example, if $b = 2$, the Halton sequence starts with

$$\frac{1}{2}, \frac{1}{4}, \frac{3}{4}, \frac{1}{8}, \frac{3}{8}, \frac{5}{8}, \frac{7}{8}, \frac{1}{16}, \ldots.$$

Intuitively, each new number is placed to fill one of the large remaining gaps.

One of the reasons that low-discrepancy sequences are important in practice is illustrated by a theorem due to Koksma and Hlawka, which states that

$$\left| \int_{[0,1]^k} f(x)\,dx - \frac{1}{n} \sum_{i=1}^{N} f(x_i) \right| < V(f) D_N^{(k)},$$

where $V(f)$ denotes the variation of the function f. Note that this gives a *deterministic* error bound for the approximation. This is in contrast to the classical Monte Carlo method using pseudo-random numbers, which gives a probabilistic error bound. In some vague sense, Monte Carlo with low-discrepancy numbers constitutes a mixture of classical numerical integration techniques and Monte Carlo simulation.

7.2.4 Sobol Numbers

One popular class of low-discrepancy sequences are *Sobol numbers*. We give a brief outline of how they (or rather a finite version of them) can be generated efficiently on a computer.

For what follows, it is convenient to define the bitwise addition modulo 2 of the binary representation of two non-negative integers n and m by $n \oplus m$. In other words, if

$$n = \sum_{i=0}^{k} a_i 2^i, \quad a_i \in \{0, 1\}$$

and

$$m = \sum_{i=0}^{k} b_i 2^i, \quad b_i \in \{0, 1\},$$

then

$$n \oplus m = \sum_{i=0}^{k} c_i 2^i, \quad c_i \in \{0, 1\},$$

where $c_i = a_i + b_i \pmod 2$.

We can use this to define the Gray code $G(i)$ for $i \geqslant 0$ by $G(i) = i \oplus \lfloor i/2 \rfloor$. Note that $\lfloor x \rfloor$ denotes the largest integer not exceeding x. So, we have

$$G(0) = 0, \quad G(1) = 1,$$
$$G(2) = 3, \quad G(3) = 2,$$
$$G(4) = 6, \quad G(5) = 7,$$
$$G(6) = 5, \quad G(7) = 4,$$
$$\vdots \qquad \vdots$$

The Gray code gives a one-to-one map from $\{0, 1, \ldots, 2^k - 1\}$ onto itself, and it has the property that $G(i)$ and $G(i+1)$ differ by exactly one bit in their binary representation. These are the only two properties that we will make use of.

Let $p_1(x), p_2(x), \ldots$ denote the sequence of irreducible polynomials over the field with two elements, ordered by non-decreasing degree. So, we have

$$p_1(x) = x + 1,$$
$$p_2(x) = x^2 + x + 1,$$
$$p_3(x) = x^3 + x + 1,$$
$$p_4(x) = x^3 + x^2 + 1,$$
$$\vdots$$

We write

$$p_k(x) = \sum_{l=0}^{q_k} a_l^{(k)} x^{q_k - l}.$$

For an n-dimensional Sobol sequence, we choose n of these polynomials, eg $p_1(x), \ldots, p_n(x)$. Each polynomial corresponds to one component of the n-dimensional Sobol sequence.

The kth component of the nth Sobol sequence is now given by

$$S_n^{(k)} = \bigoplus_i \frac{1}{2^i} M_i^{(k)},$$

where we sum over all indices i such that the ith digit (the coefficient of 2^i) in the binary expansion of $G(n)$ is equal to 1. The numbers $M_i^{(k)}$ are given by the recursion

$$M_i^{(k)} = \left(\bigoplus_{l=1}^{q_k} 2^l a_l^{(k)} M_{i-l} \right) \oplus M_{i-q_k}, \quad i > k,$$

with initial values, for example, set to 1 (so $M_1^{(k)} = \cdots = M_q^{(k)} = 1$). The use of the Gray code makes this efficient to implement, since $S_{n+1}^{(k)}$ can be computed from $S_n^{(k)}$ by a single "$\oplus$" operation.

Another popular class of low-discrepancy numbers are given by *Faure sequences*. Their construction is given in Tezuka [72], for example, which in general is a good reference book on low-discrepancy sequences.

7.2.5 Brownian Bridges

When the Monte Carlo method is used as described so far, barriers are monitored discretely. The probability of reaching a barrier can differ substantially between a continuous barrier and a discrete barrier, even if monitored daily. So one needs a large number of sampling dates to model a barrier option accurately, which is computationally expensive. It is, under certain circumstances, possible get away with very few sampling dates by using Brownian bridge techniques.

For this, note that it is easy (with constant market parameters) to compute the probability that a geometric Brownian bridge, ie a geometric Brownian motion that is conditional on two specific values, reaches a certain barrier level

H. See Section 1.1.16 for a derivation of the formula (1.8), such that

$$\mathbb{P}[S_u = H \text{ for some } t_1 \leqslant u \leqslant t_2 \mid S_{t_1}, S_{t_2}]$$
$$= \exp\left(\frac{-2\log(H/S_{t_1})\log(H/S_{t_2})}{\sigma^2(t_2 - t_1)}\right),$$

where σ is the local volatility in the period from t_1 to t_2, which has to be constant.

This probability can now be used in the following way. Suppose, for example, that we want to price a partial barrier with knock-out between two dates t_1 and t_2 and a payoff f depending only on S_{t_2}. We simulate only the asset price on the two dates t_1 and t_2, and then compute the probability p of the asset price breaking the barrier during the knock-out period, given S_{t_1} and S_{t_2}. We then use the payoff $(1-p)f(S_{t_2}) + pG$, where G denotes the knock-out rebate.

7.2.6 Markov Chain Monte Carlo Methods

For very high dimensional problems, eg to price options with many underlyings and with path-dependent features, it is interesting to investigate Markov Chain Monte Carlo (MCMC) methods, which were originally developed for statistical physics. For this, one does not simply produce independent price paths for the assets, as with the classical Monte Carlo method. Instead, one constructs a certain Markov chain on the space of paths, and then one samples from this Markov chain.

Popular techniques are the Metropolis–Hasting algorithm and the Gibbs sampler. Consult Gilks, Richardson and Spiegelhalter [29] as a reference and introduction. Currently, there is very little literature on MCMC methods applied specifically to finance.

7.3 Generic Monte Carlo Pricing

The flexibility of the Monte Carlo method allows one to build generic pricing tools, allowing spreadsheet users to price a large variety of European options that depend on a finite list of fixings. For this, one needs a language to describe the payoff. This language can be parsed and translated into a sequence of operations on a stack. These operations have then to be applied for each simulated path.

For example, the language could be specified in such a way that the user can enter

$$\text{``max}(S(T) - K, 0)\text{''}$$

to price a plain vanilla call, or

$$\text{``max}\big(\max(S(t0), S(t1), S(t2), S(t3), S(t4), S(t5))- S(t5), 0\big)\text{''}$$

for a discrete lookback option with six sampling points.

The complete application consists of three relatively independent pieces:

- a parser that takes the string entered by the user and translates it into a sequence of operations on a stack;

- an efficient implementation of a stack, together with all necessary operations on this stack;

- a Monte Carlo engine that generates asset price paths according to the distribution assumed and executes the sequence of operations generated by the parser using these prices to compute the payoff for each price path.

7.4 Hybrid Monte Carlo

We can also use the Monte Carlo method to price products in our hybrid model.

Suppose that the present value of some payoff depends on the stock price S_{t_i}, the cash bond B_{t_i} and the short rate r_{t_i} at given times $0 < t_1 < \cdots < t_n$. Conditional on $\mathcal{F}_{t_{i-1}}$ (ie with given numbers $S_{t_{i-1}}$, $B_{t_{i-1}}$ and $r_{t_{i-1}}$), the three random variables

$$\log \frac{S_{t_i}}{S_{t_{i-1}}}, \qquad \log \frac{B_{t_i}}{B_{t_{i-1}}}, \qquad \Lambda_{t_i} r_{t_i} - \Lambda_{t_{i-1}} r_{t_{i-1}}$$

are jointly normally distributed. Recall that $\Lambda_t = \exp\left(\int_0^t \lambda_s \, ds\right)$. Expressions for all first and second moments can be found in the chapters on the hybrid model.

In our implementation, we never explicitly make use of θ_t. This means that we have to re-express the integrals

$$\int_{t_1}^{t_2} \theta_s B_1(s, t_2) \, ds \quad \text{and} \quad \int_{t_1}^{t_2} \theta_s \Lambda_s \, ds,$$

which we need to compute expectations, in a form using market data directly.

Specifically, we have

$$\int_{t_1}^{t_2} \theta_s B_1(s, t_2) \, ds = \log \frac{P(0, t_1)}{P(0, t_2)} - f(0, t_1) B_1(t_1, t_2) + C + \tfrac{1}{2} B_2(t_1, t_2),$$

with C as in equation (1.25), and

$$\int_{t_1}^{t_2} \theta_s \Lambda_s \, ds = \\ \Lambda_{t_2} f(0, t_2) - \Lambda_{t_1} f(0, t_1) + \int_0^{t_2} \sigma_s^2 \Lambda_s B_1(s, t_2) \, ds - \int_0^{t_1} \sigma_s^2 \Lambda_s B_1(s, t_1) \, ds.$$

This means we are able to sample from the joint distribution for S_{t_i}, B_{t_i} and r_{t_i}.

Using the forward measure It is often very convenient to replace the measure $\mathbb{P}$ by a forward measure $\mathbb{P}_T$. This makes it possible (if all the payments are at time T) not to simulate B_T, since discounting is done by multiplying with $P(t, T)$.

The chapters on the hybrid model contain formulas for the means and variances of

$$\log \frac{S_{t_i}}{S_{t_{i-1}}} \quad \text{and} \quad \Lambda_{t_i} r_{t_i} - \Lambda_{t_{i-1}} r_{t_{i-1}}$$

under $\mathbb{P}_T$. The correlation is (as also are the variances) unaffected by the change from $\mathbb{P}$ to $\mathbb{P}_T$.

7.5 Monte Carlo for American Options

The straightforward Monte Carlo approach cannot directly be used to price American or Bermudan options. The reason is that, to make an exercise decision for a specific price path at a specific time, one needs to know the holding value of the option, ie the discounted expected value from one time period ahead.

There have been several attempts to cope with this problem. We will present two approaches here. Note that American options are treated as Bermudan options with a high density of exercise dates, eg daily.

7.5.1 Tilley's Method

One of the earliest attempts was given by Tilley [73]. There are more effective methods now available, but it is still useful to outline this simple method as an illustration. The main drawback of Tilley's method is the large computer memory requirement.

We start by simulating N price paths as in the basic method. However, we now store all those paths. The algorithm works backwards in time through the exercise dates $t_m > t_{m-1} > \cdots > t_0$ and computes for each date and each path a 'preliminary' option value and an exercise decision (a Boolean variable that says whether to exercise or not). Only the exercise decision will be used later.

At a given exercise date $t_i < t_m$ (the situation at maturity is trivial) we order the N asset prices, and then bundle them in N^α groups of $N^{1-\alpha}$ prices each, where α is some constant with $0 < \alpha < 1$. For simplicity, we assume that N is chosen such that N^α and $N^{1-\alpha}$ are both integers.

The idea is that the price paths in the same group have a similar holding value, which can be approximated by averaging the option values at time t_{i+1} and discounting. We now make an exercise decision for each path in the group based on this estimated holding value. This gives us an option value at time t_i for each path as the maximum of inner value and estimated holding value. We store this value and continue proceeding backwards until we reach t_0. We now compute the option payoff for each path by using the discounted inner value at the earliest exercise date that we have determined.

The usual averaging now gives us the option value. Since both N^α and $N^{1-\alpha}$ approach infinity if N does, it is intuitive and can be proved that this value converges to the correct theoretical value.

So far, it is possible to have inconsistent exercise decisions between paths in different groups, even for a simple plain vanilla American option. This means that a call, say, is exercised at a lower spot and not exercised at a higher spot, because the paths belong to distinct groups. This can be fixed by modifying the exercise decision accordingly, which can speed up convergence.

7.5.2 The Method of Broadie and Glasserman

The second American Monte Carlo method we want to present is due to Broadie and Glasserman [9].

We start by discretising the asset price process using a non-recombining random tree with b branches per node. For this, we first simulate b asset prices for t_1. Then we generate for each of them b prices for t_2, ie a total of b^2 prices. Continuing like this, we arrive at a total of b^m paths, ending at the maturity date t_m.

We now compute to estimates V and W for the price of the American option. The estimation V will be biased high, while W will be biased low.

We then proceed backwards through the tree, and, at maturity, we initialise the estimators by the inner value. For all other nodes, we have to compute the inner value (which is easy) and an estimated holding value, in order to be able to make an exercise decision. For the holding value for V, we simply take the discounted average of V at the next time step following the b branches. The new V is then the maximum of the inner value and this holding value.

For W, we proceed as follows. Let $W^1_{t_{i+1}}, \ldots, W^b_{t_{i+1}}$ denote the estimators for date t_{i+1} for the b branches emanating from the node under consideration. We then compute, for each $k = 1, \ldots, b$, a value A_k in the following way. If the inner value at t_i is bigger than the discounted average

$$\text{df} \cdot \frac{1}{b-1} \sum_{j \neq k} W^j_{t_{i+1}},$$

where df denotes the relevant discount factor, then we set A_k equal to the inner value. If this is not the case, we set A_k equal to $\text{df} \cdot W^k_{t_{i+1}}$ instead. Now the new value for W is given by the average

$$\frac{1}{b} \sum_{k=1}^{b} A_k.$$

This method can be extended for multi-asset options. It also works well with control variate techniques.

7.5.3 Average Strike Reset Options

We now present the valuing of a specific option with early exercise features, for which the Monte Carlo method is combined with a finite grid method.

An *average strike reset* option gives its holder the right to exercise at pre-specified times, as with a standard Bermudan option, but, if the option is not exercised, the strike will be reset to some fixed factor α times the average of the underlying price over certain days.

Formally, we have n exercise dates

$$0 < t_1 < \cdots < t_n,$$

Monte Carlo Methods

and n_i averaging dates

$$t_{i-1} \leq s_1^i < \cdots < s_{n_i}^i \leq t_i$$

corresponding to the exercise date t_i, with $i = 1, \ldots, n-1$.

At each date t_i, the holder has the choice to exercise with the strike K_i previously set for the current period (so, for a call, $(S_{t_i} - K_i)^+$ is paid) or to hold the option with a new strike $K_{i+1} = \alpha \cdot (S_{s_1^i} + \cdots + S_{s_{n_i}^i})/n_i$ (except, of course, at maturity t_n).

The initial strike K_1 is specified in the option contract.

Pricing methodology For notational simplicity, we concentrate in the following on the case of a call. We denote the option value at time t_i for the case that it is not exercised at t_i and that the strike is set to K_{i+1} by

$$V_i(S_{t_i}, K_{i+1}).$$

We can now determine recursively the option price $V_0(S_0, K_1)$ by

$$V_{n-1}(S_{t_{n-1}}, K_n) = e^{-R_{t_{i-1},t_i}} \cdot \mathbb{E}\left[(S_{t_n} - K_n)^+\right]$$

and

$$V_{i-1}(S_{t_{i-1}}, K_i) = e^{-R_{t_{i-1},t_i}} \cdot \mathbb{E}\left[\max\left((S_{t_i} - K_i)^+, V_i(S_{t_i}, K_{i+1})\right)\right], \qquad (7.1)$$

with

$$K_{i+1} = \alpha \cdot \frac{1}{n_i} \sum_{j=1}^{n_i} S_{s_j^i}.$$

From this, it follows by induction that each V_i is a homogeneous function in the spot S and the strike K, so

$$V_i(S_{t_i}, K_{i+1}) = S_{t_i} \cdot V_i(1, K_{i+1}/S_{t_i}).$$

Pricing algorithm We now describe the pricing algorithm for $V_0(S_0, K_1)$, which uses a finite grid method together with a Monte Carlo technique.

For each t_i ($i < n$), we fix a list $k_1^i, \ldots, k_N^i$ of potential values for K_{i+1}/S_{t_i}. We want to compute a corresponding list of option values $v_1^i, \ldots, v_N^i$ with $v_j^i = V_i(1, k_j^i)$. We do this by using equation (7.1) and working backwards recursively through $i = n-1, n-2, \ldots, 1$. The case $i = n-1$ is special and can be handled either by a standard Monte Carlo method or, better still, by the plain vanilla closed formula.

We now generate sample values for $S_{s_1^i}, \ldots, S_{s_{n_i}^i}$ and for K_{i+1}/S_{t_i}. From the lists $k_1^i, \ldots, k_N^i$ and $v_1^i, \ldots, v_N^i$, we determine (an approximation of) $V_i(1, K_{i+1}/S_{t_i})$ by linear interpolation.

So we can compute

$$\max\left((1 - k_j^{i-1})^+, V(1, K_{i+1}/S_{t_i})\right) \qquad (7.2)$$

for each $j = 1, \ldots, N$.

Generating a large number of these sample paths and evaluating and averaging over (7.2), we get an approximation for

$$\mathbb{E}\big[\max\big((1 - k_j^{i-1})^+,\ V(1, K_{i+1}/S_{t_i})\big)\big].$$

By discounting, we arrive at a value for v_j^{i-1} for each $j = 1, \ldots, N$.

Chapter 8
A Partial Differential Equation Solver

The fundamental stochastic differential equation (SDE) for a single asset process is given in Section 1.2.1. In this equation, the volatility parameter is a deterministic function of time only. If the market were to price options according to that model, then implied volatilities for traded options would vary with maturity but would be the same for all strikes at a given maturity. It is well known that this is not the case and that, generally in equity markets, implied volatility for index options rises as the strike decreases.

One approach to accounting for this "smile" in implied volatility is to posit an asset process obeying the stochastic differential equation

$$\frac{dS_t}{S_t} = (r_t - d_t)\,dt + \sigma(S, t)\,dW_t,$$

in which the instantaneous volatility of the asset depends on S and t. The volatility in this context is referred to as "local volatility" when it is required to emphasise that it has this spot dependence in addition to time dependence.

The application of Itô's lemma to the function $P(S, t)$ representing the price of an option on the asset S results in a version of the Black–Scholes partial differential equation (PDE) in which volatility is explicitly dependent on the asset price S and time t:

$$\frac{\partial P}{\partial t} + (r_t - d_t)S\frac{\partial P}{\partial S} + \tfrac{1}{2}S^2\sigma^2(S, t)\frac{\partial^2 P}{\partial S^2} = r_t P. \tag{8.1}$$

Note that the quantities $\hat{\sigma}(K, T)$ and $\sigma(S, t)$ are quite distinct. The former is a property of a particular traded option; the latter is a property of an asset price process. Local volatility is not a market observable, in contrast to implied volatility, and must therefore be obtained by a process of calibration to observed market prices. The purpose of this exercise is not, of course, to value European options but rather to value other option types in the context of the market information provided by European option prices. For example, while a European option of arbitrary strike and maturity could clearly be valued using a volatility interpolated from observed implied volatilities, it is far from clear which volatility it is appropriate to use in the Black–Scholes model for pricing a barrier option, ie whether it is $\hat{\sigma}(K, T)$, $\hat{\sigma}(H, T)$, or some other value. Intuitively, the price of a path-dependent option will sample the entire $\sigma(S, t)$ ($t < T$) surface, inasmuch as it is an expectation over all possible paths, unless some region of (S, t)-space is forbidden (ie has zero probability). As explained below, local volatility is obtained from the full set of observed implied volatilities, and consequently the price of a barrier option will be determined, in this approach, not by a single volatility or even by a single volatility term structure, but by the whole implied volatility surface, thus rendering obsolete the question of which volatility to use for pricing.

While many of the option valuation formulas presented here proceed directly from the stochastic differential equation, this chapter outlines a solution method for equation (8.1), based on a lattice discretisation of the (S, t)-space. We shall follow the approach of standard texts, eg [68], for the discretisation scheme, then introduce the boundary conditions of particular interest to pricing a range of barrier options, and finally discuss calibration of $\sigma(S, t)$.

8.1 Discretisation of the PDE

After time reversal $t \to -t$, equation (8.1) becomes

$$\frac{\partial P}{\partial t} = (r_t - d_t)S\frac{\partial P}{\partial S} + \tfrac{1}{2}S^2\sigma^2(S, t)\frac{\partial^2 P}{\partial S^2} - r_t P, \tag{8.2}$$

and the problem of option valuation given the exercise value at maturity becomes a well-posed initial boundary value problem, with, for example,

$$P(S, t = 0) = (S - K)^+.$$

We require the solution to this equation at the evaluation date $t = T$ in reversed time or at $t = 0$ in the original time coordinates.

The transformation
$$x = \ln(S/S_0)$$
linearises (8.2) to yield

$$\frac{\partial P}{\partial t} = [r_t - d_t - \tfrac{1}{2}\sigma^2(S, t)]\frac{\partial P}{\partial x} + \tfrac{1}{2}\sigma^2(S, t)\frac{\partial^2 P}{\partial x^2} - r_t P, \tag{8.3}$$

with, of course, the correspondingly transformed initial condition $P(x, 0)$.

We define the computational mesh to be a regular lattice of points (x_i, t_j) in the region $\{0 \leqslant t \leqslant T,\ L \leqslant x \leqslant H\}$, the positions of the x boundaries to be determined by the problem. The spacing of the points is uniform in x-space:

$$x_i = x_0 + i\,\delta x, \quad i = 0, 1, \ldots, N,$$
$$t_j = j\,\delta t, \quad j = 0, 1, \ldots, M,$$

in which δx and δt are constants to be determined.

The finite difference method for solving PDEs is constructed by using Taylor expansions around each point in the computational mesh. This results in a discretisation of the PDE of first- or higher-order accuracy, which may be solved as a system of linear difference equations. The nature of the scheme is governed by the chosen approximations for the partial derivative terms. Second-order accurate schemes are typical. How the derivative terms are approximated determines the nature of the resulting numerical scheme. The approach here employs the Crank–Nicholson scheme.

The Crank–Nicholson implicit method regards the PDE as being satisfied at the point $(x_i, t_{j+\frac{1}{2}})$, using a central difference for the time derivative and substituting the arithmetic mean of the central differences at the j and $j+1$ time steps for the space derivatives. This results in the discrete approximations for the terms in the partial differential equation:

$$P(x,t)\big|_{i,j+\frac{1}{2}} \simeq \tfrac{1}{2}(u_{i,j} + u_{i,j+1}),$$

$$\frac{\partial P(x,t)}{\partial t}\bigg|_{i,j+\frac{1}{2}} \simeq \frac{1}{\delta t}(u_{i,j+1} - u_{i,j}),$$

$$\frac{\partial P(x,t)}{\partial x}\bigg|_{i,j+\frac{1}{2}} \simeq \frac{1}{4\delta x}(u_{i+1,j+1} - u_{i-1,j+1} + u_{i+1,j} - u_{i-1,j}),$$

$$\frac{\partial^2 P(x,t)}{\partial x^2}\bigg|_{i,j+\frac{1}{2}} \simeq \frac{1}{2\delta x^2}(u_{i+1,j+1} - 2u_{i,j+1} + u_{i-1,j+1} + u_{i+1,j} - 2u_{i,j} + u_{i-1,j}),$$

in which P is the exact solution of the PDE and u is the exact solution of the difference equation that results from substituting the finite difference approximations into the PDE (8.3):

$$\frac{1}{\delta t}(u_{i,j} - u_{i,j+1}) = \frac{\eta}{4\delta x}(u_{i+1,j+1} - u_{i-1,j+1} + u_{i+1,j} - u_{i-1,j})$$

$$+ \frac{\sigma^2}{4\delta x^2}(u_{i+1,j+1} - 2u_{i,j+1} + u_{i-1,j+1} + u_{i+1,j} - 2u_{i,j} + u_{i-1,j})$$

$$- \tfrac{1}{2}r_t(u_{i,j} + u_{i,j+1}). \tag{8.4}$$

Collecting terms results in the set of linear simultaneous equations

$$\alpha u_{2,j+1} + \beta u_{1,j+1} = \kappa u_{2,j} + \lambda u_{1,j} + \xi u_{0,j} - \gamma u_{0,j+1},$$

$$\alpha u_{i+1,j+1} + \beta u_{i,j+1} + \gamma u_{i-1,j+1} = \kappa u_{i+1,j} + \lambda u_{i,j} + \xi u_{i-1,j},$$

$$\beta u_{N-1,j+1} + \gamma u_{N-2,j+1} = \kappa u_{N,j} + \lambda u_{N-1,j} + \xi u_{N-2,j} - \alpha u_{N,j+1},$$

in which

$$\alpha(x,t) = \sigma^2 \delta t + \eta\, \delta t\, \delta x,$$

$$\beta(x,t) = -2\sigma^2 \delta t - 4\delta x^2 - 2r\, \delta t \delta x^2,$$

$$\gamma(x,t) = \sigma^2\, \delta t - \eta\, \delta t\, \delta x,$$

$$\kappa(x,t) = -\alpha,$$

$$\lambda(x,t) = 2\sigma^2 \delta t - 4\delta x^2 + 2r\, \delta t\, \delta x^2,$$

$$\xi(x,t) = -\gamma.$$

Since the simultaneous equations result in the solution of the unknown values at time $j+1$, the Crank–Nicholson discretisation is described as an example of an *implicit* scheme. In many problems the matrix elements are independent of time, but our model regards η as time-dependent and σ as both spot- and time-dependent, and this simplification does not apply.

We may write the above equations in matrix form as

$$A\boldsymbol{u}_{j+1} = z,$$

in which the matrix A is tridiagonal. The matrix equation can be solved without inversion of the matrix, by successive elimination.

The Crank–Nicholson discretisation may be placed in a more general framework by considering a weighted average for the x-derivatives rather than a simple arithmetic average. This treatment leads to the stability conditions for the discretisation and to the relationship between Crank–Nicholson, explicit and fully implicit schemes.

Consider for the discretisation at the point $(i\delta x, (j+\tfrac{1}{2})\delta t)$ the approximation

$$\left.\frac{\partial P(x,t)}{\partial x}\right|_{i,j+\tfrac{1}{2}} \simeq \frac{1}{2\,\delta x}[\theta(u_{i+1,j+1} - u_{i-1,j+1}) + (1-\theta)(u_{i+1,j} - u_{i-1,j})]$$

and the analogous weighted average for $\partial^2 P/\partial x^2$. The case $\theta = 0$ gives an explicit expression for $u_{i,j+1}$ in equation (8.4) when the finite differences are substituted: no simultaneous equations arise and the method is then said to be an explicit scheme. The case $\theta = 1$ gives a fully implicit scheme, and $\theta = \tfrac{1}{2}$ corresponds to Crank–Nicholson.

The importance of this is that it may be shown that the equations are unconditionally stable and convergent for $\tfrac{1}{2} \leqslant \theta \leqslant 1$, but for $0 \leqslant \theta < \tfrac{1}{2}$ we require

$$\frac{\delta t}{(\delta x)^2} \leqslant \frac{1}{\sigma(S,t)^2(1-2\theta)}. \tag{8.5}$$

It follows from this that if, for example, we double the number of x-mesh points from N to $2N$ in order to obtain the solution on a finer mesh, we must also quadruple the number of t-mesh points, so that the problem scales as N^3. The fact that Crank–Nicholson does not suffer from this severe stability constraint, having $\theta = \tfrac{1}{2}$, is the main reason why it is chosen here. None the less, it remains true that too great a value of δt will leave the solution far from convergence: we cannot get something for nothing. (It is also worth pointing out that the Crank–Nicholson can suffer from spurious oscillations, especially when the spot is very close to a barrier.)

We can now show the relationship between the finite difference methods discussed here and the tree method of pricing options given in Chapter 6. If we consider the $\theta = 0$ explicit discretisation, then we can write equation (8.4) as

$$u_{i,j+1} = \frac{1}{1+r\,\delta t}(p_u u_{i+1,j} + p_m u_{i,j} + p_d u_{i-1,j}),$$

where

$$p_u = \frac{\sigma^2 \delta t}{2\,\delta x^2} + \frac{\eta\,\delta t}{2\,\delta x},$$

$$p_m = 1 - \frac{\sigma^2 \delta t}{\delta x^2},$$

$$p_d = \frac{\sigma^2 \delta t}{2\,\delta x^2} - \frac{\eta\,\delta t}{2\,\delta x}.$$

It is clear that this is now the same as the trinomial tree formulation (except here we are working in reverse time). The multiplication factor discounts from one node to the next, and the variables (p_u, p_m, p_d) can be interpreted as the probabilities of moving to $(u_{i+1,j}, u_{i,j}, u_{i-1,j})$ from $u_{i,j+1}$. These probabilities have the usual properties: they sum to one,

$$p_u + p_m + p_d = 1,$$

and they are consistent with the drift and volatility of the underlying process,

$$\mathbb{E}[u_{i,j+1}] = p_u(u_{i,j} + \delta x) + p_m u_{i,j} + p_u(u_{i,j} - \delta x)$$
$$= u_{i,j} + \eta\,\delta t,$$

$$\mathbb{E}\big[(u_{i,j+1} - \mathbb{E}[u_{i,j+1}])^2\big] = p_u(\delta x - \eta\,\delta t)^2 + p_m(\eta\,\delta t)^2 + p_u(\delta x + \eta\,\delta t)^2$$
$$= \sigma^2 \delta t,$$

where we have only kept terms of $O(\delta t)$. The stability condition above translates to a statement that the transition probabilities always remain positive.

This shows that the explicit PDE method is equivalent to the trinomial tree approach. It can also be shown that the implicit PDE method is equivalent to a generalised discrete stochastic process where the asset price can jump to an infinite number of possible future values. This is also true for the Crank–Nicholson discretisation, except that this approach provides an unbiased estimator for the variance.

8.2 Boundary Conditions

The payoff function, eg $(S_T - K)^+$, acts as an initial condition in reversed time. However, it is clear from the matrix equation given above that conditions at the boundaries $i = 0$ and $i = N$ are required. Boundary conditions commonly exist in two forms. In the first of these, the value of the solution is known on the boundary; this boundary condition is applicable to the case of single and double barriers when the boundary value of the solution is equal to the rebate, if any, paid when the barrier is hit. In cases where the boundary is open, no barrier being present, we impose a condition on the x-derivative of the solution at the open boundary.

8.2.1 Dirichlet Boundary Conditions

In this case the boundary condition takes the form

$$P(x_b, t) = f(x_b, t),$$

in which x_b is the position of the boundary. This form is appropriate in cases where there is a well-defined boundary fixed by the nature of the problem, such as with single and double barriers. Thus for a knock-out barrier that pays a rebate G immediately when hit:

$$P(x_b, t) = G.$$

If the payment of this rebate is delayed until maturity, the boundary condition becomes time-dependent:

$$P(x_b, t) = G \exp(-R_{T-t,t}).$$

In the case of knock-in barriers, the boundary condition is again Dirichlet in nature, but instead of being defined by a function known from the definition of the option, it is given by the solution of an associated problem, evaluated at that boundary. Explicitly, in the notation of Chapter 3:

$$C_{in} : f(x_b, t) = C(x_b, t),$$
$$C^{in} : f(x_b, t) = C(x_b, t),$$
$$C^{out}_{in} : f(x_b, t) = C^{out}(x_b, t),$$
$$C^{in}_{out} : f(x_b, t) = C_{out}(x_b, t),$$
$$C^{in}_{in} : f(x_b, t) = C(x_b, t).$$

In every case, to summarise, the boundary condition to be applied at the knock-in barrier is the solution of the corresponding problem obtained by removing the knock-in condition.

8.2.2 Neumann Boundary Conditions

The Neumann boundary condition is of the form

$$\frac{\partial P}{\partial x}(x_b, t) = f(x_b, t)$$

and is appropriate for cases where the boundary is open, notably the case of vanilla options. In the particular case of the vanilla call, we have at the upper and lower boundaries

$$\left.\frac{\partial C}{\partial S}(x, t)\right|_{x=+\infty} = 1 \quad \text{and} \quad \left.\frac{\partial C}{\partial S}(x, t)\right|_{x=-\infty} = 0,$$

which corresponds to the statement that, as the spot tends to infinity, the delta for the call tends to 1, and, as the spot tends to 0, the delta tends to 0. Similarly, for digital options, the derivatives are zero on both boundaries.

8.2.3 Lattice Generation

Knowledge of the boundary conditions for the solution of the PDE equation allows us to generate an appropriate computational mesh.

As mentioned earlier, the spacing of the grid points is uniform in both x- and t-space. The t-space step size is therefore given by $\delta t = T/M$. Specification of the x-space domain, and therefore the x-space interval length, depends on where the boundary conditions are applied. For Dirichlet (closed) boundary conditions, the value of the solution is known at a fixed x-space point, eg at the barrier level for barrier options, and we therefore know x_b.

The situation for Neumann boundary conditions is complicated by the fact that it is not possible to implement the condition at infinity. Here we take the approach of generating an x-space domain that captures the statistically significant portion of the complete domain. This is achieved by setting x_b to be Z standard deviations away from the expected stock price at maturity,

$$x_{\min} = \mathcal{M}_T - \tfrac{1}{2}\sigma^2 T - Z\sigma\sqrt{T},$$
$$x_{\max} = \mathcal{M}_T - \tfrac{1}{2}\sigma^2 T + Z\sigma\sqrt{T},$$

for a lower and upper open boundary condition respectively. Here we take the cautious view that σ is the maximum volatility up to maturity. We found that a value of $Z = 4$ provided very accurate results and is equivalent to saying that the (risk-neutral) probability that the final stock price is outside the PDE mesh is about 0.005%.

The specification for the x-space boundary is given in Table 8.1. In the table, B_L and B_U are the lower and upper barrier levels respectively. Now that we have a well-defined x-space domain, the mesh step size is given by

$$\delta x = \frac{x_{\max} - x_{\min}}{N}.$$

A useful technique for improving the convergence and accuracy of the PDE approach is to be able to place certain reference values on the grid, eg a strike level. This can be achieved by modifying the x-space step size.

Table 8.1

	x_0	x_N
No barriers	$x_{\min}$	$x_{\max}$
Lower single barrier	$\ln(B_L/S_0)$	$x_{\max}$
Upper single barrier	$x_{\min}$	$\ln(B_U/S_0)$
Lower and upper barrier	$\ln(B_L/S_0)$	$\ln(B_U/S_0)$

8.2.4 Greeks

Option sensitivities can easily be obtained from the resulting PDE calculational mesh. The delta and gamma are obtained by using the first- and second-order differential operators on the evaluation date time slice. Similarly, theta is known because the price is available at the first two time slices. These greeks are obtained with no further calculations required. However, the remaining sensitivities have to be calculated on a perturbation basis, which will inevitably require the PDE to be solved again. We shall return to the subject of hedging later in this chapter.

8.3 Moving Barriers

We may extend the treatment of barrier options by considering cases in which the barrier level is not constant in time: $x_b = g(t)$. Our approach will be to transform the resulting problems into the constant-barrier case already addressed, defining in each case an appropriate transformation Q. We will treat the two cases of barriers whose level varies linearly and exponentially with time.

8.3.1 *Single Exponential Barrier*

The simplest case is that of a single barrier whose time dependence is exponential. The initial-value problem must again be solved, now subject to the boundary condition applied at

$$B(t) = B(0)e^{at}.$$

We define the transformation

$$q = Q(S) = Se^{-at}\frac{B(T)}{B(0)},$$

which transforms the exponential barrier into a constant barrier at a level $B(T)$. An application of Itô's lemma to the process q easily shows that q satisfies the process

$$dq = (r_t - d_t - a)\,dt + \sigma\,dW.$$

Thus we need only to solve the original SDE with an adjusted drift and a constant barrier.

8.3.2 *Single Time-Varying Linear Barrier*

A barrier whose level varies linearly with time,

$$B(t) = B(0) + at,$$

can be approximated within the context of a single time step $[t_j, t_{j+1}]$ of duration δt by an exponential, ie

$$B(t) = B(t_j)e^{a_j(t-t_j)},$$

with a_j chosen so that the exponential approximation matches the true linear barrier at t_j and t_{j+1}. As before, the transformation

$$q_t = Q(S_t) = S_t e^{-a_j(t-t_j)}\frac{B(T)}{B(t_j)}$$

defines a new process q and transforms the (locally exponential) barrier onto a flat barrier at $B(T)$. As above, it results in a change of drift and the process q

has drift term $\eta - a_j$ in the time interval. Thus the transformation required to flatten the piecewise exponential approximation to the linear time-varying barrier requires us to solve the original PDE with an adjusted drift.

8.3.3 Double Time-Varying Linear Barriers

Consider two time-varying barriers $u(t)$ and $l(t)$, with $u(t) > l(t)$. We may make the linear transformation

$$q = \frac{S - l(t)}{u(t) - l(t)} u_T + \frac{u(t) - S}{u(t) - l(t)} l_T$$

$$= \frac{U_T - l_T}{u(t) - l(t)} S + \frac{u(t) l_T - l(t) u_T}{u(t) - l(t)},$$

$$l_T = l(T),$$

$$u_T = u(T).$$

Applying the transformation to the barriers $u(t)$ and $l(t)$ results in constant barriers at the levels $l(T)$ and $u(T)$. The Itô formula can be applied to the above transformation to obtain the SDE satisfied by q, and it emerges that again we have to solve the original equation with a modified drift and, in this case, also a modified volatility.

8.3.4 Complex Barrier Conditions

An extension of the scheme given thus far is to consider functional forms for the barrier that are different in different intervals of time. Above, we considered the cases of a single linear barrier, a single exponential barrier and double linear barriers. We may also consider a class of barrier options whose barriers are not continuous functions but, within each region of time, are drawn from the above cases, plus the case of no barriers (the Neumann condition at each open boundary). We call these regions of time *subdomains*: the boundary conditions take a different functional form within adjacent subdomains or are discontinuous between them.

Barrier options having such barrier conditions may be defined, and these may be addressed by dividing up the region $0 \leqslant t \leqslant T$ into subdomains and separately solving within each. Working back from maturity to evaluation date, the initial condition for subdomain n is obtained from the solution at the earliest time of the subsequent subdomain.

It will be rare for the x-meshes to match up exactly in adjacent time domains, since it is clear that the boundary conditions determine the solution grids. Thus, for this approach to work, an interpolation method must be implemented to export the solution from one subdomain into the next. It is found that linear interpolation in log space is suitable. Cases will arise where $u_{i,0}$ in the nth subdomain lies outside the solution range $[u_{0,M}, u_{N,M}]$ of the $(n+1)$th subdomain. In this case, $u_{i,0}$ is set equal to the appropriate extreme value in the completed subdomain, corresponding to extrapolation at a constant level.

8.4 Range and Fade-In Options

It is possible within the PDE framework to approach options of the type that accrue a payoff profile conditional on the asset remaining within a range defined by two barriers. This section deals with three examples having varying payoff profiles, the same examples for which closed-form solutions are presented in Sections 3.2 and 2.4.6.

8.4.1 *Range Options*

The *range* option is a structure that pays a fixed coupon for each day the asset remains within a range defined by two barriers. It can be seen as accruing a payoff profile that is independent of the spot at maturity: a fixed cashflow. A variant of this option type has at least one "exploding" barrier: hitting such a barrier causes the payoff to stop accruing permanently, although in contrast to knock-out barriers it does not extinguish the option entirely.

Considering first the range option with two non-exploding barriers, we may write the payoff as

$$X = \frac{1}{n}\sum_{i=1}^{n} 1_{\{S_{t_i} \in [a,b]\}}.$$

We may approach the range option with non-exploding barriers as a sequence of one-day options each contributing an additional coupon conditional on the value of the asset at the end of that day. Consider the first of these (working in reverse time): it has initial condition

$$P_1(S, 0) = \begin{cases} 1/n & \text{if } L < S < H, \\ 0 & \text{otherwise}, \end{cases} \quad (8.6)$$

in which n is the number of days between evaluation date and maturity, on each of which the asset is observed once in order to determine whether the coupon is accrued. The boundary conditions are of the Neumann type:

$$\left.\frac{\partial P}{\partial S}\right|_{x \to \infty} = 0, \quad \left.\frac{\partial P}{\partial S}\right|_{x \to 0} = 0.$$

Its solution domain is $t = 0$ to $t = 1$. The second one-day option exists between $t = 1$ and $t = 2$ and its initial condition is the solution $P_1(S, 1)$ to the first option, plus a second term of similar form to equation (8.6):

$$P_2(S, 1) = \begin{cases} (1/n)e^{-r_1 \cdot 1} & \text{if } L < S < H, \\ 0 & \text{otherwise}, \end{cases}$$

in which r_1 is the discounting rate for $t = 1$ to $t = 0$. The boundary conditions at $S = 0$ and $S = \infty$ are unchanged. The complete range option is priced by iterating this scheme over all n dates in the range period.

In the case of a range option with an upper exploding barrier, the payoff X at maturity can be written as

$$X^u = \sum_{i=1}^{n} 1_{\{S_{t_i} \in [L,H],\, \forall u \in [0,t_i]\,:\, S_u < H\}}.$$

Range options with exploding barriers are approached using a similar strategy but different boundary conditions. In particular, if one or both barriers are of the exploding type, the boundary condition at that barrier is changed to one of Dirichlet type $P(B, t) = 0$.

8.4.2 Fade-In Options

Fade-in options are similar to range options except that the maximum payoff at maturity is not a fixed cash amount but rather the payoff of a standard call or put. Thus the call payoff X at maturity is written as

$$X = \frac{(S_T - K)^+}{n} \sum_{i=1}^{n} 1_{\{S_{t_i} \in [L,H]\}},$$

ie the payoff is the call payoff multiplied by the fraction of the option's lifetime for which the asset remains within the range.

The approach to pricing fade-in options in the PDE framework is similar to that for range options, in that a sequence of suboptions is used, each of whose initial values consists of the solution of a previous suboption plus a contribution conditional on the asset price being within the range.

As before, we may write the initial condition for the first option, which runs from $t = 0$ to $t = t_1$, in reverse time, as

$$P_1(S, 0) = \begin{cases} (1/n)(S_0 - K)^+ & \text{if } L < S < H, \\ 0 & \text{otherwise,} \end{cases}$$

with boundary conditions of Neumann type:

$$\left.\frac{\partial P}{\partial S}\right|_{x \to \infty} = 0, \qquad \left.\frac{\partial P}{\partial S}\right|_{x \to 0} = 0.$$

In parallel with the solution for the fade-in option, we construct for all $t \in [0, T]$ the solution to the plain vanilla call: denote this solution $C(S, t)$. The Neumann boundary conditions for this solution are, of course,

$$\left.\frac{\partial P}{\partial S}\right|_{x \to \infty} = 1, \qquad \left.\frac{\partial P}{\partial S}\right|_{x \to 0} = 0. \tag{8.7}$$

For the second suboption in the sequence, the initial condition is, as before, the solution to the first suboption at $t = 1$, plus an additional term

$$P_2(S, t_1) = \begin{cases} (1/n)C(S, t_1) & \text{if } L < S < H, \\ 0 & \text{otherwise,} \end{cases}$$

with the Dirichlet boundary conditions as above. The solution is evolved backwards towards evaluation date, adding similar terms at each fade-in date t_i.

8.4.3 Fade-In Barrier Options

The range and fade-in options have in common the fact that a payoff is accrued through time according to the fraction of the lifetime of the option during which the underlying asset remains within the corridor. The only difference between the two options is the nature of the payoff, in one case a fixed cash flow and in the other a standard option. We may therefore extend the same accrual mechanism to other payoffs, in particular to a barrier option that, of course, pays out conditionally on hitting the barrier.

The approach to pricing the fade-in barrier within a PDE framework is exactly as for the fade in, except for the nature of the solution, which is evolved in parallel with the primary option solution. In this case, we construct for all $t \in [0, T]$ the solution to the appropriate barrier option. The boundary conditions for this solution have a Dirichlet condition replacing one of the two open-boundary Neumann conditions (8.7). Thus, for an up-and-out call, for example, the boundary conditions are

$$P(H, t) = 0, \qquad \left.\frac{\partial P}{\partial S}\right|_{x \to 0} = 0,$$

in which the open upper boundary has been replaced by a fixed boundary at which the option value is known.

Given the secondary solution – that of the plain barrier option – the fade-in barrier follows in the same way as before, with additional components of the secondary solution being added into the primary solution at each fade-in date t_i, conditional on the position of the lattice point.

8.5 American Options

The extension of the PDE method to the pricing of American options requires the solution of a free-boundary problem, and here we follow the widely used approach of Brennan and Schwartz [8] applied to our discretisation method.

Following Section 2.2, the price of an American put at time t is given by

$$P(S_t, t) = \sup_{\tau \in \Phi_{t,T}} \mathbb{E}\left[e^{-R_{t\tau}}(K - S_\tau)^+ \right],$$

where $\Phi_{t,T}$ is the set of all possible stopping times τ.

It will not be optimal to exercise the option at time t if $P(S_t, t) > K - S_t$, whereas exercise will occur when $P(S_t, t) = K - S_t$. This naturally leads to the definition of a continuation region $\mathcal{C}$ and a stopping region $\mathcal{S}$:

$$\mathcal{C} = \{(x, t) \in (0, \infty) \times [0, T] \mid P(x, t) > x - K\}$$
$$\mathcal{S} = \{(x, t) \in (0, \infty) \times [0, T] \mid P(x, t) = x - K\}.$$

The line separating these two regions defines the optimal exercise boundary

$S^*(t)$. It can be shown [55] that on C the price of the American option satisfies the Black–Scholes PDE (8.1) subject to the following boundary conditions (in reverse time)

$$P(x, t)|_{t=0} = 0, \quad x \geq K,$$

$$P(x, t)|_{x=\infty} = 0,$$

$$P(x, t)|_{x=S^*(t)} = K - S^*(t),$$

$$\left.\frac{\partial P}{\partial S}(x, t)\right|_{x=S^*(t)} = -1.$$

Each iteration in the solution of this PDE is broken up into two steps. In the first step, we ignore the early exercise feature and simply do a European-style iteration. In the second step, at the new nodes, we check and replace these values if the exercise value is greater. In essence we implement the American feature at each time step.

In more detail, the steps are as follows. We first calculate the auxiliary values at the next time step, $p_{i,j+1}$ ($i = 0, \ldots, N$), subject to the boundary conditions $p_{0,j+1} = K$ and $p_{N,j+1} = 0$ by solving the PDE. We know the American option values $P_{i,j}$ and can calculate the values at the next time step using

$$P_{i,j+1} = \max(K - S_i, p_{i,j+1}).$$

The optimal exercise boundary node is then given by

$$\max(i \mid P_{i,j+1} = K - S_i)$$

for $S^*(j+1)$.

8.6 Discrete Dividends

So far, we have only considered an asset process that has a continuously compounded dividend yield. In this section we extend these results for the PDE-based approach to include also discrete dividend payments [75, 76].

Whenever an equity process is modelled, there are different approaches to incorporate discrete dividends (see Section 1.3.2). Here we follow Method I from that section:

$$\tilde{S}_t = S_t - D_t, \tag{8.8}$$

where $\tilde{S}_t$ is lognormal and

$$D_t = \sum_{t \leq t_i < T} e^{-r_{t_i}(t_i - t)} \delta_i.$$

For a European option, this amounts to replacing the spot S_0 by $S_0 - D_0$. The appropriate SDE is therefore

$$\frac{d\tilde{S}_t}{\tilde{S}_t} = (r_t - d_t) dt + \sigma(\tilde{S}_t, t) dW_t, \quad t \in [0, T]. \tag{8.9}$$

We note that here the underlying process volatility $\sigma(\tilde{S}_t, t)$ is now a function of price and time, and is assumed known.

Here we work directly with the Black–Scholes PDE from the modelling equation for $\tilde{S}$. After applying time reversal and the further change of variables $\tilde{x} = \ln(\tilde{S}/\tilde{S}_0)$, this may be written as

$$\frac{\partial p}{\partial t}(\tilde{x}, t) = \tfrac{1}{2}\sigma^2 \frac{\partial^2 p}{\partial \tilde{x}^2}(\tilde{x}, t) + \eta \frac{\partial p}{\partial \tilde{x}}(\tilde{x}, t) - rp(\tilde{x}, t), \quad t \in [0, T]. \tag{8.10}$$

The option payoff function now becomes an initial condition. In the case of a plain vanilla call option, this is

$$p(\tilde{S}, T) = \big(K - S(\tilde{S}, T)\big)^+, \quad \tilde{S} \in [L, H].$$

At $t = T$, the present value D_t of the future dividend stream is zero, so we have $S(\tilde{S}, T) = \tilde{S}$. Thus the initial condition for the PDE (8.10) in the case of a plain vanilla call is simply

$$p(\tilde{S}, T) = (K - \tilde{S})^+, \quad \tilde{S} \in [L, H]. \tag{8.11}$$

8.6.1 Boundary Conditions

The PDE (8.10) above is defined on "dividend space" $\tilde{S}$ with coordinates $(\tilde{S}, t) \in \mathbb{R}_+ \times \mathbb{R}_+$, while financial contracts are defined on "stock space" S with coordinates $(S, t) \in \mathbb{R}_+ \times \mathbb{R}_+$. It follows that the correct initial boundary value problem (IBVP) for a given contract is found by converting the defining boundary conditions into $(\tilde{S}, t)$-coordinates. A given Dirichlet condition in stock space

$$p(x_b, t) = f(x_b, t), \quad t \in [0, T],$$

now becomes the corresponding Dirichlet condition

$$p(\tilde{x}_b, t) = f\big(x_b(\tilde{x}_b, t), t\big), \quad t \in [0, T]. \tag{8.12}$$

A Neumann condition in stock space applied at $x = +\infty$, such as the one encountered in the case of the plain vanilla call, translates simply to the dividend space as

$$\left.\frac{\partial p}{\partial \tilde{S}}(\tilde{x}, t)\right|_{\tilde{x}=+\infty} = 1, \quad t \in [0, T].$$

The application of the above boundary conditions is made more difficult by the discontinuous nature of the function D_t across dividend payments. Dividend payments will result in discontinuous changes in boundary position within the dividend space $\tilde{S}$ as a result of the correspondence (8.8). In addition to these discontinuous changes in boundary position, there will be an exponential drift of the position of the boundary position in $\tilde{S}$-space as a result of the discounting in time of the forward dividend stream. Thus, fixed boundaries in stock space become drifting boundaries within dividend space, and moving boundaries in stock space acquire additional motion via drift. Below, we outline the solution strategy employed to accommodate this.

8.6.2 Solution Strategy

We calculate the price for the above contract by computing a rolling series of

IBVPs, starting at maturity (ie at $t = t_N$) and moving back in time on a dividend by dividend basis. A new IBVP is defined at a given dividend payment only in the case where the contract involves boundaries whose position S_b is defined in stock space as less than $+\infty$. That is, "open" Neumann-like conditions applied at infinity are unaffected by dividend payments. For the case of boundaries defined at some position S_b in stock space, each dividend payment will result in a discontinuous change to the boundary's position in dividend space, and the generation of a new IBVP for the next period between dividends. Specifically, if a sequence of dividends $d_{t_0}, d_{t_1}, \ldots, d_{t_N}$ are paid on the underlying stock at forward times $t_0, t_1, t_2, \ldots, t_N$ and where we define time t_M to be the maturity, then for the first period in reverse time $[t_{N-1}, t_N]$, the IBVP in dividend space, denoted by IBVP$_0$, will be

$$p(\tilde{S}_b, t) = f(S_b - D_t), \quad t \in [t_{N-1}, t_N],$$

and with initial condition given by (8.11). Thus, the effect a dividend paid at maturity has on the boundary condition is that it is now applied at the new position $\tilde{S}_b(0) = S_b - d_{t_N}$. We denote the solution to IBVP$_0$ by $DS_0(\tilde{S}, t)$ for $t \in [t_{N-1}, t_N]$. At time t_{N-1}, the solution of IBVP$_0$ is used to initialise the second initial boundary value problem, IBVP$_1$, where now the initial condition becomes

$$p(\tilde{S}, t_{N-1}) = \begin{cases} DS_0(\tilde{S}, t_{N-1}) & \text{for } \tilde{S} \in [\tilde{S}_b(t_{N-1}), H], \\ DS_0(\tilde{S}_b(t_{N-1}), t_{N-1}) & \text{for } \tilde{S} < \tilde{S}_b(t_{N-1}). \end{cases}$$

Here $t \in [t_{N-1}, t_{N-2}]$ and where we have assumed the barrier at S_b to be a lower barrier, with H representing the position of the upper boundary. The boundary condition at the lower boundary is simply (8.12). Note that the above initial condition flat interpolates out the t_{N-1} boundary condition prior to the t_{N-1} dividend payment as initial data to the IBVP$_1$ problem in the region where no $\tilde{S}$ solution is defined. This process is then repeated, giving for the initial boundary value problem i, IBVP$_i$, the initial data

$$p(\tilde{S}, t_{N-i}) = \begin{cases} DS_{i-1}(\tilde{S}, t_{N-i}) & \text{for } \tilde{S} \in [\tilde{S}_b(t_{N-i}), H], \\ DS_{i-1}(\tilde{S}_b(t_{N-i}), t_{N-i}) & \text{for } \tilde{S} < \tilde{S}_b(t_{N-i}), \end{cases}$$

where $DS_{i-1}(\tilde{S}, t)$ is the solution to IBVP$_{i-1}$ and $t \in [t_{N-i}, t_{N-i-1}]$. Upon completion of this sequence of IBVPs, the final solution $DS_{N-1}(\tilde{S}, t)$ at $t = t_0$ is transformed back into stock space using (8.8) to give the final option price.

8.6.3 Treatment of Moving Boundaries

Movement of the boundary position between dividend payments can come about either through the intrinsic motion of the barrier through the contract definition in "stock space" or by the drift imposed by the transformation (8.8) moving to dividend space $\tilde{S}$. The approach we take here is to incorporate the boundary drift into the natural (ie contract-defined) motion of the boundary on a time step by time step basis. Thus, in the case of a linear barrier, the additional exponential drift of the barrier as a result of the dividend is incorporated into the barrier's motion over the time step, with the appropriate

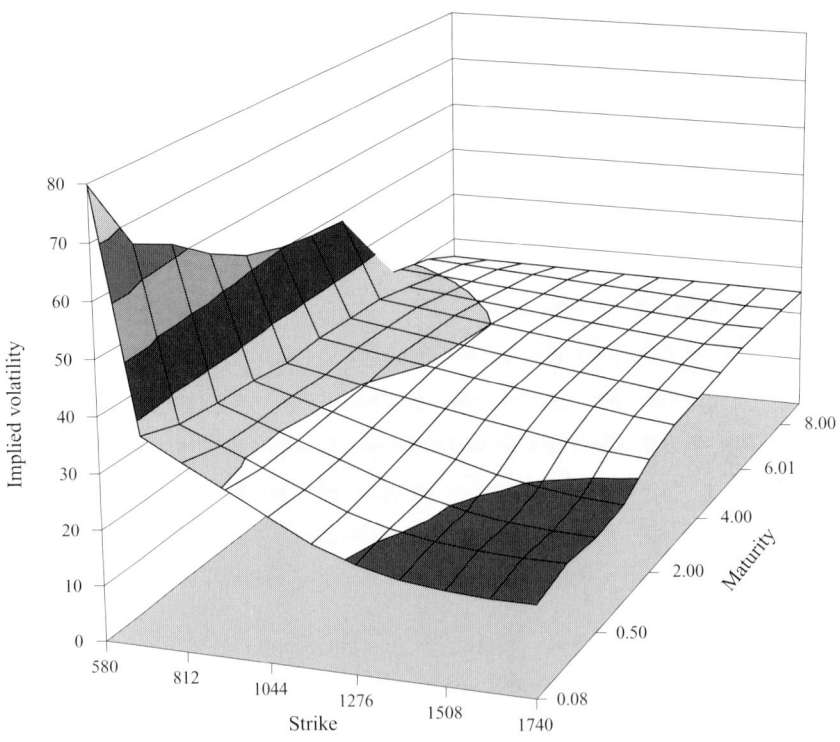

8.1 Implied volatility surface of DAX equity index.

transform being applied to "flatten" the barrier motion. Exponential barrier motion is treated in a similar manner.

8.7 Model Calibration

Liquid European call and put option prices are available in the market for a range of values of the strike price K_i and maturity T_j. It is a well-observed fact that these prices are not consistent with the Black–Scholes formulation. The Black–Scholes equation assumes a single constant volatility parameter $\tilde{\sigma}$, and implies that the asset follows a lognormal process with this volatility regardless of the asset level at future times. The market prices in fact indicate an implied volatility surface $\tilde{\sigma}(K_i, T_j)$ for some spot price S_0 at $t = 0$, and this generally is a skew surface in equity markets.[1] In Figure 8.1, we show such a volatility skew; the skew is more pronounced for shorter maturities.

Generally, the implied volatility is higher for lower strikes, with the skew becoming less pronounced at longer maturities. A possible explanation for this is the market's aversion to downward jumps.

The approach taken here is to go one step beyond the Black–Scholes and Merton models, and to assume that the instantaneous (local) volatility is a known (deterministic) function of the asset price S_t and time t. We do not assume any stochastic behaviour of volatility. In making the local volatility a function of the asset level, we lose all the nice features of the Black–Scholes

---

[1] If we denote the observable market price of a plain vanilla European call by $C(K, T; S = S_0, t = 0)$, then the implied volatility surface $\tilde{\sigma}(K_i, T_j)$ is defined by solving $C(K, T; S = S_0, t = 0) = BS(K, T; \tilde{\sigma})$ for each pair (K_i, T_j).

model of asset prices; however, we do produce a *single* asset process that is capable of correctly pricing all input market prices for all strikes and maturities. The PDE approach described above allows us to model this underlying process. This then leads naturally to a consistent pricing scheme for other structured and exotic options in the presence of our smile surface.

Calibration of the PDE solver in this approach requires the determination of the local volatility function $\sigma(S, t)$ such that the plain vanilla call price returned by the PDE solver for a particular strike K_i and maturity T_j agrees with the call prices available in the market.

Thus local volatility $\sigma(S, t)$ can be obtained from the implied volatility surface $\tilde{\sigma}(K, T)$ using implied diffusion theory [21, 22, 12, 1]. Here we work in a continuous-time setting and extend the work of Dupire [22] to non-zero interest rates and dividends (independently derived by Andersen and Brotherton-Ratcliffe [1]).

8.7.1 Implied Diffusion Theory

We start by assuming that at time t we can observe a continuum of European call option prices at all possible strikes $K > 0$ and at all maturities $T > t$. The value $C(K, T)$ of these market options can be written in terms of the underlying process density $\phi_T(S)$:

$$C(K, T) = B(t, T) \int_0^\infty (S - K)^+ \phi_T(S)\, dS,$$

with $B(t, T) = \exp(-\int_t^T r_u\, du)$. Differentiating twice with respect to K gives[2]

$$\phi(K, T; S, t) = \frac{1}{B(t, T)} \frac{\partial^2 C(K, T)}{\partial K^2}, \qquad (8.13)$$

where we now show the dependence on the option prices (K, T) in the density function and where

$$\phi(K, T = t; S, t) = \delta(S - K).$$

That is, the density for an option expiring on its evaluation date t is just a delta-function in the strike centred on the spot price. Figure 8.2 shows the implied risk-neutral probability density of the asset at several maturities, resulting from the implied volatility surface displayed in Figure 8.1. It should be remembered that this market-implied density is *not* the actual stock price distribution but a risk-adjusted distribution.

It can be seen more clearly from Figure 8.3 that, for short maturities, the distribution is approximately lognormal. However, for longer maturities, the market's (risk-neutral) probability becomes more skewed and in general has fat tails. The fat tail is more pronounced for lower spots and is a reflection of the increased probability given to market crashes over the lognormal models (such as the Black–Scholes model). There is also a tendency for the density function to be bimodal.

The problem now is to use the density to determine the diffusion defined by

---
[2] After integrating by parts, $C(K, T) = B(t, T) \int_K^\infty \int_y^\infty \phi_T(y)\, dy\, dS$ and the result follows trivially.

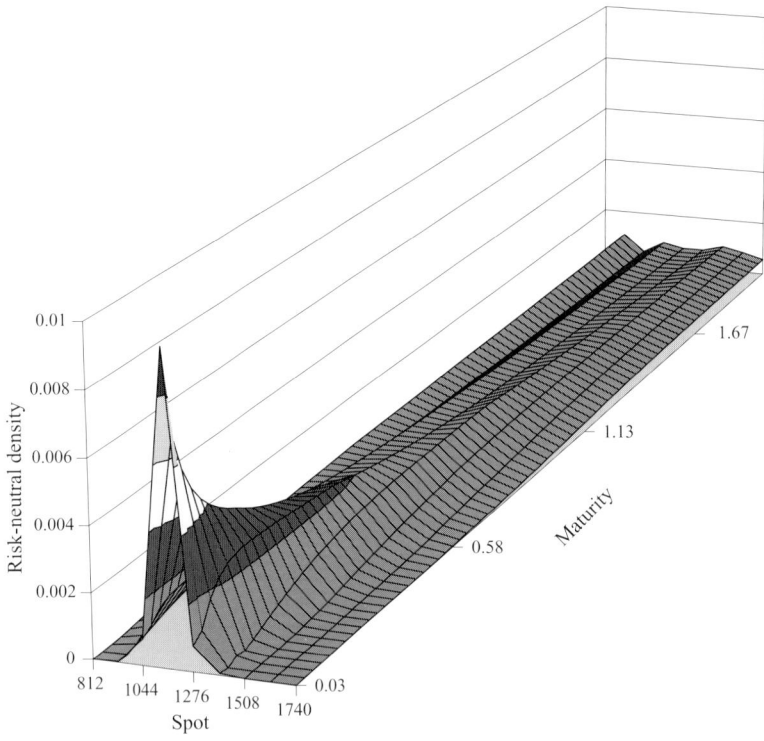

8.2 Implied risk-neutral density.

the process SDE. The solution to the converse problem of determining the density from the SDE is well known, where, under certain restrictions on the drift and volatility, the density $\phi(K, T; S_0, t)$ satisfies the forward Kolmogorov equation:

$$\frac{\partial \phi}{\partial T} - \frac{\frac{1}{2}\partial^2 (K^2 \sigma^2(K, T)\phi)}{\partial K^2} + \frac{\partial (K(r_T - d_T)\phi)}{\partial K} = 0. \tag{8.14}$$

Determining the diffusion from the distribution is more difficult. It is possible to show that two distinct diffusions can share the same distribution [21].

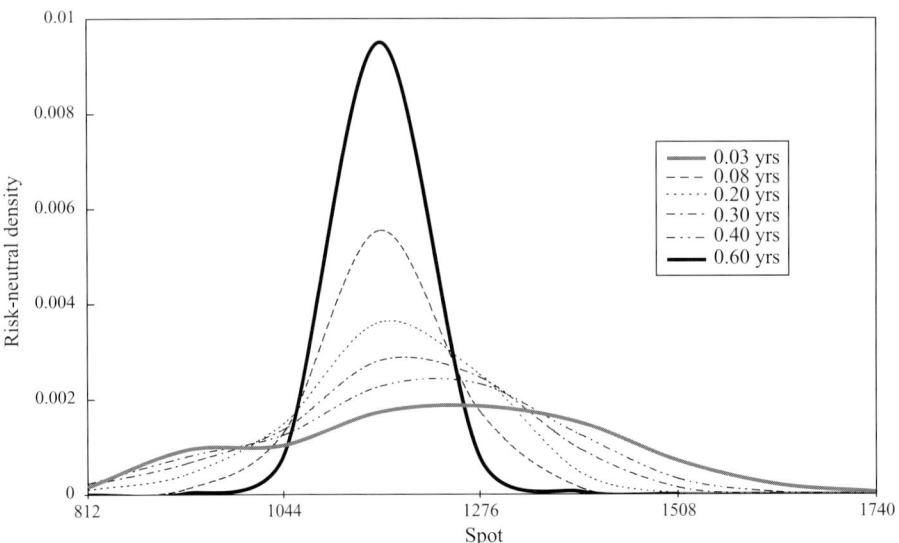

8.3 Implied risk-neutral density of various maturities.

However, limiting the collection of diffusions to those that are *risk neutral* makes it possible to associate a unique diffusion with some density $\phi(K, T; S_0, t)$. Proceeding formally, substitution of (8.13) into the first term in (8.14) gives

$$\frac{\partial \phi}{\partial T}(K, T; S_0, t) = \frac{1}{B(t, T)}\left(r_T \frac{\partial^2 C(K, T)}{\partial K^2} + \frac{\partial^2}{\partial K^2}\frac{\partial C(K, T)}{\partial T}\right).$$

Then integrating equation (8.14) twice over K, while holding T constant, gives

$$\tfrac{1}{2} K^2 \sigma^2(K, T) \frac{\partial^2 C(K, T)}{\partial K^2}$$
$$= \frac{\partial C(K, T)}{\partial T} + d_T C(K, T) + (r_T - d_T) K \frac{\partial C(K, T)}{\partial K} + \alpha(T) K + \beta(T),$$

where $\alpha(T)$ and $\beta(T)$ are unknown functions of T and where we have used the result

$$K \frac{\partial^2 C(K, T)}{\partial K^2} = \frac{\partial}{\partial K}\left(K \frac{\partial C(K, T)}{\partial K} - C(K, T)\right).$$

In the limit, as $K \to \infty$, we assume that all terms involving C vanish. This is a reasonable assumption, which is based on the fact that C does tend to zero and that we expect the functions to be regular. This forces the functions $\alpha(T)$ and $\beta(T)$ to be identically zero. Therefore

$$-\frac{\partial C}{\partial T} + (d_T - r_T) K \frac{\partial C}{\partial K} + \tfrac{1}{2} K^2 \sigma^2(K, T) \frac{\partial^2 C}{\partial K^2} = d_T C. \tag{8.15}$$

Notice the similarity between this equation (valid for European call options) and the general Black–Scholes PDE equation (8.1) (valid for any option). In fact we have a *duality* present, and it can be shown that our original problem can be solved in a dual market. In this dual market, time is reversed, the strike price becomes the underlying with the asset price as the strike, and the roles of interest rates and dividend yields are reversed.[3]

We now have our final result, which relates the local volatility surface $\sigma(K, T)$ to the market option prices $C(K, T)$:

$$\sigma^2(K, T) = \frac{2\dfrac{\partial C}{\partial T} + d_T C + (r_T - d_T) K \dfrac{\partial C}{\partial K}}{K^2 \dfrac{\partial^2 C}{\partial K^2}}. \tag{8.16}$$

The model developed here is complete (allowing arbitrage-free pricing and hedging) and assumes a continuum of market option prices across both strike and maturity.

In practice the market prices may exhibit arbitrage opportunities, and in any implementation one should check that the following no-arbitrage conditions

---

[3] Thus K follows the process $dK_t/K_t = (d_t - r_t) dt + \sigma(K, t) d\tilde{W}_t$, with t being backward time.

hold (for call option prices):

$$\frac{\partial C(K,T)}{\partial K} < 0 \quad \text{and} \quad \frac{\partial^2 C(K,T)}{\partial K^2} \geq 0.$$

The first condition should always be true for a European call option, while the second simply states the the risk-neutral probability density should always be positive. We also have the condition

$$\frac{\partial C(K,T)}{\partial T} > 0.$$

No-arbitrage portfolio dominance arguments [1] imply that the numerator of equation (8.16) is always positive. The above conditions then imply that $\sigma^2(K,T) \geq 0$, and therefore that the local volatility is real.

Our expression for the local volatility is in terms of the call prices and their derivatives, all of which are available by direct market observation. However, this market information is held (and often quoted) in the form of implied volatility surfaces. It is therefore computationally more efficient to work directly from this surface. The above equation can be written in terms of the implied volatility surface by making use of the Black–Scholes equation for the call price in terms of the implied volatility:

$$\sigma^2(K,T) = \frac{2\frac{\partial \hat{\sigma}}{\partial T} + \frac{\hat{\sigma}}{T-t} + 2(r_T - d_T)K\frac{\partial \hat{\sigma}}{\partial K}}{K^2 \left[\frac{\partial^2 \hat{\sigma}}{\partial K^2} + \frac{1}{(T-t)K^2\hat{\sigma}} + \frac{d_+}{\hat{\sigma}K\sqrt{T-t}} \frac{\partial \hat{\sigma}}{\partial K} + \frac{d_+ d_-}{\hat{\sigma}} \left(\frac{\partial \hat{\sigma}}{\partial K}\right)^2 \right]}, \quad (8.17)$$

where

$$d_\pm = \frac{\ln[S_t/K] + \mathcal{M}_{tT}}{\hat{\sigma}\sqrt{T-t}} \pm \tfrac{1}{2}\hat{\sigma}\sqrt{T-t}.$$

We can now look at the limiting cases of our approach. For the special case of a flat implied volatility surface, the local volatility is also flat and equal to the implied volatility,

$$\sigma^2 = \hat{\sigma}^2,$$

and we recover the Black–Scholes model. When the implied volatility surface has a term structure and no strike dependence, the local volatility is given by

$$\sigma^2(T) = \hat{\sigma}^2(T) + 2(T-t)\hat{\sigma}(T)\frac{\partial \hat{\sigma}(T)}{\partial T},$$

the instantaneous forward volatility. Alternatively, the implied volatility is the time average of the local volatility:

$$\hat{\sigma}^2(T) = \frac{1}{T-t}\int_t^T \sigma^2(u)\,du = \sigma_{tT}^2.$$

This recovers Merton's extension of the Black–Scholes framework.

Consider an implied volatility surface with a term structure and skew. Intuitively, we would expect the implied volatility to be a weighted average of

A Partial Differential Equation Solver

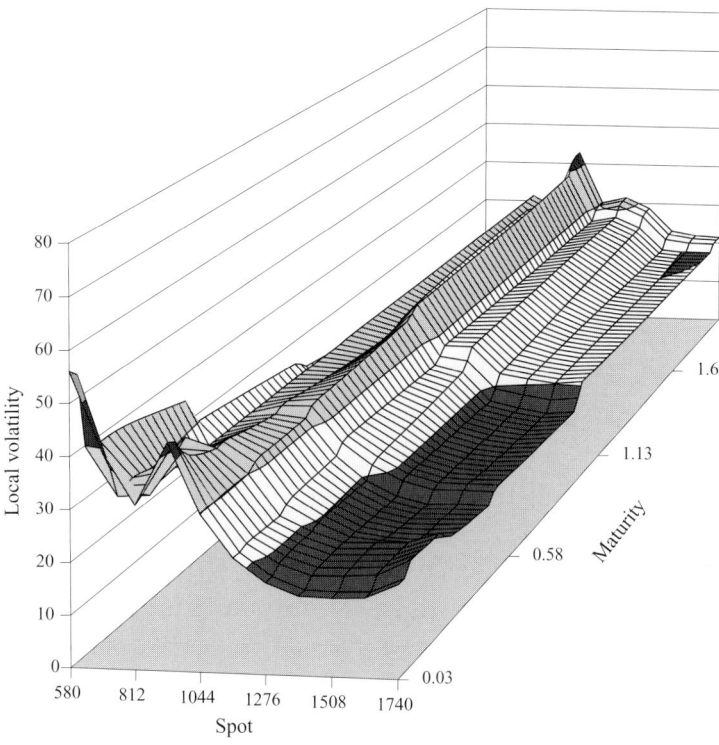

8.4 Local volatility surface.

the local volatility, averaged over all possible times and asset levels. Dupire [21] has shown that this is indeed the case and that the weighting scheme is roughly proportional to the Brownian bridge density, $\mathbb{P}[S_t = x \mid S_T = K]$.

Figure 8.4 shows the local volatility surface generated from the previous implied volatility surface (Figure 8.1), based on an appropriate numerical approximation for the implied volatility derivatives.

8.7.2 Hedging

Now that our model captures the volatility smile effect, we can investigate its effect on our hedging strategies. It turns out that the implied deltas and gammas can deviate quite considerably from the Black–Scholes results, and hedging in the Black–Scholes environment can result in significant "overhedging". This can be seen by looking at the following approximation for delta:

$$\Delta(S) = \frac{C(S + \delta S) - C(S - \delta S)}{2 \, \delta S}.$$

Because of the volatility skew, $C(S + \delta S)$ in our model is smaller than in the Black–Scholes model (because our volatility at $S + \delta S$ is smaller than the Black–Scholes volatility, which is constant). Similarly, $C(S - \delta S)$ is larger, and hence the price difference and hedging ratios are smaller in our model, by as much as 20% (see Figure 8.5). This effect obviously becomes more pronounced with higher skews.

Similarly, we can see that the gammas are bigger for in-the-money and smaller for out-of-the-money calls (see Figure 8.6). This might result in increased transaction costs required to maintain a hedge.

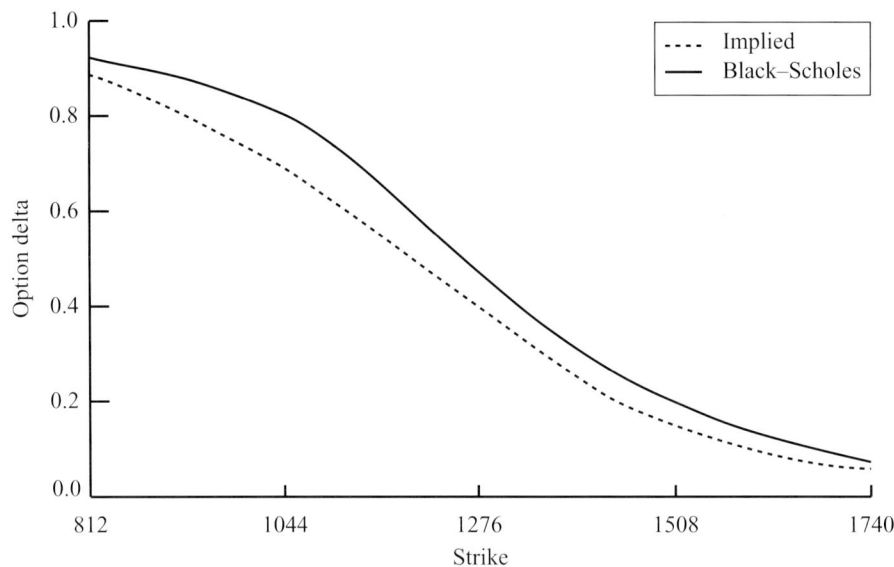

8.5 Delta for a call option compared with Black–Scholes.

Our proposed model of local volatility is a static model and assumes that the asset follows the implied process. In reality the local volatility surface will be dynamic and, as such, will result in our delta hedge becoming less effective. However, we can adopt a vega position that will allow us to capture some of this dynamic volatility. For a particular exotic option, this hedge can be achieved by shifting the implied volatility of one of the European options, regenerating the local volatility surface and repricing the exotic. This gives the vega sensitivity to that option. If this is done for all the European options used in the implied volatility surface, we can get a global vega.

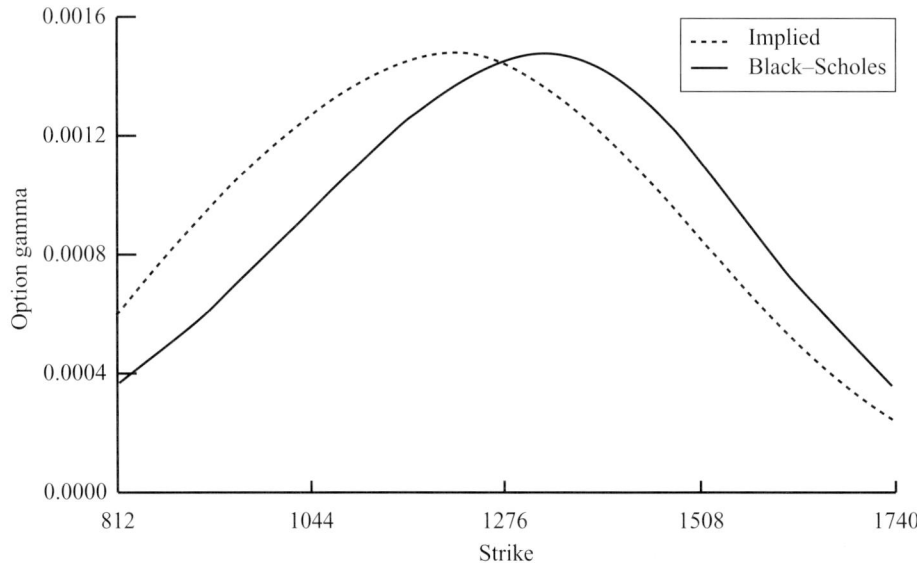

8.6 Gamma for a call option compared with Black–Scholes.

8.7.3 Calibration Strategy

The first decision is to choose a spanning set (S_i, T_k) of spots and dates over which the local volatility matrix will be constructed. The sampling points chosen define values on the local volatility surface. A sensible choice for the spot range must include all the values of significance for the products being priced from that local volatility surface. Examples of significant points would include barrier levels, strike levels, and so on. In practice, a pragmatic choice for the spot range is simply the range of strikes present in the definition of the implied volatility surface. The question of the date resolution for the local volatility surface is less clear, but is easiest to see in the simple case of an implied volatility that just contains a time structure. (We also assume that implied volatility is interpolated bilinearly in variance and that the local volatility surface is interpolated bilinearly.) In this case, the correct local volatility at each date is just the instantaneous forward volatility from the implied volatility surface taken at that date. Between date fixings in the implied volatility, the forward volatility is flat. Given that bilinear interpolation is used between date fixings in the local volatility surface, at least some additional dates must be provided between implied volatility fixing dates in order that the local volatility surface resolves the flat forward structure. Normally two additional dates between each pair of implied volatility dates should be sufficient to capture the forward structure.

Algorithmically, two special cases can be separated out. In the case of a flat volatility then, local volatility is just identically the flat implied volatility. Likewise for pure time structures, the local volatility may be determined directly from the forward volatility. Calibration is often required for the case where the implied volatility has a strike-dependent structure. In order to employ equation (8.17) to determine the local volatility, an interpolation and extrapolation method in strike K and maturity T of the implied volatility matrix $\tilde{\sigma}_{ij}$ is required. This should ensure the smoothness of the derivative terms in equation (8.17), which are computed numerically. Practical difficulties are encountered for realistic volatility smiles, where, for short-dated maturities far-in-the-money, the curvature term in the denominator gets vanishingly small, which can lead to numerical instabilities. Arbitrage opportunities present in the implied surface can also result in complex local volatilities. In effect, this approach tries to calculate numerically first- and second-order derivatives of a surface defined by a discrete number of points, where each of these points is prone to noise. It therefore appears that the key ingredient to a successful implementation of this approach is the interpolation and extrapolation methods used.

Experience suggests that the fit produced by local volatility surfaces generated by equation (8.17) for at-the-money strikes is good, but far-from-the-money results can deteriorate. With careful choices of the interpolation and extrapolation method used, realistic implied volatility surfaces regenerate the input European prices to within 10 basis points for everything except far-from-the-money options at long maturities.

For this reason, additional fitting may be required, with the Dupire surface equation (8.17) acting as a first approximation. The following procedure can easily be implemented, giving accurate results.

Suppose that the surface can be adjusted by three parameters α, β and γ, with

$$\delta\sigma(S_i, T_k) = \alpha(S_i - S_0) + \beta(T_k - T_0) + \gamma,$$

where S_0 is the spot and T_0 the evaluation date. The quantity $\delta\sigma(S_i, T_k)$ is the amount added to the local volatility surface at spot S_i and time T_k. Automatic adjustment of the surface can be performed. This consists of a constrained minimisation of the sum-of-squares distance function

$$F(T_k) = \sum_{i=1}^{N}[P_{\text{PDE}}(S_i, T_k) - P_{\text{BS}}(S_i, T_k)]^2, \qquad (8.18)$$

which is the square of the difference between the PDE-based price and the Black–Scholes price summed over a collection of spot values for some chosen maturity T_k. Additional constraints may also be placed on this problem, such as that PDE prices must fall within the bid–ask spread.

For example, we can choose three equally spaced spot values across the defined spot range. The local surface is fitted at a set of calibration dates, representing the maturities at which the PDE prices *must* match the Black–Scholes prices. The surface fitting works forward in maturity starting from the shortest date out to the longest. For a complete surface calibration, this set of dates would be those defining the implied volatility surface. In practice, applying this sort of least-squares approach is computationally expensive, and so it makes sense to fit as few dates as possible. For each calibration date, a two-parameter perturbation of the local volatility surface is performed for all surface dates prior to the calibration date, up to but not including any earlier calibration date. That is,

$$\delta\sigma(S_i, T_k; \alpha, \gamma) = \alpha(T_k)(S_i - S_0) + \gamma(T_k),$$

where

$$\sigma(S_i, T_k) \longrightarrow \sigma + \delta\sigma \quad \text{for all } T_{\text{cal0}} < T_k \leqslant T_{\text{cal1}}.$$

Here $T_{\text{cal0}} < T_{\text{cal1}}$ are two consecutive calibration dates. After each successive perturbation, the distance function (8.18) is recalculated, where the prices are computed from the option evaluation date up to the given calibration date.

To summarise, this approach generates the local volatility surface from the pure market prices and then, if necessary, calibrates this surface to provide a better fit back to the market prices. Noise and arbitrage opportunities in the market prices cause these inaccuracies. An alternative approach would be to start with a smooth implied volatility surface that is consistent with the market prices (to within, say, the bid–ask spread). This can be achieved by fitting a parametric form to the surface before applying the implied diffusion method. Because the surface is a smooth known function, the model calibration process is very quick and accurate (all the derivatives are known functions). However, this poses the question of what parametric form to fit.

8.7.4 Barrier Example

The local volatility surface that has been generated, which is consistent with the market prices, can now be used to price exotic options. Consider a one-year

Table 8.2

Barrier level	PDE price	Implied volatility (%)	Volatility range (%)
580	151.49	21.5	25.0–56.8
700	151.24	24.0	25.0–33.8
810	149.45	25.1	25.0–31.6
930	141.16	25.5	25.0–29.4
1050	107.87	25.5	25.0–27.2

knock-out at-the-money call option ($S_0 = 1160$, $K = 1160$, $r = 6\%$, $N = 150$, $M = 100$). Table 8.2 shows the option's price for several barrier levels, as calculated with the local volatility surface in the PDE approach and the resulting Black–Scholes implied volatility. The last column shows the range of implied volatilities between the strike and barrier level. Pricing these options within the Black–Scholes framework requires the selection of a single volatility, usually within this range. An important point to notice is that the Black–Scholes volatility implied by the PDE price is not always in this range, and in fact there are situations [1] where either no or two Black–Scholes implied volatilities are consistent with the PDE price.

8.7.5 Alternative Models

In this section we briefly summarise a group of models that provide a method of generating the local volatility surface within the deterministic volatility framework.

Implied tree models [66, 17, 22] use a discrete approximation to the risk-neutral continuous process in the form of a binomial–trinomial tree. The regular tree is distorted so as to capture all the market option prices. This distortion is achieved by carefully manipulating the local branching probabilities. Checks must be performed to make sure that these probabilities do not become negative.

In the spirit of the implied tree model, but applied to finite difference methodologies, is a forward induction approach [12]. Option prices for each point on the finite difference grid are calculated. With this PDE method, we have an operator that allows us to evolve the values from one time slice to the next. This evolution operator depends on the diffusion process, namely on the unknown local volatility at the nodes in the time slice. Given that there are N x-space points, there will then be N unknown local volatilities. The forward induction method uses the fact that for each time slice we know the value of N options. Therefore, at each time step, we can solve these N equations with N unknown local volatilities and thereby iteratively generate the full local volatility surface.

All the methods require interpolation and extrapolation from the discrete set of market prices available. An alternative method exists [46] that does not make this restriction but instead finds, by a minimisation procedure, the local volatility function such that the solutions of the Black–Scholes equation fall between the corresponding bid–ask market quotes. There is generally not enough information to determine $\sigma(S_t, T)$ (which should depend continuously on the market data) uniquely. So a secondary constraint is placed on the minimisation function, which ensures the smoothness of $\sigma(S_t, T)$. A gradient descent procedure can be used to perform this minimisation.

The approach we have taken here is to assume that the instantaneous volatility is a deterministic function of the asset level and time (ie it is not random). Alternative models are available that assume that the volatility is itself a stochastic process [33]. One of the main advantages that the *deterministic volatility* models have over the majority of *stochastic volatility* models is one of completeness. This then provides a secure foundation for pricing and hedging all options.

Section 9.3 describes another method for capturing the volatility smile effect. This skew model is based on an Edgeworth expansion of a probability distribution around the lognormal distribution and allows for the modelling of skewness and kurtosis.

Chapter 9
Further Modelling Issues

9.1 Calibration of the Extended Vasicek Model

In this section we present a method for calibrating the parameters of the extended Vasicek interest rate model.

The method assumes as input a set of market rates (money market and/or swap market rates) as well as a set of flat volatilities of caps and/or swaptions. The market prices for these instruments are obtained using the Black model [6], as described in Sections 5.2 and 5.1 respectively.

The goal is to determine the parameters of the Vasicek model so that the difference between the market prices and the corresponding prices using the Vasicek model is minimal.

The model Recall that in the generalised Vasicek model the short rate process follows the equation

$$dr_t = (\theta_t - \lambda_t r_t)\, dt + \sigma_t\, dW_t,$$

with the following deterministic time-dependent parameters: mean reversion rate λ, mean reversion level η and volatility σ. Given λ and σ, the parameter θ can be chosen to provide an exact fit of a given yield curve. We assume the parameters λ and σ to be piecewise constant, ie

$$\lambda_t = \lambda_m, \quad t \in (s_{m-1}, s_m], \quad m = 1, 2, \ldots, M,$$

$$\sigma_t = \sigma_n, \quad t \in (t_{n-1}, t_n], \quad n = 1, 2, \ldots, N.$$

General approach We assume a set of market instrument prices (caps and/or swaptions) to be given. We know how to compute these prices from flat Black implied volatilities.

Now, given piecewise constant parameters λ and σ, the Vasicek prices of caps and swaptions can be given in closed form (see Sections 5.2 and 5.1).

As a goal function for the minimisation process, we use a suitable function

$$U(\lambda_m, \sigma_n : m = 1, 2, \ldots, M; n = 1, 2, \ldots, N) = F(V_i, B_i : i = 1, 2, \ldots, I),$$

where B_i denotes the market price of the ith market instrument and V_i denotes the corresponding Vasicek price calculated assuming the piecewise constant parameters with $s_0 = t_0 = 0$ as well as $s_M = t_N = \infty$. The result of the calibration process depends on the choice of the following:

- the function U;
- the market instruments B_i ($i = 1, 2, \ldots, I$);
- the model dates s_m ($m = 1, 2, \ldots, M$) and t_n ($n = 1, 2, \ldots, N$).

We discuss these items in the following sections.

Utility function The function F defining the utility function U via

$$U(\lambda_m, \sigma_n : m = 1, 2, \ldots, M; n = 1, 2, \ldots, N) = F(V_i, B_i : i = 1, 2, \ldots, I)$$

has to satisfy:

- $F \geqslant 0$;
- $F = 0$ if and only if $V_i = B_i$ ($i = 1, 2, \ldots, I$);
- F is convex in V_i ($i = 1, 2, \ldots, I$).

In addition, the parameters may be restricted by *a priori* bounds $\Lambda_-, \Lambda_+, \Sigma_-, \Sigma_+$, ie

$$\Lambda_- \leqslant \lambda_m \leqslant \Lambda_+, \quad \text{and} \quad \Sigma_- \leqslant \sigma_n \leqslant \Sigma_+.$$

As an example, one could choose a quadratic utility function

$$F(V_i, B_i) = \sum_{i=1}^{I} (V_i - B_i)^2$$

and bounds

$$\Lambda_- = 0.00001, \quad \Lambda_+ = 10, \quad \Sigma_- = 0, \quad \Sigma_- = 100\%.$$

Market instruments The choice of market instruments should depend on the option to be priced. Two issues have to be considered:

- It is desirable to include market instruments that are similar to the product to be priced. For example, if the option is an outperformance equity bond, the bond price distribution at maturity of the option implied by the Vasicek model should be close to the distribution implied by market prices for bond options with similar maturity and bond expiry. This corresponds to the definition of a time-dependent volatility based on plain vanilla options in the equity derivatives world.

- The Black volatilities for caps depend on the fixed level. Similarly, the Black swaption volatilities depend on the strike. In order to capture this "smile", either market instruments with different strikes should be included in the calibration process or an at-the-money strike should be used. This corresponds to the smile in the equity derivatives world.

One possibility is to use at-the-money caplets covering the time interval needed for the specific product, as well as swaptions in the case of products with a longer maturity.

Market dates The number of model dates defines the number of free parameters in the calibration process.

In order to define a well-posed optimisation problem, the number of any subset of parameters should always be less than or equal to the number of market instruments influenced by these parameters.

Given a market instrument, some parameters will influence the price more than others. Generally, the mean reversion λ_t for small t has less effect on the short rate distribution at T than the mean reversion for t close to T. For the short rate volatility σ_t, the effect for small t is greater than that for large t.

Generally, optimisation with two time-independent parameters λ and σ is fast and more stable. If time-dependent parameters are needed, we suggest that only the volatility σ be allowed to be time-dependent.

9.2 Basket and Asian Underlyings

In this section we will develop lognormal approximations to two non-lognormal distributions. The first of these is the arithmetic average of observations on a single asset at times t_i ($i = 1\ldots m$); the second is the weighted sum of observations $S_T^1, \ldots, S_T^n$ on n assets at time T. In each case, the approximating distribution will be chosen so that its first two moments match those of the target distribution.

If this approximation is good, it will allow us to make the following assumption: that we may approximately price an option on an average or basket by applying the standard algorithm to the corresponding approximating lognormal process.

9.2.1 Asian Underlyings

We develop in this section the lognormal approximation to the arithmetic average A of an asset process S,

$$A = \frac{1}{m}\sum_{i=1}^{m} S_{t_i}, \tag{9.1}$$

by means of a lognormal random variable B. In addition, given a second average

$$\tilde{A} = \frac{1}{n}\sum_{j=1}^{n} \tilde{S}_{t_j} \tag{9.2}$$

on a second asset process $\tilde{S}$ with lognormal approximation $\tilde{B}$, we calculate the correlation of B and $\tilde{B}$, given the correlation of S and $\tilde{S}$.

The moments of an average process Following the approach of Turnbull and Wakemann [74], let X be a standard normally distributed random variable and let

$$B = \exp(\alpha + \beta X)$$

denote the lognormal random variable approximating A. The parameters (α, β)

of B can be chosen so that the first two moments of A and B coincide, ie $\mathbb{E}[A] = \mathbb{E}[B]$ and $\text{Var}[A] = \text{Var}[B]$. Since

$$\mathbb{E}[B^m] = e^{m\alpha + m^2\beta^2/2}, \tag{9.3}$$

we have to choose

$$\alpha = \sqrt{\log\left(\frac{\text{Var}[A]}{\mathbb{E}[A]^2} + 1\right)},$$

$$\beta = 2\log\mathbb{E}[A] - \tfrac{1}{2}\log(\text{Var}[A] + \mathbb{E}[A]^2).$$

To calculate the moments of A, let

$$R_i = S_{t_i}/S_{t_{i-1}}, \quad i = 1, 2, \ldots, n,$$

with $t_0 = 0$. The moments $(r_i^m, m = 1, 2, \ldots)$ of R_i are given by

$$r_i^m = \mathbb{E}[R_i^m]$$
$$= \exp\bigl(\tfrac{1}{2}m^2 \Sigma^2_{t_{i-1}t_i} + m\Gamma_{t_{i-1}t_i}\bigr)$$

(see equation (9.3)). We now define $(L_i, i = 1, 2, \ldots, n)$ recursively by

$$L_{n+1} = 1$$
$$L_i = 1 + R_i L_{i+1}, \quad i = n, n-1, \ldots, 2.$$

Then

$$A = \frac{1}{n}\sum_{i=1}^n S_{t_i} = \frac{1}{n} S_{t_1} \sum_{i=1}^n \prod_{j=2}^i R_j$$
$$= \frac{S_0}{n} R_1 L_2.$$

The moments $(l_i^m, m = 1, 2, \ldots)$ of L_i can be calculated recursively by

$$l_n^m = \mathbb{E}[(1 + R_n)^m]$$
$$= \sum_{j=0}^m r_n^j \begin{bmatrix} m \\ j \end{bmatrix},$$

$$l_i^m = \mathbb{E}[(1 + R_i L_{i+1})^m]$$
$$= \sum_{j=0}^m r_i^j l_{i+1}^j \begin{bmatrix} m \\ j \end{bmatrix} \quad (i = n-1, \ldots, 2).$$

We thus obtain the moments of $(n/S_0)A = R_1 L_2$. In particular, we have

$$\mathbb{E}[A] = \frac{S_0}{n} r_1^1 l_2^1,$$

$$\text{Var}[A] = \frac{S_0^2}{n^2} r_1^2 l_2^2,$$

and hence Σ_T^2 for $T \geqslant t_n$. Note that this treatment also provides higher moments.

Further Modelling Issues

The correlation of two averages Let A and $\tilde{A}$ denote the two averages (9.1) and (9.2). Then, using the above results, we obtain the approximations

$$A \approx B = e^{\alpha + \beta X},$$

$$\tilde{A} \approx \tilde{B} = e^{\tilde{\alpha} + \tilde{\beta}\tilde{X}},$$

with X and $\tilde{X}$ denoting two standard normally distributed random variables. We assume the assets S and $\tilde{S}$ to be instantaneously correlated with the parameter ρ_t, ie the driving Brownian motions W and $\tilde{W}$ of S and $\tilde{S}$ respectively satisfy $d\langle W_t^i, W_t^j \rangle = \rho_t^{ij} dt$. The aim of this section is to find a correlation ρ_{avg} of X and $\tilde{X}$ such that the correlation of A and $\tilde{A}$ is equal to the correlation of B and $\tilde{B}$.

It is straightforward to show that

$$\rho_{\text{avg}} = \text{Corr}[X, \tilde{X}] = \frac{1}{\beta\tilde{\beta}} \log\left(\text{Corr}[A, \tilde{A}] \sqrt{e^{\beta^2} - 1} \sqrt{e^{\tilde{\beta}^2} - 1} + 1\right). \quad (9.4)$$

Hence, it remains to calculate $\text{Corr}[A, \tilde{A}]$. We have

$$\mathbb{E}[S_s \tilde{S}_t] = F_s \tilde{F}_t \exp\left(\int_0^{s \wedge t} \rho_u \sigma_u \tilde{\sigma}_u \, du\right),$$

with $F_s = \mathbb{E}[S_s]$ and $\tilde{F}_t = \mathbb{E}[\tilde{S}_t]$.

For the averages, we obtain

$$\mathbb{E}[A\tilde{A}] = \frac{1}{mn} \sum_{i=1}^m \sum_{j=1}^n \mathbb{E}[S_{s_i} \tilde{S}_{t_j}]$$

$$= \frac{1}{mn} \sum_{i=1}^m \sum_{j=1}^n F_{s_i} \tilde{F}_{t_j} \exp\left(\int_0^{s_i \wedge t_j} \rho_t \sigma_t \tilde{\sigma}_t \, dt\right).$$

To obtain the average correlation in (9.4), we apply the definition

$$\text{Corr}[A, \tilde{A}] = \frac{\mathbb{E}[A\tilde{A}] - \mathbb{E}[A]\mathbb{E}[\tilde{A}]}{\sqrt{\text{Var}[A] \text{Var}[\tilde{A}]}}$$

and the moments calculated earlier in this section.

9.2.2 Lognormal Approximation of a Basket

In this section a lognormal approximation to the basket price at time t is developed.

The basket price at time t is given by the weighted sum

$$\sum_{i=1}^n w_i S_t^i,$$

where $w_1, \ldots, w_n$ are weights.

We wish to find a new lognormal variable

$$S_t = F_t \exp\left(\sigma W_t - \tfrac{1}{2}\sigma^2 t\right)$$

such that

$$\mathbb{E}[S_t] = \mathbb{E}\left[\sum_{i=1}^{n} w_i S_t^i\right] \quad (9.5)$$

and

$$\text{Var}[S_t] = \text{Var}\left[\sum_{i=1}^{n} w_i S_t^i\right]. \quad (9.6)$$

To satisfy (9.5), we simply need to set

$$F_t = \sum_{i=1}^{n} w_i F_t^i,$$

in which F_t^i are the forwards of the basket constituents. To satisfy (9.6), we compute

$$\text{Var}\left[\sum_{i=1}^{n} w_i S_t^i\right] = \sum_{i=1}^{n}\sum_{j=1}^{n} w_i w_j \mathbb{E}[S_t^i S_t^j] - \left(\sum_{i=1}^{n} w_i \mathbb{E}[S_t^i]\right)^2$$

$$= \sum_{i=1}^{n}\sum_{j=1}^{n} w_i w_j F_t^i F_t^j [\exp(\Psi_t^{ij}) - 1].$$

Since

$$\text{Var}[S_t] = F_t^2 (e^{\Sigma_t^2} - 1),$$

we get

$$\sigma = \sqrt{\frac{1}{t} \ln\left(\frac{\sum_{i=1}^{n}\sum_{j=1}^{n} w_i w_j F_t^i F_t^j \exp(\Psi_t^{ij})}{\sum_{i=1}^{n}\sum_{j=1}^{n} w_i w_j F_t^i F_t^j}\right)}.$$

9.3 Volatility Smile

As mentioned earlier (see Section 1.2), the Black–Scholes model assumes a constant (or time-dependent) volatility of the asset prices. It is well known that the market prices for European options do not agree with this assumption. This becomes apparent in the "volatility smile", ie the plot of implied volatility versus strike for a fixed maturity.

Different extensions to the model have been suggested to obtain a better fit to a given set of market data.

One approach assumes that the volatility in the governing stochastic differential equation is a deterministic function of time and asset price rather than of time only. Closed-form solutions will exist for very specific forms of $\sigma(S_t, t)$; however, for all other cases, prices can be obtained using a numerical scheme. This is the subject of Chapter 8.

In this section we aim at approximating the observed density of S_T as implied by market prices of European plain vanilla options with maturity T. This is done via a two-dimensional parametrisation of the space of densities. The two parameters account for the skew and kurtosis of the observed density of the return. We thus add two more degrees of freedom to the model, which allows us to approximate a given volatility smile.

We present two approaches, applying ideas presented in [41] and [15]. Moreover, we describe the implementation of an optimisation tool that calculates optimal skew and kurtosis parameters to fit a given set of market data.

9.3.1 Expansion and Option Pricing

Let F be the real distribution of S_T and let A be an approximating lognormal distribution with the same first moment. We denote the density of A by a and its kth derivative by $a^{(k)}$. With $C(A, K)$ and $P(A, K)$ denoting the Black–Scholes prices for calls and puts with strike K, we have the expansion

$$C(F, K) = C(A, K)$$
$$+ \frac{e^{-R_T}}{2!} \{\kappa_2(F) - \kappa_2(A)\} \int_{\mathbb{R}} (x - K)^+ a^{(2)}(x)\, dx$$
$$- \frac{e^{-R_T}}{3!} \{\kappa_3(F) - \kappa_3(A)\} \int_{\mathbb{R}} (x - K)^+ a^{(3)}(x)\, dx$$
$$+ \frac{e^{-R_T}}{4!} \{\kappa_4(F) - \kappa_4(A) + 3[\kappa_2(F) - \kappa_2(A)]^2\} \int_{\mathbb{R}} (x - K)^+ a^{(4)}(x)\, dx$$
$$+ \varepsilon,$$
$$= C(A, K)$$
$$+ \frac{e^{-R_T}}{2!} \{\kappa_2(F) - \kappa_2(A)\} a(K)$$
$$- \frac{e^{-R_T}}{3!} \{\kappa_3(F) - \kappa_3(A)\} a^{(1)}(K)$$
$$+ \frac{e^{-R_T}}{4!} \{\kappa_4(F) - \kappa_4(A) + 3(\kappa_2(F) - \kappa_2(A))^2\} a^{(2)}(K)$$
$$+ \varepsilon,$$

where $\kappa_i(F)$ denotes the ith cumulant of distribution F, ie

$$\kappa_1(F) = \mu_1(F),$$
$$\kappa_2(F) = \mu_2(F),$$
$$\kappa_3(F) = \mu_3(F),$$
$$\kappa_4(F) = \mu_4(F) - 3\mu_2(F),$$

with $\mu_i(F)$ denoting the ith central moment

$$\int_{\mathbb{R}} \left(x - \int_{\mathbb{R}} x f(x)\, dx\right)^i f(x)\, dx$$

of F (see [70]).

The lognormal distribution and its first derivatives are given by

$$\left.\begin{aligned} a(x) &= \left(\sqrt{2\pi}\,\sigma\sqrt{T}\,x\right)^{-1} e^{-d_2^2}, \\ a^{(1)}(x) &= -\frac{a(x)}{x}\left(1 - \frac{d_2}{\sigma\sqrt{T}}\right), \\ a^{(2)}(x) &= \frac{a(x)}{x^2}\left(2 - \frac{1}{\sigma^2 T} - 3\frac{d_2}{\sigma\sqrt{T}} + \frac{d_2^2}{\sigma^2 T}\right), \end{aligned}\right\} \qquad (9.7)$$

with

$$d_2 = \frac{\log(S_0/x) + (r - d - \tfrac{1}{2}\sigma^2)T}{\sigma\sqrt{T}}.$$

Since we assumed that the first moments of A and F coincide, the call–put parity implies

$$C(F, K) - C(A, K) = P(F, K) - P(A, K).$$

Hence the additional terms for call and put options are equal. Now, as Jarrow and Rudd [41] point out, there are different possibilities for choosing the second moment of A. We discuss two of them in the following sections.

Equating the second moment of the real distribution We use two parameters s and k to model the excess $\kappa_2(F) - \kappa_2(A)$ via

$$\partial = \kappa_2(F) - \kappa_2(A)$$

$$= -\frac{1}{\sigma\sqrt{T}}\left(s\frac{(\sigma\sqrt{T})^3}{3!} + k\frac{(\sigma\sqrt{T})^4}{4!}\right).$$

The approximate formulas read:

$$C(F, K) \approx C(A, K) + \left(\frac{\partial^2}{2!}\right)(K\sigma\sqrt{T})^2 e^{-R_T} a(K)$$

$$- \left(\frac{s}{3!}\right)(K\sigma\sqrt{T})^3 e^{-R_T} a^{(1)}(K)$$

$$+ \left(\frac{k}{4!} + \frac{s\partial}{3!}\right)(K\sigma\sqrt{T})^4 e^{-R_T} a^{(2)}(K)$$

for the call option, and

$$P(F, K) \approx P(A, K) + \left(\frac{\partial^2}{2!}\right)(K\sigma\sqrt{T})^2 e^{-R_T} a(K)$$

$$- \left(\frac{s}{3!}\right)(K\sigma\sqrt{T})^3 e^{-R_T} a^{(1)}(K)$$

$$+ \left(\frac{k}{4!} + \frac{s\partial}{3!}\right)(K\sigma\sqrt{T})^4 e^{-R_T} a^{(2)}(K)$$

for the put. The lognormal density a and its derivatives $a^{(i)}$ are defined in (9.7).

Matching the first two moments The approach described here follows Corrado and Su [15]. We define

$$\mu_2(F) = \mu_2(A).$$

This implies

$$\mu_2(F) = \mu_1(F)(e^{\sigma^2 T} - 1),$$

and hence determines the parameter σ of A. We set skewness s and kurtosis k as

$$s = \frac{\kappa_3(F)}{\kappa_2^{3/2}(F)} - \frac{\kappa_3(A)}{\kappa_2^{3/2}(A)} \quad \text{and} \quad k = \frac{\kappa_4(F)}{\kappa_2^2(F)} - \frac{\kappa_4(A)}{\kappa_2^2(A)},$$

and obtain the approximation

$$C(F, K) \approx$$
$$C(A, K) - s\frac{e^{-R_T}}{3!}\mu_1^3(e^{\sigma^2 T} - 1)^{3/2} a^{(1)}(K) + k\frac{e^{-R_T}}{4!}\mu_1^4(e^{\sigma^2 T} - 1)^2 a^{(2)}(K),$$

with μ_1 being the first moment or, in financial terms, the forward. The lognormal density a and its derivatives $a^{(i)}$ are defined in (9.7). The terms

$$\frac{\kappa_3(F)}{\kappa_2^{3/2}(F)} \quad \text{and} \quad \frac{\kappa_4(F)}{\kappa_2^2(F)}$$

describe the skewness and kurtosis of the distribution F. The corresponding approximation for the put is

$$P(F, K) \approx$$
$$P(A, K) - s\frac{e^{-R_T}}{3!}\mu_1^3(e^{\sigma^2 T} - 1)^{3/2} a^{(1)}(K) + k\frac{e^{-R_T}}{4!}\mu_1^4(e^{\sigma^2 T} - 1)^2 a^{(2)}(K).$$

9.3.2 Fitting Market Values

Let the market parameters spot, interest rate, dividend yield and volatility be fixed. Let the set $(P_1, P_2, \ldots, P_n)$ be the market prices for call (put) options with common maturity and strikes $(K_1, K_2, \ldots, K_n)$. This implies a set of implied volatilities $(IV_1(P_1), IV_2(P_2), \ldots, IV_n(P_n))$, where $IV_i(P)$ denotes the implied volatility of the option price P using strike K_i.

Now the goal is to choose parameters s and k (skewness and kurtosis) such that the prices $(Q_1(s, k), Q_2(s, k), \ldots, Q_n(s, k))$ using the four-moment model and strikes $(K_1, K_2, \ldots, K_n)$ are as close as possible to $(P_1, P_2, \ldots, P_n)$.

More specifically, we minimise the function

$$(s, k) \mapsto \sum_{i=1}^{n}(Q_i(s, k) - P_i)^2$$

or the function

$$(s, k) \mapsto \sum_{i=1}^{n}\left[IV_i(Q_i(s, k)) - IV_i(P_i)\right]^2.$$

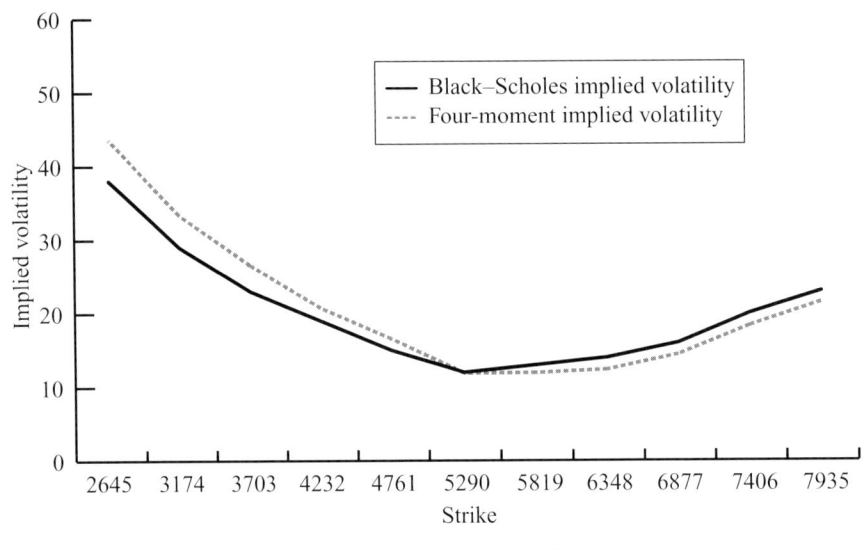

9.1 Fitted volatility smile.

Figure 9.1 shows the result of fitting to the FTSE implied volatility smile present in six-month option prices.

Chapter 10
Hedging

10.1 Hedging and Risk Management

10.1.1 *The Concept of Risk Management*

Much of the book thus far has been concerned with the mathematical derivation of formulas and algorithms to price equity derivative contracts. This compilation would be incomplete without discussing the trading of such instruments.

As trading and risk management require considerable attention to the fine detail of an individual market situation and/or transaction, there is no generic algorithm that describes risk management. Instead, we can only hope to highlight some general principles and caveats.

Evolution of derivative markets Although options existed as early as the 17th Century, when calls on the infamous tulip bulbs were traded, the quantitative approach to pricing and risk management did not start before the 1970s. Despite the fact that gut feeling always told traders to buy a couple of stocks to offset their risk when they just sold calls, the seminal work of Black and Scholes and of Merton established some crucial insights when compared with earlier option pricing formulas.

First, they made clear that in a complete market model one could replicate the payoff of an equity option perfectly by dynamically trading the underlying stock (this is called "delta hedging"). In consequence, traders should have no problem selling a call option on a stock, even if they were very bullish on that stock. Indeed, modern derivative desks do not care about directional moves of the stock but only about gamma and vega (ie the magnitude of future stock price movements, irrespective of their direction).

Second, Black and Scholes observed that in an arbitrage-free world the minimum cost of such a dynamic delta hedging strategy determines the fair price of the option. Furthermore, they gave a closed-form formula to compute the fair price of such an option based on the assumptions that the underlying stock follows a geometric Brownian motion and that interest rates, dividend yields and stock volatility remain constant.

The formula soon became a huge success with traders, as the level of volatility was the only unknown input parameter; all previous option pricing methods were dependent on the risk preferences of the option seller and hence required the input of some hard-to-know risk-aversion parameter.

With the Black–Scholes formula at hand, option trading became a lot easier and pricing essentially boiled down to the question: what is the correct volatility? Nevertheless, in the first phase, bid–ask spreads on options were wide. As implied volatility was still difficult to obtain, historical volatility had to be used to provide a guideline for the correct pricing volatility. Having sold an option, the seller was left with gamma and vega risk during the life of the option, which was expensive to lay off with other market participants, and hence remained as a residual risk with the option seller. Delta risk could, of course, be hedged in the stock as a liquid underlying and therefore did not pose a residual risk.

Thus, practical trading and pricing in this early Phase 1 (see Table 10.1) was only risk-preference free as regards delta risk. As far as the transfer of gamma and vega risk – which is involved in every option transaction – was concerned, risk preferences related to this residual risk still affected the pricing.

A lot of option markets, however, did eventually see reduced bid–ask spreads and increased liquidity. In this Phase 2, plain vanilla options can be traded in and out at low cost almost as a commodity, therefore serving as an underlying security themselves.

In such a mature market, second-generation equity derivative products, such as those discussed in this book, emerge. Here it is possible to hedge gamma and vega risks by trading the appropriate single stock or index options. However, correlation, cross-gamma or barrier exposures inherent in these second-generation equity derivative products are still impossible or very expensive to lay off, since no liquid market for those risks has developed.

As a consequence, one has to assess carefully which of the risks inherent in trading any of the second-generation products can reasonably be hedged in the existing markets. A lot of less-developed markets or markets for mid-cap stocks are, even today, more or less in Phase 1, without a liquid market for plain vanilla options. While this should not prevent a trader from selling Asian options, it might not be a good idea to start trading compound or power options in such a market.

The concept of hedging We shall now look at the mechanics of pricing and hedging on a more formal basis. As a starting point, we choose a derivative contract X being entered into. As an example, we take a 2-year American put on a basket of two stocks, so the payoff of X to the buyer at the time of

Table 10.1 Phases of derivatives market development.

	Phase 1	Phase 2
Traded underlying securities	Stocks, interest rate products	Stocks, interest rate products, plain vanilla options
Newly introduced derivative security	Calls, puts	Second-generation (barrier, Asian, lookback, compound, callable)
Hedgeable risks	Delta, rho	Delta, rho; single stock and index gamma and vega
Residual risks	Gamma, vega, dividends	Correlation, cross-gamma, dividends

exercise t would be
$$\max\left(100 - \tfrac{1}{2}(S_t^1 + S_t^2), 0\right).$$

Once this contract is entered into and exists in an equity derivative trading book, a pricing engine V and market data input $\mathcal{M}$ will be used to calculate the current mark-to-market value $V(X, \mathcal{M})$ of the position. The set of market parameters $\mathcal{M} = (\mathcal{M}_1, ..., \mathcal{M}_M)$ will consist of all model input data necessary to compute $V(X, \mathcal{M})$.

If we assume for our example that both stocks trade in US$ and that US$ is also the accounting currency of the financial institution, then we might, for example, have
$$\mathcal{M} = (r_{1y}^{US\$}, r_{2y}^{US\$}, d^1, d^2, S_0^1, S_0^2, \sigma^1, \sigma^2, \rho^{12}),$$
with

$r_{1y}^{US\$}$ = 1-year US$ risk-free rate,

$r_{2y}^{US\$}$ = 2-year US$ risk-free rate,

d^1 = dividend yield or discrete dividends for Stock 1,

d^2 = dividend yield or discrete dividends for Stock 2.

In accordance with market standard for equity derivatives and as has been done in this book, the pricing engine will return the expected value of the discounted payoff of X under the assumption that S^1 and S^2 follow geometric Brownian motion and that the exercise policy τ is optimal for the option holder. However, for $V(X, \mathcal{M})$ to be a viable pricing engine for the use by the trading and controlling functions of a financial institution, the following constraints have to be met:

1. If market values for the elements of $\mathcal{M}$ are known or can be obtained from traded instruments through calibration, these values should be used. In our example, this would be easy for $r_{1y}^{US\$}$ and $r_{2y}^{US\$}$ but probably difficult for ρ^{12} since there rarely is a market for implied correlation. In particular, if a market price for X is available, $V(X, \mathcal{M})$ should be close to this price.

2. If there is a payment or delivery under the contract X, the change in value of $V(X, \mathcal{M})$ should reflect this. In our example, this would mean
$$V(X, \mathcal{M}) \geqslant \max\left(100 - \tfrac{1}{2}(S_\tau^1 + S_\tau^2), 0\right) \tag{10.1}$$
for all times τ, as the contract is American. Equality would only hold in (10.1) if the counterparty chooses the optimal time to exercise or if $t = 2y$, the maturity of the contract. In the PDE approach to option pricing, this condition is incorporated into boundary conditions.

3. The same pricing engine V should be used to evaluate all similar contracts X throughout the financial institution.

From a practical standpoint, satisfaction of the constraints 1–3 can be seen as more important than the theoretical requirements of using an arbitrage-free and complete model.

Constraint 1 should ensure that the contract X can be sold at a price not too far from its book value, Constraint 2 guarantees that all payments required under X can be paid from its book value, and constraint 3 prevents arbitrage and the creation of artificial profits between different departments within the same financial institution.

In contrast to equity derivatives or short-term FX options, where the geometric Brownian motion is ubiquitous, there is no such standard model for interest rate options. Firms apply various one-, two- or multi-factor short rate or Heath–Jarrow–Morton yield curve models (or blends of different models), and yet usually meet the constraints 1–3. Even in the equity derivatives world, there is no general agreement on the volatility model to be used, with approaches to the skew ranging from complete market models such as

$$\frac{dS_t}{S_t} = \sigma(S, t)\, dW_t + (r_t - d_t)\, dt$$

to modelling σ as a stochastic process itself.

Even the way that the Black–Scholes formula is used by most equity derivative desks is unsatisfactory from a theoretical standpoint. Take, for example, a trader who has in his book two 1-year call options written on the same stock S. For the at-the-money call with strike 100, he uses a volatility of 20% and thus assumes a stock price dynamic of

$$\frac{dS_t}{S_t} = 0.20\, dW_t + (r_t - d_t)\, dt, \tag{10.2}$$

which also gives him the Black–Scholes delta. For the out-of-the-money call with strike 150, he uses a volatility of 30% and hence a stock price dynamic of

$$\frac{dS_t}{S_t} = 0.30\, dW_t + (r_t - d_t)\, dt, \tag{10.3}$$

from which he also gets his Black–Scholes delta. Since both equations (10.2) and (10.3) are modelling the dynamics of the same stock, they can't both be correct simultaneously, and so the trader's hedging approach based on his combined delta is inconsistent.

Once the financial institution has chosen a viable model and built a pricing engine V based on it, the actual task of risk management is straightforward. As $V(X, \mathcal{M})$ gives – by definition – the fair value of X, every change in a market parameter within $\mathcal{M}$ leads to a change in the fair value of X and hence constitutes a risk. Note that, although some parameter changes within $\mathcal{M}$ might lead to a *profit* for the financial institution, we will nevertheless refer to any change as a risk. With this definition, a *fully hedged portfolio* will have the same mark-to-market value for all possible values of input parameters $\mathcal{M}$. A *locally hedged portfolio* will, by definition, not change its mark-to-market value for small changes in the set of market parameters $\mathcal{M}$.

If we denote by Y the portfolio consisting of our derivative contract X plus all further hedge contracts entered into, then the two definitions can be formally expressed as

$$V(Y, \mathcal{M}) = \text{const} \quad \text{for all } \mathcal{M}$$

for a full hedge, and

$$\frac{\partial V}{\partial \mathcal{M}_m}(Y, \mathcal{M}) = 0 \quad \text{for all } \mathcal{M}_m \in \mathcal{M}$$

for a local hedge.

The trader could therefore construct a local hedge by simply calculating all first-order derivatives of X and then buy and sell some underlying securities that yield exactly the opposite exposure. In our earlier example of having sold the American put on two stocks S^1 and S^2, this would mean that the trader would go short the quantities

$$\frac{\partial V}{\partial S^1} \quad \text{and} \quad \frac{\partial V}{\partial S^2}$$

in stocks S^1 and S^2 respectively plus, for example, buy some two-year bonds to hedge the rho

$$\frac{\partial V}{\partial r_{2y}^{\text{US\$}}}.$$

The dividend risk $\partial V/\partial d^1$ and $\partial V/\partial d^2$ could theoretically be hedged by entering into two-year forward contracts on the stocks; as those forwards would have delta and rho themselves, the short stock and long bond quantities have to be adjusted accordingly.

In a market that has reached Phase 2, a hedge for the vegas $\partial V/\partial \sigma^1$ and $\partial V/\partial \sigma^2$ would be obtainable from buying puts on S^1 and S^2, thus leaving correlation ρ^{12} as the only unhedged market parameter in $\mathcal{M}$.

As markets move and since the higher-order derivatives of $V(X, \mathcal{M})$ do not vanish, the local hedge is not the end of the story. In a classical Black–Scholes world, there is no distinction between a local and a full hedge, because Black and Scholes assumed interest rates and volatilities (and dividends, if considered at all) to remain constant. Furthermore, Black and Scholes assumed that it was possible to readjust the delta continuously, so that gamma risk $\partial^2 V/\partial S^2$ is non-existent in this ideal world.

In reality, however, continuous trading is not possible and a stock that closed at 100 might easily open at 90 the next morning. Thus, in addition to hedging locally, a hedge of the most critical higher-order derivatives of V might be considered. In our example, these are the two gammas $\partial^2 V/\partial (S^i)^2$ ($i = 1, 2$) and and the cross-gamma $\partial^2 V/\partial S^1 \partial S^2$. Having sold the American put on the basket of S^1 and S^2, the trader might hedge his gamma by buying plain vanilla puts on stock S^1 and and on stock S^2. If the market is in Phase 2 and there is a sufficiently competitive single stock option market, doing so would be a good idea as it also approximately hedges the vegas $\partial V/\partial \sigma^1$ and $\partial V/\partial \sigma^2$. Unfortunately, the hedge would not be static as there might arise situations where it is optimal to exercise the basket put but not both single-stock puts. Hence the quantity of hedge puts might need to be changed if there is a large market move. Like correlation risk, the cross-gamma cannot usually be hedged in the market except by buying other second-generation equity derivatives.

In practice, most traders would not follow the above mechanistic hedging approach strictly. With all the bid–ask spreads involved in hedging with equity forwards or other plain vanilla equity options, it would be hard to hedge all risks completely and still have a profit left.

If the trader has a large book, he might benefit from risk reduction through diversification. With a large number of disparate equity derivative trades and a flow that is equally balanced between selling and buying, a lot of risks in the trading book will at least partially cancel out. Using this macro approach to risk-managing the whole derivative book, instead of each transaction separately, enables an institution to benefit from economies of scale, by having a large book, and also leads to prices for new trades that are heavily dependent on the overall risk situation of the trading book. If the book is, for example, heavily gamma or cross-gamma short in certain underlyings, the trader will only be willing to increase his short if he can receive very high option premiums in return. On the other hand, the trader might be willing to enter into gamma-hedging trades at mid-market prices, ie without profit, just for the sake of overall risk reduction.

For example, if the book is short a lot of gamma in a certain index due to warrant sales, the book might issue some range note or double digital warrants on this index under favourable terms, just to buy some gamma. Buying this gamma through the warrant market might still be better for the trader than buying it at expensive prices from a competitor, who might be gamma short himself. This does not mean, however, that it is a bad deal for the buyer of the range warrant: if he wants to take an outright gamma short risk, this would be an opportunity for him to buy the position at the mid-market price. Eventually, even with a large book, traders have to make limited and well-analysed bets on the direction of market parameters and leave some risks unhedged.

10.1.2 Some Mathematics of Real Life Hedging

Let us consider a market consisting of n underlyings assets. We assume as usual that the price processes of these assets through $[0, T]$ are given by

$$\frac{d\tilde{S}_t^i}{\tilde{S}_t^i} = (r_t - d_t^i)\,dt + \tilde{\sigma}_t^i\,dW_t^i, \quad n = 1, \ldots, n, \qquad (10.4)$$

where $(W^1, \ldots, W^n)$ is an n-dimensional correlated Brownian motion under $\tilde{\mathbb{P}}$, with

$$d\langle W^i, W^j\rangle_t = \tilde{\rho}_t^{ij}\,dt.$$

We use the tilde to denote true volatility $\tilde{\sigma}$ and true correlation $\tilde{\rho}$, which are future volatility and correlation of the assets through the lifetime of the option. Of course, $\tilde{\sigma}$ and $\tilde{\rho}$ are unknown to the trader today (at $t = 0$) and will in general be different from the volatilities σ and the correlations ρ that the trader will use for his pricing and hedging model. We assume for now that there are no transaction costs. If the trader pursues a dynamic and self-financing strategy of holding $\tilde{\varphi}^i$ units of asset i at time t in his portfolio (which has a total value of V_t), then his profit and loss in the time interval $[t, t + \Delta t]$ will be

$$\text{P\&L}(\tilde{\varphi}) = \int_t^{t+\Delta t} \left(\sum_{i=1}^n (d_u^i - r_u) \tilde{S}_u^i \tilde{\varphi}_u^i + r_u V_u \right) du + \int_t^{t+\Delta t} \sum_{i=1}^n \tilde{\varphi}_u^i d\tilde{S}_u^i. \quad (10.5)$$

This is clear, since:

- $\tilde{S}_u^i \tilde{\varphi}_u^i$ is the total mark-to-market value of the trader's position in asset i at time u, so his income from dividend yield will be $\int_t^{t+\Delta t} d_u^i \tilde{S}_u^i \tilde{\varphi}_u^i du$.

- His cash position at time u is

$$V_u - \sum_{i=1}^n \tilde{S}_u^i \tilde{\varphi}_u^i,$$

on which he will earn interest at a rate of r_u.

- His P&L from movements in the price of asset i is $\tilde{\varphi}_u^i d\tilde{S}_u^i$.

Note that we may even assume that $\tilde{\sigma}$ and $\tilde{\rho}$ are adapted stochastic processes.

Continuous-time hedging with incorrect volatility and correlation Now we assume that the trader wants to use the strategy $\tilde{\varphi}$ to hedge some complicated derivative X in which he is short. We assume that the trader uses the n-dimensional Black–Scholes equity model

$$\frac{dS_t^i}{S_t^i} = (r_t - d_t^i) dt + \sigma_t^i dW_t^i, \quad n = 1, \ldots, n,$$

where W is a correlated Brownian motion with $d\langle W^i, W^j \rangle = \rho^{ij} dt$. On the basis of that model, he devised a pricing engine $V(S_t^1, \ldots, S_t^n, t)$. Note that the one-asset PDE for V used earlier generalises to

$$r_t V(x, t) = \frac{\partial V}{\partial t}(x, t) + \sum_{i=1}^n (r_t - d_t^i) x^i \frac{\partial V}{\partial x^i}(x, t) + \tfrac{1}{2} \sum_{i,j=1}^n \rho_t^{ij} \sigma_t^i \sigma_t^j x^i x^j \frac{\partial^2 V}{\partial x^i \partial x^j}(x, t)$$

(10.6)

for all $t \in (0, T)$ and $x = (x^1, \ldots, x^n) \in (0, \infty)^n$, where the PDE is supplemented by the appropriate boundary conditions corresponding to X. The trader will choose $\tilde{\varphi}$ to be the delta hedging strategy based on his model V. He will therefore hold

$$\tilde{\varphi}_t^i = \frac{\partial V}{\partial x^i}(\tilde{S}_t, t)$$

units of asset i at time t and the total value of his portfolio will be $V(\tilde{S}_t, t)$.

The P&L of his strategy, including the total P&L made on the derivative X, in the time period from t to $t + \Delta t$ is therefore given by

$$\begin{aligned}
\text{P\&L} &= -V(\tilde{S}_{t+\Delta t}, t + \Delta t) + V(\tilde{S}_t, t) + \text{P\&L}(\tilde{\varphi}) \\
&= -V(\tilde{S}_{t+\Delta t}, t + \Delta t) + V(\tilde{S}_t, t) \\
&\quad + \int_t^{t+\Delta t} \left(\frac{\partial V}{\partial t}(\tilde{S}_u, u) + \tfrac{1}{2} \sum_{i,j=1}^n \rho_u^{ij} \sigma_u^i \sigma_u^j \tilde{S}_u^i \tilde{S}_u^j \frac{\partial^2 V}{\partial x^i \partial x^j}(\tilde{S}_u, u) \right) du \\
&\quad + \int_t^{t+\Delta t} \sum_{i=1}^n \frac{\partial V}{\partial x^i}(\tilde{S}_u, u) \, d\tilde{S}_u^i,
\end{aligned}$$
(10.7)

which is seen from inserting equation (10.6) along the path $x = \tilde{S}_t(\omega)$ into (10.5).

On the other hand, being the solution of PDE (10.6), the function V is sufficiently differentiable to allow the computation of $dV(\tilde{S}_t, t)$ for the process $\tilde{S}$. This gives

$$\begin{aligned}
&V(\tilde{S}_{t+\Delta t}, t + \Delta t) - V(\tilde{S}_t, t) \\
&\quad = \int_t^{t+\Delta t} dV(\tilde{S}_u, u) \\
&\quad = \int_t^{t+\Delta t} \left(\frac{\partial V}{\partial t}(\tilde{S}_u, u) + \tfrac{1}{2} \sum_{i,j=1}^n \tilde{\rho}_u^{ij} \tilde{\sigma}_u^i \tilde{\sigma}_u^j \tilde{S}_u^i \tilde{S}_u^j \frac{\partial^2 V}{\partial x^i \partial x^j}(\tilde{S}_u, u) \right) du \\
&\quad\quad + \int_t^{t+\Delta t} \sum_{i=1}^n \frac{\partial V}{\partial x^i}(\tilde{S}_u, u) \, d\tilde{S}_u^i.
\end{aligned}$$

Substituting this in equation (10.7) gives us a P&L of

$$\tfrac{1}{2} \int_t^{t+\Delta t} \left(\sum_{i,j=1}^n (\rho^{ij} \sigma_u^i \sigma_u^j - \tilde{\rho}_u^{ij} \tilde{\sigma}_u^i \tilde{\sigma}_u^j) \tilde{S}_u^i \tilde{S}_u^j \frac{\partial^2 V}{\partial x^i \partial x^j}(\tilde{S}_u, u) \right) du. \qquad (10.8)$$

This last equation gives us several insights.

- If we had $\tilde{\sigma} = \sigma$ and $\tilde{\rho} = \rho$, ie if the trader hedges using the correct future volatilities and correlations, then we would get a perfect hedge, since P&L = 0. So, in the case of continuous-time hedging, gamma only enters the picture when volatility or correlation are misspecified.

- In the one-asset case $n = 1$, the P&L impact from using the wrong model volatility σ instead of the true volatility $\tilde{\sigma}$ of the asset is

$$\text{P\&L} = \tfrac{1}{2} \int_t^{t+\Delta t} (\sigma_u^2 - \tilde{\sigma}_u^2) \tilde{S}_u^2 \frac{\partial^2 V}{\partial x^2}(\tilde{S}_u, u) \, du \qquad (10.9)$$

So, if the trader has, for example, sold a plain vanilla put or call option, then he will be gamma short, ie $\partial^2 V / \partial x^2 (\tilde{S}_u, u) < 0$. If he sold the option for a volatility of σ, ie too cheaply compared with true volatility ($\sigma < \tilde{\sigma}$), and he uses σ to mark his position, then he will lose money over the time of his delta hedge. The exact loss will depend on the stock price path; it will, however, be proportional to the difference of the squared volatilities $(\sigma^2 - \tilde{\sigma}^2)$.

On the other hand, if the option was sold at a very high volatility, ie $\sigma > \tilde{\sigma}$, and the trader chooses to mark his position at σ (not realising an immediate profit), then he will be rewarded over time according to (10.9). Note, however, that in this case the P&L will again depend on the actual stock price path; the higher the option gamma along that path, the higher the profit.

- In the general case, the difference between the covariances

$$(\rho^{ij}\sigma_u^i\sigma_u^j - \tilde{\rho}_u^{ij}\tilde{\sigma}_u^i\tilde{\sigma}_u^j)\tilde{S}_u^i\tilde{S}_u^j,$$

together with the sign of the cross-gamma $\partial^2 V/\partial x^i \partial x^j$, decides whether the error term in equation (10.8) leads to an additional profit or loss in the hedging strategy.

Depending on the structure of the derivative X, the error term can become quite large. One can show that for digital options, for example, it can actually become infinite.

Discrete-time hedging with incorrect volatility and correlation In reality, not only are $\tilde{\sigma}$ and $\tilde{\rho}$ unknown, but continuous adjustment of the hedge portfolio is impossible. We will therefore assume that $\tilde{\varphi}_t^i$ is constant throughout the time interval $[t, t + \Delta t]$ and simply denote it by $\tilde{\varphi}^i$. Thus, (10.5) becomes

$$\text{P\&L}(\tilde{\varphi}) = \sum_{i=1}^n \tilde{\varphi}^i(d_u^i - r_u) \int_t^{t+\Delta t} \tilde{S}_u^i \, du + r_u \int_t^{t+\Delta t} V_u \, du + \sum_{i=1}^n \tilde{\varphi}^i(\tilde{S}_{t+\Delta t}^i - \tilde{S}_t^i)$$

$$\approx \sum_{i=1}^n \tilde{\varphi}^i(\tilde{S}_{t+\Delta t}^i - \tilde{S}_t^i)$$

if Δt is sufficiently small that interest and dividend yields can be neglected.

Using a Taylor expansion, we obtain

$$V(\tilde{S}_{t+\Delta t}, t) = V(\tilde{S}_t, t) + \sum_{i=1}^n \frac{\partial V}{\partial x^i}(\tilde{S}_t, t)(\tilde{S}_{t+\Delta t}^i - \tilde{S}_t^i)$$

$$+ \tfrac{1}{2} \sum_{i,j=1}^n \frac{\partial^2 V}{\partial x^n \partial x^m}(\tilde{S}_t, t)(\tilde{S}_{t+\Delta t}^i - \tilde{S}_t^i)(\tilde{S}_{t+\Delta t}^j - \tilde{S}_t^j).$$

If the trader adjusts his delta at time t, ie $\tilde{\varphi}^i = \partial V/\partial x^i(t, \tilde{S}_t)$, we get for the P&L of the derivative X and the hedge

$$\text{P\&L} = -V(\tilde{S}_{t+\Delta t}, t + \Delta t) + V(\tilde{S}_t, t) + \text{P\&L}(\tilde{\varphi})$$

$$= -\frac{\partial V}{\partial t}(\tilde{S}_{t+\Delta t}, t)\Delta t - V(\tilde{S}_{t+\Delta t}, t + \Delta t) + V(\tilde{S}_t, t) + \text{P\&L}(\tilde{\varphi})$$

$$\approx -\frac{\partial V}{\partial t}(\tilde{S}_{t+\Delta t}, t)\Delta t - \tfrac{1}{2}\sum_{i,j=1}^n \frac{\partial^2 V}{\partial x^i \partial x^j}(\tilde{S}_t, t)(\tilde{S}_{t+\Delta t}^i - \tilde{S}_t^i)(\tilde{S}_{t+\Delta t}^j - \tilde{S}_t^j).$$

From (10.4), we have, for small Δt,

$$\mathbb{E}\big[\,(\tilde{S}^i_{t+\Delta t} - \tilde{S}^i_t)(\tilde{S}^j_{t+\Delta t} - \tilde{S}^j_t) \mid \tilde{S}^i_t, \tilde{S}^j_t\,\big]$$
$$= \tilde{S}^i_t \tilde{S}^j_t \mathbb{E}\big[\,\tilde{\sigma}^i(W^i_{t+\Delta t} - W^i_t)\tilde{\sigma}^j(W^j_{t+\Delta t} - W^j_t)\,\big]$$
$$= \tilde{S}^i_t \tilde{S}^j_t \tilde{\sigma}^i \tilde{\sigma}^j \tilde{\rho}^{ij} \Delta t.$$

Hence

$$\tilde{\mathbb{E}}[\,\text{P\&L} \mid \tilde{S}_t\,] = -\frac{\partial V}{\partial t}(\tilde{S}_{t+\Delta t}, t) - \tfrac{1}{2}\sum_{i,j=1}^{n} \frac{\partial^2 V}{\partial x^i \partial x^j}(\tilde{S}_t, t)\tilde{S}^i_t \tilde{S}^j_t \tilde{\sigma}^i \tilde{\sigma}^j \tilde{\rho}^{ij} \Delta t \quad (10.10)$$

for small Δt. This equation (10.10) can be interpreted as follows.

- If the trader only adjusts his delta hedging strategy at discrete points in time, then the expected P&L of this hedging strategy is not zero but proportional to the length of the time period between two adjustments.

- Usually, eg for a call or put option X, the theta $\partial V/\partial t$ is negative, so that $-\partial V/\partial t$ actually represents income to the trader. The expected P&L in discrete-time delta hedging is thus the sum of the theta earned on the short position in the derivative X minus a "slippage" or "hedging cost" term.

- If the true correlation is zero, $\tilde{\rho}^{ij} = 0$, the cross-gamma $\partial^2 V/\partial x^i \partial x^j$ is irrelevant for the hedging cost term.

- Considering PDE (10.6) for V, where the terms containing r and d^i are negligible for short time periods, we obtain

$$\frac{\partial V}{\partial t}(x, t) + \tfrac{1}{2}\sum_{i,j=1}^{n} \frac{\partial^2 V}{\partial x^i \partial x^j}(x, t)\rho^{ij}\sigma^i \sigma^j x^i x^j = 0.$$

For the trader, this means that the time decay of the option (theta $= \partial V/\partial t$) should exactly offset the hedging cost term, according to theory. Since the equation is linear, it should also be expected that for large portfolios the total gamma exposure would be matched by the theta. However, with term structure in interest rates, volatilities and correlations, this is not necessarily correct.

- Using the PDE (10.7) for V and again neglecting those terms that depend on interest and dividend yield, we can rewrite (10.10) as

$$\mathbb{E}[\,\text{P\&L} \mid \tilde{S}_t\,] = \tfrac{1}{2}\sum_{i,j=1}^{n} \frac{\partial^2 V}{\partial x^i \partial x^j}(t, \tilde{S}_t)\tilde{S}^i_t \tilde{S}^j_t (\sigma^i \sigma^j \rho^{ij} - \tilde{\sigma}^i \tilde{\sigma}^j \tilde{\rho}^{ij})\Delta t,$$

and so if the trader were to use true volatility and true correlation in his model, ie $\sigma^n = \tilde{\sigma}^n$ and $\rho^{nm} = \tilde{\rho}^{nm}$, then he would have

$$\mathbb{E}[\,\text{P\&L} \mid \tilde{S}\,] = 0.$$

Hence, in this situation, the fact that he only hedges at discrete points in time and not continuously would increase his risk, but, apart from this, it would on average not lead to other economic costs, as $\mathbb{E}[\,\text{P\&L} \mid \tilde{S}\,] = 0$.

Hedging

Example Let us again assume that the trader has sold an American put X on a basket of two stocks, as already considered in Section 10.1.1. If the trader uses the discrete-time delta hedging strategy and adjusts his portfolio according to his option deltas at the beginning of the short time interval $[t, t + \Delta t]$, then he can expect the following P&L at $t + \Delta t$:

$$\mathbb{E}[\text{P\&L} \mid \tilde{S}_t] =$$
$$-\theta \Delta t - \tfrac{1}{2}\Gamma_{11}(\tilde{S}_t^1)^2(\tilde{\sigma}^1)^2 \Delta t - \tfrac{1}{2}\Gamma_{22}(\tilde{S}^2)^2(\tilde{\sigma}^2)^2 \Delta t - \Gamma_{12}\tilde{S}_t^1 \tilde{S}_t^2 \tilde{\sigma}^1 \tilde{\sigma}^2 \tilde{\rho}^{12} \Delta t, \quad (10.11)$$

with

$\theta = \dfrac{\partial V}{\partial t}(\tilde{S}_t^1, \tilde{S}_t^2, t)$ as the theta of the American put;

$\Gamma_{ii} = \dfrac{\partial^2 V}{(\partial x_i)^2}(\tilde{S}_t^1, \tilde{S}_t^2, t)$ as the two stock gammas of the American put ($i = 1, 2$);

$\Gamma_{12} = \dfrac{\partial^2 V}{\partial x_1 \partial x_2}(\tilde{S}_t^1, \tilde{S}_t^2, t)$ as the cross-gamma of the American put.

To use some specific numbers, let us assume that both stocks trade at a price of $\tilde{S}_t^1 = \tilde{S}_t^2 = \$100$, and the trader uses a daily readjustment of his delta hedge, so $\Delta t = 1$ day. Suppose that his option theta is $-\theta \Delta t = \$20,000$; so, if the market does not move, the trader earns \$20,000 on his position. With ΔS^i being the movement in asset S^i to the next day, his P&L is given by

$$\text{P\&L} = \$20,000 - \tfrac{1}{2}\Gamma_{11}(\Delta \tilde{S}^1)^2 - \tfrac{1}{2}\Gamma_{22}(\Delta \tilde{S}^2)^2 - \Gamma_{12}\Delta \tilde{S}_t^1 \Delta \tilde{S}_t^2.$$

Note that the movement in the stock enters squared, so, with $\Gamma_{11} = 1000$, $\Delta S^1 = \$20$ and $\Delta S^2 = 0$, the trader will lose

$$-\tfrac{1}{2} \cdot \$1000 \cdot 20^2 = -\$200,000$$

from the \$20 move in stock S^1. With $\Delta S^2 = \$40$, he will lose

$$-\tfrac{1}{2} \cdot \$1000 \cdot 40^2 = -\$800,000$$

from a move twice as large.

Note also that the cross-gamma time Γ_{12} does not carry the factor $\tfrac{1}{2}$; so, when comparing gamma and cross-gamma in absolute terms, one should remember that the latter carries twice the weight. Let us assume that $\Gamma_{12} = 1000$, so that whenever stock S^1 increases by \$1 the option delta in stock S^2 will increase by 1000 shares and vice versa for stock S^2. So, when $\Delta S^1 = \$20$ and $\Delta S^2 = \$20$, the P&L impact of the cross-gamma time will be

$$-\Gamma_{12}\Delta S^1 \Delta S^2 = -\$400,000.$$

To look at the expected P&L, we know from the PDE (10.6) with all interest rate and dividend yield terms neglected

$$\$20,000 = -\theta \Delta t = \tfrac{1}{2}\Gamma_{11}(\tilde{S}^1)^2(\sigma^1)^2 + \tfrac{1}{2}\Gamma_{22}(\tilde{S}^2)^2(\sigma^2)^2 + \Gamma_{12}\tilde{S}_t^1 \tilde{S}_t^2 \sigma^1 \sigma^2 \rho^{12}.$$

So, if the true volatility and correlation of the stock equals the volatility and correlation at which the book is marked (ie $\sigma^1 = \tilde{\sigma}^1$, $\sigma^2 = \tilde{\sigma}^2$, $\rho^{12} = \tilde{\rho}^{12}$), then from equation (10.11) it follows that $\mathbb{E}[\text{P\&L} \mid \tilde{S}_t] = 0$, and that means the trader will on average exactly lose his theta of \$20,000 from market movements. If the trader is lucky and true volatility is below, then he should on average lose less than \$20,000. In the extreme case of $0 = \tilde{\sigma}^1 = \tilde{\sigma}^2 = \tilde{\rho}$, equation (10.11) shows that

$$\mathbb{E}[\text{P\&L} \mid \tilde{S}_t] = \$20{,}000.$$

On the other hand, if true market volatility and correlation are higher than in the trading model (ie $\tilde{\sigma}^1 > \sigma^1$, $\tilde{\sigma}^2 > \sigma^2$, $\tilde{\rho}^{12} > \rho^{12}$), then he will lose an average according to

$$\mathbb{E}[\text{P\&L} \mid \tilde{S}_t] = \tfrac{1}{2}\Gamma_{11}(\tilde{S}_t^1)^2[(\sigma^1)^2 - (\tilde{\sigma}^1)^2] + \tfrac{1}{2}\Gamma_{22}(\tilde{S}_t^2)^2[(\sigma^2)^2 - (\tilde{\sigma}^2)^2]$$

$$+ \Gamma_{12}\tilde{S}_t^1 \tilde{S}_t^2(\sigma^1\sigma^2\rho^{12} - \tilde{\sigma}^1\tilde{\sigma}^2\tilde{\rho}^{12})$$

$$< 0. \qquad (10.12)$$

If the true volatility exceeds the volatility at which the trader marks his position, then he has two alternatives: either he can take the pain now by adjusting σ and ρ and take a corresponding loss on the short vega, or he can take the pain slowly by leaving σ and ρ as they are and expect to realise a daily loss according to (10.12) in his delta hedging activity. The situation cannot be improved by hedging with other options instead of doing a pure delta hedge, because in order to buy other options the trader now has to pay the higher volatility $\tilde{\sigma}$ in the market, so he would realise an immediate loss if he marks those options at σ. Even if the derivative X is long term and the trader hedges by buying short-term options at a volatility of $\tilde{\sigma}$ for a hedge (also marking short-term volatility at $\tilde{\sigma}$), he is not necessarily better off. The trader's book is now gamma flat, but he still has to pay theta every day, because the theta income of \$20,000 from the long-term option will be lower than the theta cost of holding the short-term hedge recorded at $\tilde{\sigma} > \sigma$. If short-term volatility stays at $\tilde{\sigma}$ for the entire lifetime of X and the trader is forced to roll his short-term option hedge again and again, then he will have paid the same money through his negative overall theta than he would have lost otherwise through delta hedging according to (10.12). His only hope is that short-term volatility in the future will be below σ, so that he then can gamma hedge X cheaply through short-term options.

This reasoning shows that the cost of immediately adjusting the trader's volatility σ to the correct level just reflects the average cost of holding the position gamma flat through buying short-term options in the future. As equation (10.12) shows, the expected loss in delta hedging is particularly high if the trader has a large gamma position in those assets for which the vega mismatch $\sigma^2 - \tilde{\sigma}^2$ is large. This shows a relationship between volatility levels, gamma and vega risk.

10.2 Pricing and Hedging European Options Under Transaction Costs

The standard Black–Scholes model assumes that the underlying can be traded continuously without any transaction costs. This assumption allows a perfect delta hedge of an option position by replicating the payoff in the underlying stock market. Thus, Black and Scholes arrived at a unique preference-free price and a riskless hedging strategy.

Real stock markets have transaction costs, meaning bid–ask spreads as well as proportional fees and provisions. Under transaction costs, perfect option replication is impossible. Therefore, we concentrate on computing a hedging strategy that reduces the expected error in hedging an option position by holding a number of stocks that typically differ from the Black–Scholes delta. Technically this means that a utility function measuring the "badness" of the hedging error is introduced and subsequently a minimisation procedure over all possible strategies is performed. The Black–Scholes partial differential equation for option pricing is therefore replaced by a Hamilton–Jacobi–Bellman equation (see Section 10.2.1), which describes this optimal control problem.

The solution of this problem results in a bid and an ask call (put) option price for the same given volatility. The proposed hedging strategy under transaction costs works as follows. For each stock price, two deltas are calculated. The graphs of these two deltas divide the state space into a buy, a sell and a do-nothing region. Only if a stock price movement leads to leaving the do-nothing region should stocks be bought or sold so that the number of stocks in the hedge portfolio equals the respective delta.

The result of the above hedging strategy has been benchmarked in Monte Carlo simulations against the *naive* hedging strategy of using the Black–Scholes delta and buying or selling stocks as soon as this Black–Scholes delta changes by, say, more than 5 or 10%. In all cases considered, the naive strategy led to a hedging error whose distribution had higher variance and mean than the proposed hedging strategy. The higher the transaction costs, the more clearly the naive strategy was beaten. However, the numerical computations required by the proposed strategy are rather involved, so that the proposed strategy seems to be particularly useful if the bid–ask spread is about 1.5% or higher.

10.2.1 *The Model for Transaction Costs*

We now give a brief sketch of a transaction cost model.

Risk aversion of the hedger is measured by a *utility function* $\mathcal{U}: \mathbb{R} \to \mathbb{R}$, which is assumed to be concave and increasing.

We use the following notation. Trading strategies are denoted by π, while the set of all admissible (self-financing) trading strategies is called $\mathcal{T}$. Further, y^π and B^π denote the number of shares and the cash held in the hedge portfolio. Finally, the cash value of the shares is $c(y_t, S_t)$, while Φ is the portfolio value.

One obtains

$$c(y_t, S_t) = \begin{cases} (1+\lambda)y_t S_t & \text{if } y_t < 0, \\ (1-\mu)y_t S_t & \text{if } y_t \geq 0, \end{cases}$$

with λ and μ as the proportion of transaction costs for buying and selling respectively.

The value function is defined as

$$V(s, B, y, S) = \sup_{\pi \in \mathcal{T}} \mathbb{E}\left[\mathcal{U}\left(\Phi(T, B_T^\pi, y_T^\pi, S_T)\right)\right],$$

where T is, as usual, the option maturity date. Option prices and deltas can be computed from the value function.

Equations governing trading The following equations govern trading:

$$dB_t = rB_t\,dt - (1+\lambda)S_t\,dL_t + (1-\mu)S_t\,dM_t,$$

$$dy_t = dL_t - dM_t,$$

$$dS_t = S_t(\alpha\,dt + \sigma\,dW_t),$$

with r, α, σ constant, W_t Brownian motion, and L_t and M_t as the total number of shares bought or sold respectively.

Hamilton–Jacobi–Bellman equation We restrict our attention to the following (smaller) class of trading strategies $\mathcal{T}'$ of the form

$$L_t = \int_s^t l(\xi)\,d\xi \quad \text{and} \quad M_t = \int_s^t m(\xi)\,d\xi,$$

with $l(\xi)$ and $m(\xi)$ positive and bounded by $k < \infty$. The processes $l(\xi)$ and $m(\xi)$ describe buying and selling of shares during the time interval $[s, t]$.

This gives the Hamilton–Jacobi–Bellman equation for V

$$\max_{0 \leq l, m \leq k}\left\{\left(\frac{\partial V}{\partial y} - (1+\lambda)S\frac{\partial V}{\partial B}\right)l - \left(\frac{\partial V}{\partial y} - (1-\mu)S\frac{\partial V}{\partial B}\right)m\right\}$$

$$+ \frac{\partial V}{\partial s} + rB\frac{\partial V}{\partial B} + \alpha S\frac{\partial V}{\partial S} + \tfrac{1}{2}\sigma^2 S^2\frac{\partial^2 V}{\partial S^2} = 0$$

for $(s, B, y, S) \in [0, T] \times \mathbb{R} \times \mathbb{R} \times [0, \infty)$.

The *proposed strategy* is characterised by the three cases:

$$\frac{\partial V}{\partial y} - (1+\lambda)S\frac{\partial V}{\partial B} \geq 0 \quad \text{and} \quad \frac{\partial V}{\partial y} - (1-\mu)S\frac{\partial V}{\partial B} > 0,$$

where the maximum is attained by $m = 0$ and $l = k$, which means that k shares are bought;

$$\frac{\partial V}{\partial y} - (1+\lambda)S\frac{\partial V}{\partial B} < 0 \quad \text{and} \quad \frac{\partial V}{\partial y} - (1-\mu)S\frac{\partial V}{\partial B} \leq 0,$$

where the maximum is attained by $l = 0$ and $m = k$, which means that k shares are sold; and

$$\frac{\partial V}{\partial y} - (1+\lambda)S\frac{\partial V}{\partial B} \leq 0 \quad \text{and} \quad \frac{\partial V}{\partial y} - (1-\mu)S\frac{\partial V}{\partial B} \geq 0,$$

where the maximum is attained by $l = m = 0$, which means nothing is done.

Hedging

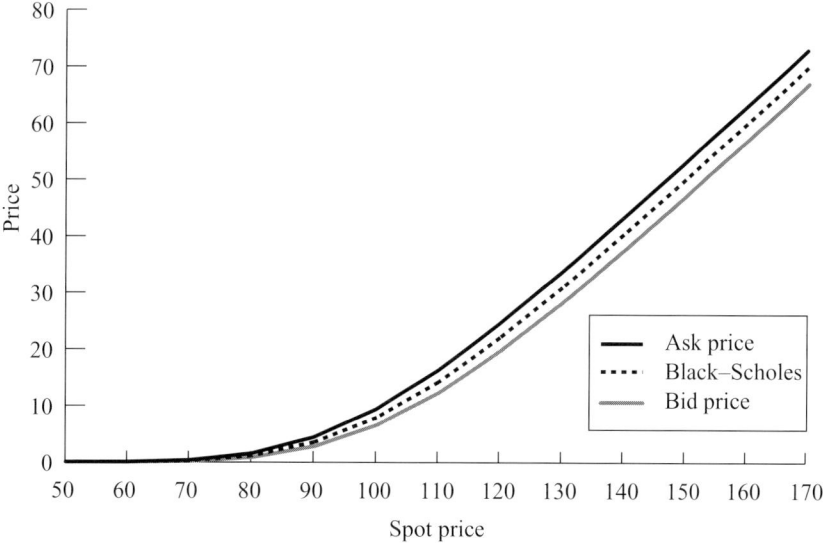

10.1 Call price under transaction costs.

The state space is thus divided into three different regions:

- the *buy* region;
- the *sell* region;
- the *no-transaction* region.

10.2.2 Numerical Results

In the accompanying figures, the case $\mu = \lambda = 1\%$ is considered for a one-year call with a strike of 100. Volatility is taken to be 20% and, for simplicity, interest rates and the dividend yield are set to zero. The utility function is assumed as $\mathcal{U}(x) = -e^{-x}$.

Figure 10.1 shows the resulting bid and ask option prices for a range of spot values. Figures 10.2 and 10.3 show the two deltas for long and short call

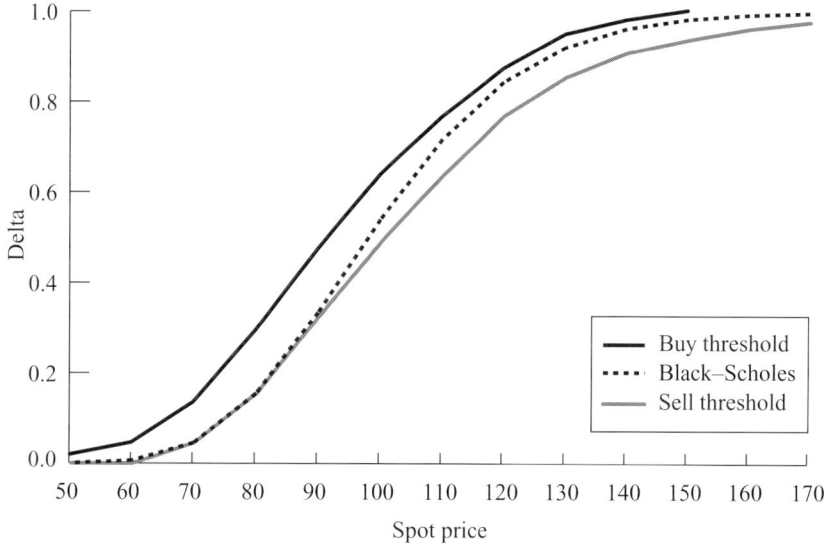

10.2 Hedging a short call position under transaction costs.

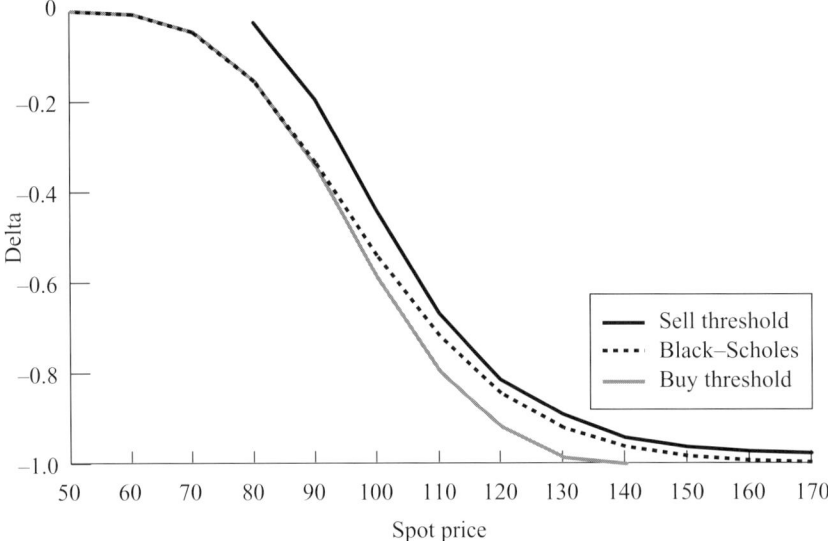

10.3 Hedging a long call position under transaction costs.

positions, respectively. These last two figures show the division of the space into the three regions in which the proposed hedging strategy requires that hedge be bought, sold or no transaction carried out.

10.3 Hedging of Specific Products

We shall now discuss hedging strategies for some of the equity and fixed income derivatives already discussed in earlier chapters. As the general philosophy and mathematics of hedging is discussed in Section 10.1, we shall focus here on the very specific risks involved in some particular products.

10.3.1 Static Hedging

If X is the payoff of some complex path-dependent equity derivative on one underlying that should be hedged, the idea of static hedging would be to find a partition $0 \leqslant t_0 < t_1 < \cdots < t_n \leqslant T$ and real numbers α_i, K_i and β_i such that

$$X = \sum_{i=0}^{n} e^{-R_{t_i}} [\alpha_i (S_{t_i} - K_i)^+ + \beta_i S_{t_i}].$$

Finding such a decomposition would mean that the payoff of X can be expressed as a linear combination of European calls and forward contracts, or alternatively (by put–call parity) of European calls and puts. For a payoff $f(S_T)$ depending on the asset price at maturity T only, this can always be achieved since

$$f(x) = \int_0^\infty f''(u)(u-x)^+ \, du,$$

$$= \int_0^\infty f''(u)(x-u)^+ \, du \qquad (10.13)$$

While finding such a decomposition may not be too difficult for certain payoffs X, it will usually be too costly to trade all the plain vanilla instruments in the market. Nevertheless, finding such a decomposition provides some helpful insights. First, it gives a good feel for the actual risks involved in X; and, second, it shows exactly how the volatility smile present in the plain vanilla options should affect the price of X. If the trading book at a financial institution is organisationally split into an exotics book and a plain vanilla flow book, the static hedge can even be used in practice. Thus, once the initial static hedge is set up, the exotic book requires less attention and the plain vanilla risk is managed out of the flow book.

10.3.2 Digital Options

We assume a notional of 1200 to be paid if the asset exceeds a level of $K = 2500$ at maturity. If the maturity value of the asset is lower, then the contract is worthless. This payoff can be hedged using a "call spread", consisting of a long and a short call with different strikes, the long strike being lower than 2500 and the short strike higher. Assuming that the hedge is executed using listed products, the width of the call spread depends on the liquidity of the derivatives market. If there exist options O_1, O_2 and O_3 with the same maturity and strikes $K_1 = 2460$, $K_2 = 2500$ and $K_3 = 2540$, a portfolio of 30 options O_1 and -30 options O_2 has a cashflow that is dominated by the "over the counter" (OTC) product. The hedge can be seen in Figure 10.4.

Similarly, a position of $30 O_2 - 30 O_3$ dominates our product. In practice, financial institutions often sell digital options to clients and charge a price close to the call spread used to hedge the position.

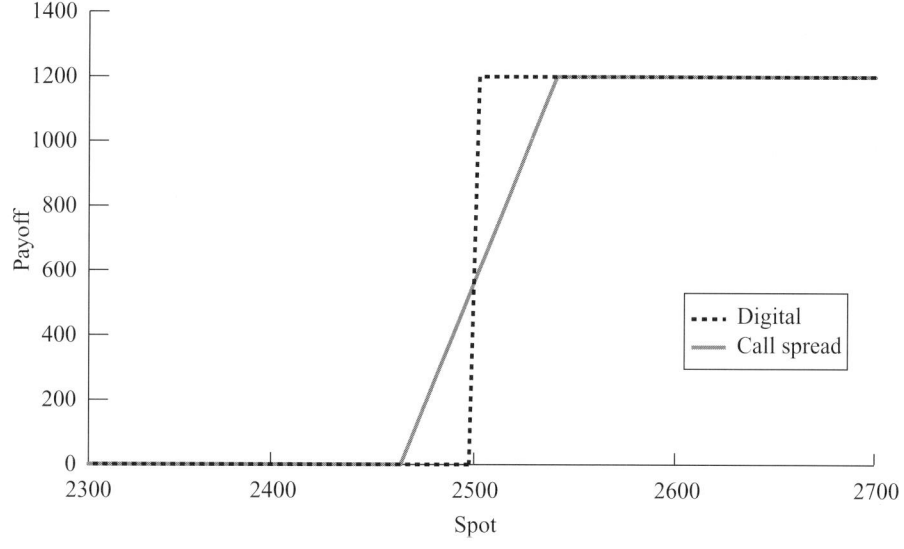

10.4 Digital hedge.

10.3.3 Chooser Options

An early example of building a static hedge is the European chooser option with maturity T, where the buyer can decide at an earlier time t if the option should be a put or a call (with the same maturity T and strike K: see

Section 3.4). The static hedge is as follows. When one chooser option is sold, the trader will buy one plain vanilla call with strike K and maturity T plus $e^{-D_t T}$ plain vanilla puts with strike $Ke^{-N_t T}$ and maturity t. If the buyer of the chooser decides at time t that he wants a call, nothing has to be done. If he decides that he wants a put, the put with maturity t will be exercised and the proceeds are used to trade the call with maturity T and strike K into a put with the same maturity and strike. This rather academic way of setting up a hedge will lead to hedging errors if the implied volatilities for the two options above disagree. In practice, a dynamic hedge of the form

$$w_1(s)C(S_s, s, K, T) + w_2(s)P(S_s, s, K, T)$$

is more common. The two weights are calculated as

$$w_1(s) = \mathbb{P}\big[\,C(S_t, K, t, T) > P(S_t, K, t, T)\,\big] \quad \text{and} \quad w_2(s) = 1 - w_1(s).$$

These probabilities change because of stock movement and changes of other market parameters, making it necessary to readjust the hedge. This hedge can be applied to more general chooser options.

10.3.4 Compound Options

We now want to hedge a compound option that requires the user to pay a compound strike K' at time t if he wants to receive a call with strike K and maturity T. This product is comparable with a European option with strike K' and maturity t. At maturity t, its "payoff" X is based on an option price rather than the asset, ie

$$X = \max\big(C(S_t, t, T, K) - K', 0\big).$$

To hedge this position, we remark that, at every time $s < t$, there is a value S^* such that

$$C(S^*, t, T, K) = K'.$$

The time dependence of S^* is due to changing market conditions. We choose a

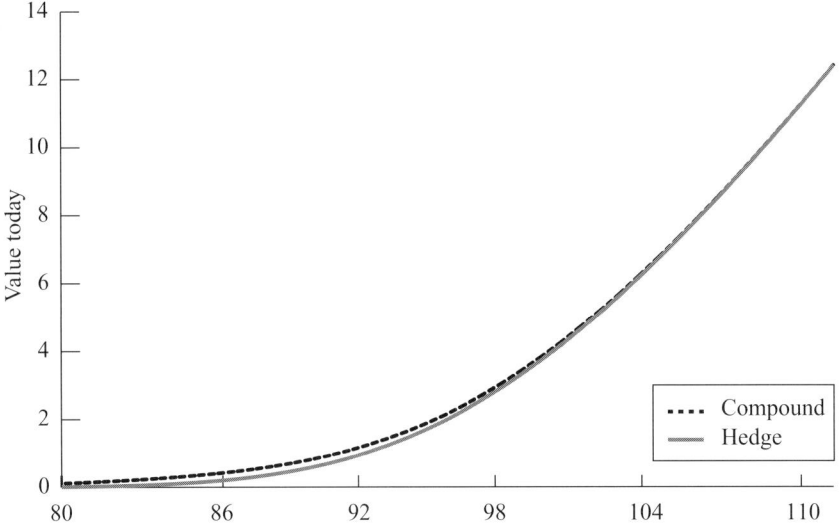

10.5 Compound option hedge value today.

Hedging

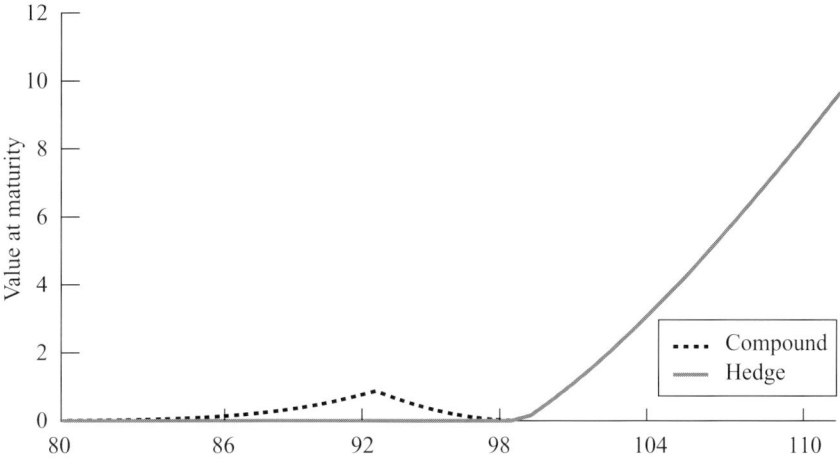

10.6 Compound option hedge value at maturity (compound) date.

portfolio consisting of a long call with strike K and maturity T, long D put spreads with strikes S^* and $S^* - K'/D$ and maturity t, and a short bond paying K' at t. The value D is the delta of $C(S_t, t, T, K)$ at $S_t = S^*$. We remark that this portfolio dominates the compound option. At time t, the two values agree for $S_t \geqslant S^*$. For $S_t < S^*$, the compound option is worthless, while, owing to the concave shape of the call price, the portfolio has a positive value. The factor D ensures that the slopes of the inner values of the portfolio and the compound option agree. See Figures 10.5 and 10.6.

10.3.5 Power and Powered Options

Equation (10.13) shows that a static hedge for a powered option with payoff

$$\left(\max(S_T - K, 0)\right)^\alpha$$

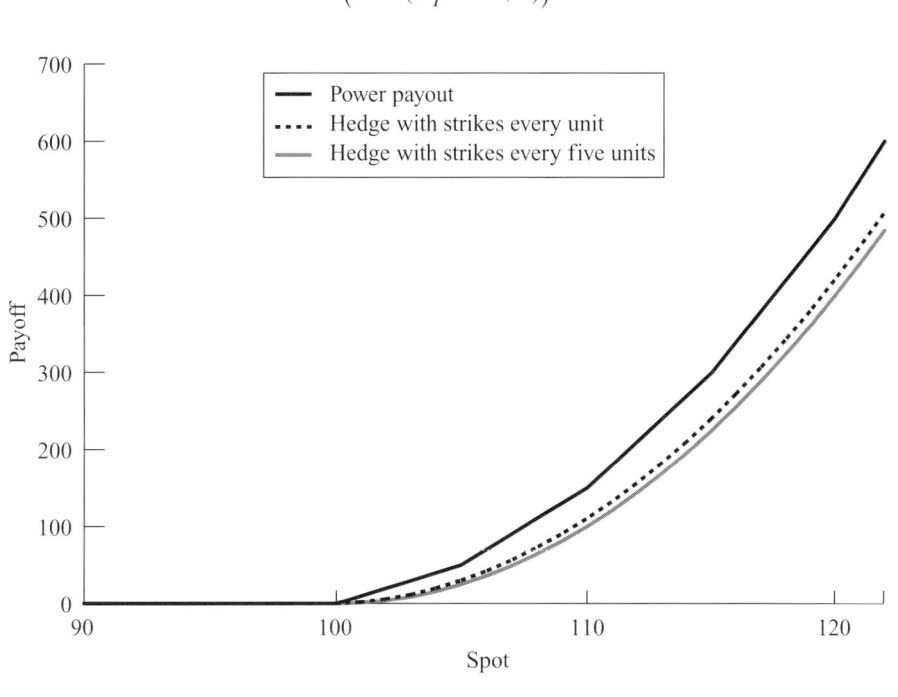

10.7 Power option hedge.

is given by a weighted sum of plain vanilla calls with strikes

$$K_i = K + i\Delta, \quad i = 0, 1, 2, \ldots,$$

for some interval delta. The weight of the strike K_i-option is given by

$$w_i = \alpha(\alpha - 1)(i\Delta)^{\alpha-2}\Delta.$$

In the special case of $\alpha = 2$, we have $w_i = 2\Delta$. Figure 10.7 shows the hedge with a Δ of 1 and 5.

Since the sequence of weights does not decrease for $\alpha \geqslant 2$, the effect of a volatility smile will be high for large exponents α.

10.3.6 Equity Swaps

We now show how to hedge equity swaps with exposure to different yield curves as well as exchange rate movements. For the definition of equity swaps, we refer the reader to Section 2.1.3. Quanto and composite products are described in Section 1.3.4.

Hedging a quantoed equity swap Executing the delta hedge for the quantoed equity swap requires holding positions in both the underlying equity and the foreign currency. In particular, it is necessary to ensure that interest is paid on the cash positions held in domestic/foreign currency according to the domestic/foreign interest rate curve. Furthermore, there will be a substantial risk resulting from possible shifts in the difference between the domestic and foreign interest rate curve. This risk can be reduced by taking positions in forward rate agreements (FRAs), interest rate futures or interest rate swaps in both currencies.

Hedging a composite equity swap When a composite equity swap is entered at t_0 (with the financial institution as payer of equity performance), it sets up a hedge in the following way.

- Borrow the amount $N_0 e$ in the underlying currency for a fixed rate until t_N, where e is the exchange rate connecting notional and underlying currencies.

- The desk uses the amount raised to buy $(N_0 e/S_0)\exp(-D_N)$ contracts of underlying equity at t_0, which are sold again at t_N.

- It enters as floating rate payer at Libor minus the stipulated spread into an interest rate swap on the fixed notional amount N_0 in foreign currency. The desk does not receive coupon payments, but only one bullet payment B in foreign currency at the end of the swap at t_N.

- We sell forward at t_N the foreign currency amount of $N_0 + B$.

For all payments that have to be made to the client at the dates $t_1, \ldots, t_N$, the desk borrows the necessary foreign currency amounts until t_N. All interest payments that are paid by the client at these dates are used for the floating rate payments under the interest rate swap (which, of course, is not exact). The desk does not change any of the transactions above.

Hedging

For $n = 1$, this hedge is exact. For multi-period swaps, it offers a static way to reduce the risk considerably.

10.3.7 Swaptions

When pricing a swaption using Black's model (see Section 5.1), we assume that the underlying swap rate is lognormally distributed. The swaption is then essentially an option on the swap rate, and therefore one should delta hedge the position with forward swap rates. This is done using forward starting swaps. For example, if you are long a five-year into five-year payer swaption, then the obvious hedging vehicle would be five-year forward five-year receiver swaps.

However, this is not always practical, and in reality we do not price using the Black model. Table 10.2 gives a delta hedge portfolio that mainly consists of selling five-year and buying 10-year treasury rates. This hedge is achieved by finding the risk-equivalent position in the market instruments (sometimes called yield curve hedging) and is based on pricing the swaption taking full account of convexity correction effects.

Table 10.2 Delta hedge for a 5 into 5 payer swaption at 6%.

	Yield curve		Risk-equivalent position	
	Rates	DV01	Swaption	Hedge
3 day	4.8%	1	0	0.08
1 month	5.6%	2	0	0.01
Dec 98	9477	25	1	0.03
Mar 99	9521	25	1	0.03
June 99	9529	25	0	0.02
Sept 99	9531	25	0	0.02
2 year	5.0%	188	4	0.02
3 year	5.0%	275	6	0.02
4 year	5.1%	358	8	0.02
5 year	5.1%	437	26510	60.66
6 year	5.2%	512	291	0.57
7 year	5.2%	582	346	0.59
8 year	5.3%	649	406	0.63
9 year	5.4%	711	470	0.66
10 year	5.4%	770	−46231	−60.02
12 year	5.5%	878	0	0.00
15 year	5.6%	1017	0	0.00
20 year	5.6%	1198	0	0.00
25 year	5.7%	1334	0	0.00
30 year	5.7%	1436	0	0.00
Total			−18187	
Parallel shift up (0.5 bp)			−18173	
Parallel shift down (−0.5 bp)			−18154	
Average of parallel shifts			−18163	
Average − Sum			23	

10.3.8 Barrier Options

In this section we present some ideas on hedging barrier options. If a pure delta hedge is applied to a barrier option, one encounters an exploding delta effect close to the barrier. This can make the option difficult and expensive to hedge. Figure 10.8 shows an example of the dependence on spot of the gamma of a barrier option.

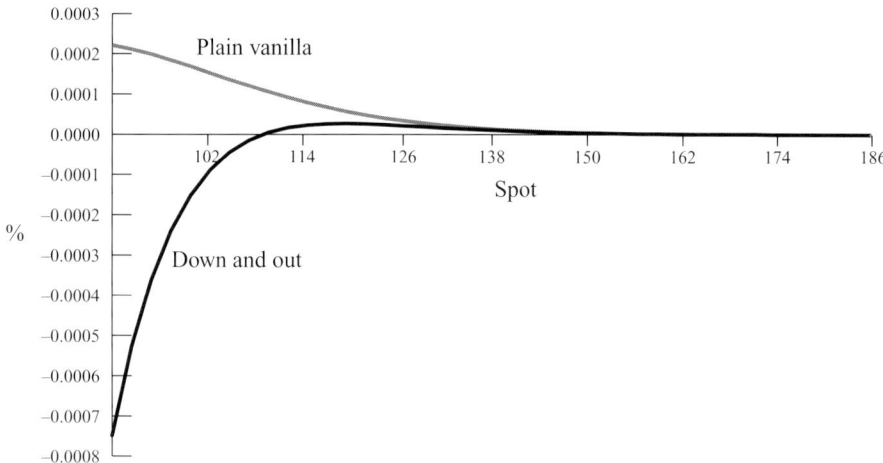

10.8 Gamma of a barrier option and plain vanilla.

Fortunately, one can also employ static hedges, ie hedges that do not need continuous readjustment, by using plain vanilla options. Note that we will use the notion of *static* hedge in a slightly extended way, since we will allow for adjustments in the hedging portfolio. However, these adjustments do not have to be done in a dynamic fashion.

We will throughout this section assume that the drift of the asset price process is zero, ie that the risk-free interest equals the dividend yield. Even though this is a severe restriction, it allows very instructive *exact* static hedges. These ideas can still be used when the drift is non-zero to construct approximate hedges.

All options considered in this section will have the same maturity T. We write $C(K)$ and $P(K)$ for the values of a plain vanilla call and put with strike K.

A simple situation For illustration, let us first present the idea in the situation where the strike of the barrier option equals the barrier level. Say we want to hedge a short position in a down-and-in call with strike K and barrier level H.

In this situation we can find a simple static hedge, which is perfect in the Black–Scholes framework. We will hedge by buying a put of strike K and we will retain this put as long as we do not hit the barrier. If the barrier is never hit during the life of the option, both the down-and-in call as well as the plain vanilla put expire worthlessly. On the other hand, if the barrier is hit at any time, we will immediately sell the put and buy a call with strike K from the proceeds. This call will now hedge the knocked-in call perfectly. The crucial point is that the put we sell and the call we buy are of exactly the same value under our assumptions. This follows simply from call–put parity.

Indeed, since we have perfectly replicated the down-and-in call with the plain vanilla put, their prices must match. So, our argument gives us not only a replication strategy but also the fair value.

Call–put symmetry The classical call–put parity implies that a European call and put have the same value when their strike is at the forward, and they have the same maturity T. Note that we assume for simplicity that there are no dividends. We write this equality as

$$C(F_T) = P(F_T).$$

This equality actually holds under a much generalised assumption on the asset price model used, since it follows from simple arbitrage considerations.

If one remains within the Black–Scholes framework, this call–put equality can be generalised to a situation where the two strikes differ. We then have *call–put symmetry*

$$\frac{C(K_C)}{\sqrt{K_C}} = \frac{P(K_P)}{\sqrt{K_P}},$$

where $K_C K_P = F_T^2$, ie the geometric mean of the two strikes K_C and K_P is equal to the forward F_T. This symmetry can be rephrased as

$$P(K) = \frac{K}{F} C\left(\frac{F^2}{K}\right).$$

The general situation Let us consider a down-and-out call, now with strike K and potentially different barrier level H, and with maturity T. We assume $H \leqslant K$. We start setting up a replication portfolio consisting of a long plain vanilla call with strike K. If at any time the barrier H is reached, the down-and-out call becomes worthless. Since we assume no drift, the forward is then equal to H. So call–put symmetry assures us that the value of our call with strike K is equal to K/H puts with strike H^2/K, ie

$$C(K) = \frac{K}{H} P\left(\frac{H^2}{K}\right).$$

Therefore, we initially go short K/H puts with strike H^2/K for our initial hedge portfolio. So, whenever the barrier is hit, our hedge portfolio has value zero, as has the down-and-out call. If the barrier is never hit, the put expires worthlessly, since $H \leqslant K$ implies $H^2/K \leqslant K$, and therefore the plain vanilla call covers exactly the payoff of the down-and-out call.

Thus, in any event, we have a perfect replicating portfolio. This shows that

$$C_{\text{out}}(K) = C(K) - \frac{K}{H} P\left(\frac{H^2}{K}\right),$$

and as a side-effect we have a quite instructive proof of a special case of our general barrier pricing formula.

10.3.9 Lookback Options

As discussed in the previous sections of this chapter, a mechanical delta hedge is often not the most feasible method of hedging an exotic option position. Other strategies have to be employed. Indeed, the presence of a liquid market for plain vanilla options gives us a richer set of instruments for our hedge.

In this section we want to consider the case of a floating strike lookback option, and we want to present one simple possible approach to an approximate hedge. We will consider the case of a put. Recall that the payoff is given by

$$\left(\max_{0 \leqslant u \leqslant T} S_u - S_T\right)^+.$$

We also assume that we sold this put, so we have a short position.

Our hedging strategy is based on buying and selling straddles. A long *straddle* position is a combination of two long plain vanilla options, a call and a put, both with the same strike and the same maturity. The common strike and maturity of these two options are called the strike and maturity of the straddle.

For illustration, assume that the underlying stock trades at a price of 100. We fix a rehedge threshold, eg at a level of four, implying increasing rehedge levels of 104, 108, 112, ... Initially, we buy an at-the-money straddle with the same maturity date as the lookback option. As soon as the first rehedge level 104 is reached, we sell the at-the-money straddle and buy a new straddle with a strike of 104 instead. We hold on to this position until we reach another rehedge level. See Figure 10.9 for an illustration.

Suppose that the maximal asset price during the life of the option is 138. Then we will at maturity hold a straddle with strike 136. In this case, no matter what the final asset price is, the payoff of the straddle will not differ by more than two. In general, the payoff of the straddle and the payoff from the lookback will never differ by more than the rehedge threshold. This means that there is an upper bound on the discrepancy between the amount we have to pay out to the lookback option holder and the amount we get from the hedging straddle. However, this hedge is not self-financing. The amount we receive when selling a straddle does not agree with the amount we have to pay to purchase a new straddle with a higher strike. This difference is not too big under reasonable circumstances.

In practice, one wants to put a time restriction on rehedging, eg by having a maximum dealing frequency of one transaction per day. In addition, one does not choose the straddle symmetrically, ie each straddle consists of one call but a fewer number of puts. This is done in such a way that the delta of the portfolio is zero.

It is important to choose the rehedge threshold carefully. If it is too small, transaction costs become a burden. On the other hand, if the threshold is too large, the option replication loses in accuracy. A reasonable value depends very much on the circumstances and can be found with the help of a Monte Carlo hedging simulation.

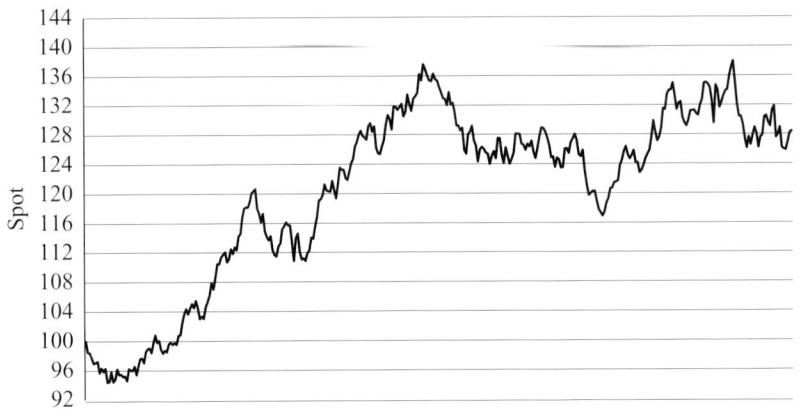

10.9 Hedging lookbacks with straddles.

10.3.10 Convertible Bonds

A convertible bond is a true hybrid security, as it combines a fixed income security with an equity option. Unlike a corporate bond that is issued with a covered warrant attached to it, with the convertible bond the equity optionality cannot be separated from the pure bond.

From the standpoint of hedging, the first decision to make concerns the choice of model from Section 6.5. There is a trade-off to be made between using the single-factor tree, which models only the equity component stochastically, and the two-factor tree, which models random equity and bond price movements. While the two-factor model is certainly a superior model of the market, it incurs overheads in respect of speed of computation, software development and parameter estimation. Thus it can be a good idea to use the more sophisticated two-factor model only for pricing, and, when it comes to hedging, traders are frequently satisfied with a single-factor tree as a fast way to compute position value and greeks. Exceptions to this rule are convertibles with long maturities and call or put rights where the embedded optionality is sensitive to random fluctuations of the yield curve. Another area for the use of two-factor trees are markets with high yield curve volatilities, eg emerging markets. In those markets, however, the volatility of the underlying stock is usually even greater and credit risk is an issue. In general, one can therefore state that a single-factor tree is often sufficient and that a three-factor tree, ie a tree where a two-factor model is used to describe the random movement of the yield curve, will usually only be of academic interest. Of practical importance, however, is the question of whether credit risk and the volatility skew of the underlying, especially for convertible bonds with embedded call or put rights, are modelled properly.

Having chosen a model, the trader should make sure not only that the convertible bond but also all the hedge products are valued using the same model approach, particularly as far as the treatment of volatility skew and stochastic interest rates is concerned. Using a consistent model for the entire position avoids unnecessary mismatches.

As convertible bonds are an area where each security has it own special features, a close look at the conditions of issue is essential. We shall therefore discuss hedging problems only in a general fashion with respect to the different features of convertible bonds described in Section 6.5.1.

Let us start with a convertible bond that carries only a coupon c and a conversion feature with a constant conversion ratio. Just as an American call option is not exercised prematurely unless the stock is about to pay a significant dividend, such a convertible security is likely only to be converted (if ever) at the latest possible time. Premature exercise would typically only occur before the last dividend, as the bond holder would give up all future coupon payments plus the time value of the conversion option, ie the downside protection offered by the convertible bond as opposed to pure equity. The static hedge would consist in buying calls with maturity T and a strike of R/α_T against a short position in the convertible bond. The interest rate exposure of the total position can then be hedged by entering into swap contracts with maturity T.

If the convertible bond is endowed with an issuer call feature, things are more complicated. Let us assume that the issuer can call the bond at any time t during the time interval $[s, T]$ for an amount C_t, unconditional on the stock price. If the stock price at time s is high, the issuer will announce his call, thus forcing bond holders to convert immediately. In practice, the issuer will place an announcement in the relevant financial press to the effect that he will redeem all unconverted bonds in, for example, 30 days at the value of C_t. If the stock trades sufficiently high, ie if $S_t \alpha_t > C_t$, bond holders would rather convert before the end of the 30-day period than accept being paid the lower redemption value. Thus, the issuer call usually serves as a means to force early conversion if the stock price is high. However, one should bear in mind that even with a low stock price the issuer might still exercise the call if interest rates at time t are significantly lower than at the time of issue. This is because the issuer will also call the bond if C_t is below the present value of the pure bond, irrespective of the stock price level at time t. As a result, the issuer call feature may shorten the lifetime of the convertible to s if S_s is high; otherwise it effectively acts as a cap on the total value of the convertible during the time interval $[s, T]$. To hedge a short position in such a convertible instrument, one has to buy as many calls as one might have to deliver stocks under the convertible bond. The maturity of those calls should be chosen according to the expected lifetime of the convertible bond. The expected lifetime is the risk-neutral expectation of the time until the convertible will be called, converted or redeemed; it can be calculated in the same tree that is used to evaluate the convertible bond (this would be a "thick tree": see Section 6.5.2). As a general rule, the expected lifetime will fall when the stock price goes up or when interest rates go down. The option hedge should correspond to the expected lifetime of the convertible bond. Thus, an initial hedge for the short convertible could be to buy calls with maturity T and to roll these calls to the shorter maturity s when the underlying stock price rises significantly. As soon as the issuer announces to call the bond at time t, the convertible is just equivalent to the sum of a fixed cashflow of C_t plus a plain vanilla call with maturity t and strike price C_t/α_t. A feature that is usually difficult to hedge with issuer calls are trigger conditions that have to be met before the issuer may announce his call. As an example, such a trigger condition might allow "the issuer to call the bond if the stock trades above 130% of the initial conversion price on 20 out of 30 trading day". This feature introduces a path dependency and is therefore difficult to model exactly in a tree. Usually a Brownian bridge is applied or the condition is, for example, rephrased as $S_t > 135\%$ of initial price on call date t. The trader therefore has to make a manual adjustment to the model if the stock was, for example, trading above 130% of initial conversion price for the last 15 days.

In addition, the trader might also consider the use of swaptions to hedge the interest rate optionality that comes with the issuer call. This is even more interesting if convertible bond holders also have a put right. Bond holders will exercise this put right if P_t is higher than the value of holding or converting the bond, so that P_t provides an absolute floor for the value of the convertible. If the stock price is low, the convertible bond will be barely more than just a pure bond, so the put right provides a protection against rising interest rates for the bond holder. The trader who is short the convertible bond should therefore consider the purchase of a put option on a bond with coupon c and maturity T if the underlying stock price has fallen significantly.

Convertible bonds where the currency of the notional differs from the currency in which the underlying stock is traded can usually be treated as composite options. The decision to convert the bond would depend on the value of the stock seen as an American Depository Receipt (ADR) in notional currency. The hedge therefore consists of options on the ADR or options on the stock in its native currency plus an accompanying FX hedge. Problems with this approach may arise if, for example, trigger conditions are still formulated in the native currency of the underlying stock. This is the case if we consider, for example, a US$ convertible that can be exchanged into a British stock, where the issuer call is linked to the trigger condition that the stock must trade above £100. In this case, treatment as a composite option is not possible and a tree that models sterling stock price and sterling/US$ exchange rate separately is required.

Finally, credit risk can be a problem. The problem will be especially pronounced if the convertible has the stock of the issuer itself as underlying. In this case, the protection offered by the bond floor of the convertible could turn out to be shallow when it is needed the most, ie when the company is in financial distress and the stock price is low. The correlation between the stock price and the default risk of the company can in principle be used to hedge credit exposure by shortening the stock, but such a hedge is very delicate. As an alternative, the trader might consider buying credit protection for the convertible through a default swap if the stock trades far-out-of-the-money.

Chapter 11
Implementation Issues

11.1 The Context of a Model Library

The mathematical treatment of equity derivatives presented in previous chapters leads naturally to software implementations of the resulting algorithms. Applications whose functionality includes a component that performs pricing or risk management of derivatives will require results calculated by these implementations. The large number of models, and their widely differing requirements for input parameters, indicates that some framework is needed in which to embed the core algorithm implementations, and with which to pass data to them and return results to the calling environment.

In this chapter we describe such a framework intended to satisfy a number of requirements, namely that it:

1. support an extensive range of pricing models, including those for multi-currency, multi-underlying structures;

2. support integration of further models, yet to be specified;

3. be reusable in several applications;

4. insulate calling applications from the volatility of pricing software;

5. enable general features, applicable to many models, including quanto, composite, forward start and "meta-underlyings";

6. perform common processing in common code, easing the burden on the core model implementations.

The first two of the above requirements, amounting to a decision to create a general-purpose rather than a special-purpose library, are perhaps the most severe and have the most far-reaching influence on the solution. Much of the complexity of a framework for model integration arises from this requirement; much simpler solutions suffice if it is known in advance that only support for simple derivatives, in particular single-underlying, single-currency structures, will ever be required.[1]

The library discussed here does not attempt to model the entire process of

---

[1] Here, and elsewhere in this chapter, the simplicity or complexity of a security does not refer to the algorithm used to price it. Complexity rather refers to the amount of data required first to define the security and only then to price it. Commonly, however, securities which are complex in this sense are also algorithmically complex.

trading and booking derivative securities: it is envisaged as a provider of calculated prices and hedge parameters to calling applications, which might include systems for trading and risk management, not as a complete application in its own right. In particular, the aim of writing a library capable of serving a number of applications – the third requirement above – rules out any explicit dependence on data sources such as a database or real-time data feeds. The calling application is therefore taken as the only source of data available to the library: all data used by the library are passed to it through its public interface.

Worthwhile though the aim of reuse is, there can be little doubt that the implementation of a general pricing library suitable for use in a number of applications rules out tailoring it for any particular one, with the optimisations in performance or functionality that would result to the benefit of that one application. Although this is a perfectly real disadvantage in using a common library in several applications, the advantage of implementing the library only once, with the cost saving and consistency of results between applications that result from this, generally outweighs this in an environment where more than one application exists that needs these calculations.

In an application concerned with many types of derivative security, such as a derivatives trading and risk-management system, it will often be the case that new option types are routinely introduced in new software releases or that enhanced models are introduced for pricing already-supported securities. If this process each time incurs extensive re-engineering of the system, costs are high and the situation is clearly untenable. It is this that leads to requirement 4 above, that the existence of an integrated framework or subsystem dealing specifically with pricing serves also to insulate calling applications from excessive change resulting from changes to existing models or the introduction of new ones. If developers of pricing algorithms produce individual packages for each model, each with a different interface, then application developers will in any event be forced to construct an insulating layer between the pricing software and the rest of the application.

Many functions of such a pricing library are common to many or all models, notably the support of general features including forward start, quanto and composite. These are shown in Sections 1.3.3 and 1.3.4 to be achievable by reusing the model implementations while providing them with suitably modified parameters. This functionality is naturally supported in common code and not reimplemented in each model. Other functions common to all models are the purely data-handling functions, including accepting requests from client applications for particular calculations, and presenting the results or reporting errors. Also in this category are functions such as managing the set of data available to the library and presenting the required subset to the models.

Finally, software problems routinely admit of a large number of solutions, and it will often be a matter of judgement which is superior, particularly if we accept that requirements will evolve over time in ways that cannot easily be predicted. This chapter therefore presents one particular solution and attempts to separate design principles from implementation details. The choice of programming language is not central to the issues discussed here, but where references to one must be made, or where the terminology used suggests a particular language, this will be C++.

11.2 Library Interface Design

Identification of the major classes in the problem domain is one of the initial tasks in object-oriented analysis. In the application domain of equity derivative pricing, the principal candidate classes are those representing market data and those representing options and their underlyings. More generally, we note that application domain objects divide naturally into two distinct categories. The first category captures information needing to be observed regularly in the market, which might be modelled with stochastic processes, such as interest rates, spot prices and implied volatilities. The second category captures information that, once defined, changes rarely if ever and can be taken to be "true by definition", such as definitions of particular options, lists of holidays observed by a particular market, and so on.

The first category, referred to here as *market data* classes, includes yield curves, implied volatility surfaces, market prices, dividends, correlation matrices, exchange rates, credit rating information, and much more. The second category, *static data*, consists mainly of definitions of options and equity underlyings.

The public interface discussed here consists of a number of classes that capture the data in the problem and a single function at global scope, the latter called by applications to request calculations.

As will be seen in the following section, the application programming interface (API) of the library must allow an indefinite number of objects of each class to be passed to the library. It follows that each of the classes defined will have to incorporate an identifier enabling objects of that class to be retrieved from container classes as necessary. It is found that most objects are identified either by the name of a currency or the name of an equity. There are exceptions to this: exchange rates are identified by pairs of currency names; and correlation matrix elements are identified by pairs of identifiers that may themselves be names of equities, currencies or exchange rates.

Design principle: Market data and static data are not mixed in the same object. It is a data modelling error for a security object to contain members that capture market data needed to price it. A security object contains the parameters defining that security. Observations of market data exist independently of the particular option being priced.

11.2.1 *Capturing Market Data*

This section provides an outline of the principal classes needed to capture market-observable data generally needed for pricing derivatives. Each description gives:

- the type of market information captured by the class;
- the principal methods that the class provides to enter this information;
- the principal methods that the class provides to later access or process it.

In the following descriptions, it is assumed that each class provides a method for naming the object and retrieving that name. As indicated below, some classes perform some processing on the entered data in response to a request, whereas others simply store and retrieve.

Dividends This class captures the information concerning the dividends paid by a share: their amounts and dates.

The market observables for a Dividends object are the announced or estimated dividends for the share. A method is therefore required for populating the object with as much dividend data as is available. In addition, it is common to make assumptions about dividend growth and a method is provided to specify this.

The commonest requests made of a dividend object by client code (ie mainly model implementations) are to enumerate the dividends in date order and to calculate the present value of dividends falling between two dates. If dividends are assumed to be replicated with some growth rate, then arbitrarily many dividend dates and amounts can be supplied in response to the former request; otherwise the sequence will terminate when the data are exhausted and the request for a further dividend will fail. This behaviour is naturally handled by an Iterator object, a pattern for which is given by Gamma *et al* [24].

Yield curve This class captures information pertaining to a deterministic model of interest rates.

The market observables are quoted rates for instruments including deposits, futures and swaps. The object provides a method for registering these instruments into the yield curve: auxiliary classes may be defined that capture the required data for each instrument type and provide methods for calculating the dates and amounts of the corresponding cashflows, for use in the yield curve stripping procedure.

The principal access methods return a discount factor or equivalent zero-coupon rate between two given dates. An internal calculation is required to extract the discount curve from market rates; this can often profitably be delayed until the discount factor request is made ("lazy evaluation": see [53]).

The procedure for calculating the discount curve from market rates is not given here.

Repo curve This class captures the rates available to borrow or lend a stock. Although their meaning is quite different and they have a mapping to an equity rather than a currency, the functionality required of the class is similar to that provided by the Yield Curve class.

Market price This is a particularly simple market data class that stores the market price of a named instrument. The instrument may be an equity, option, bond, or anything else. Market prices are, of course, required for underlyings as the spot parameter S of the model implementations, but they are also required as the target price for implied volatility calculations.

Correlation This class captures correlation data between pairs of processes, more precisely, the correlation between the logarithmic returns of these processes. For our present purposes, these processes include equities, exchange rates and short rates in the Vasicek model of interest rates.

The class provides a method for entering correlation between named processes and methods for extracting single matrix elements or submatrices. Internally, the class can exploit the symmetry of correlation matrices to provide matrix elements not explicitly entered. In addition, the correlation between an exchange rate and some other process implies the correlation involving the inverse exchange rate: the class can also provide this. Sparse population of the matrix might be required. An alternative to having a single class storing correlation data between all types of process would be to have a derived class for each combination, which might provide more type safety. It is natural to associate with this class the checks necessary to ensure that the matrix is valid.

The optimum storage structure for the data depends on the size and sparseness of the matrix. The obvious choices are, first, a square matrix and a list of names of the processes involved and, second, a list of small elements, each of which names the two correlated processes and holds the value of the correlation. Symmetry can, of course, be used to reduce storage requirements, at the cost of a little more complexity in the retrieval method.

Exchange rates This class captures spot exchange rates. It is structurally similar to the correlation class in that the individual data items are real numbers associated with two identifiers – in this case currencies.

There are obvious code reuse opportunities, even though the symmetry properties differ from the case of correlation.

Exchange rate volatility Quanto and composite option pricing requires exchange rate volatility, modelled for simplicity as a single number. With this simplification, the data structure is identical to that for exchange rates, although with differing symmetry properties. Again, code reuse is clearly possible.

Implied volatility surface This class captures all implied volatility data observable for a given underlying. Although traded option prices might be regarded more strictly as market observables, it proves convenient to regard the corresponding implied volatilities as primary data, since they are more stable than raw prices.

The class provides a method for entering the implied volatility for a given strike and maturity. Client code can request volatility at some date and strike not present in the entered data, so that some interpolation algorithm must be implemented, such as bilinear interpolation in the variance $\sigma^2 t$. This particular algorithm implies that instantaneous forward volatilities are constant between maturities in the matrix.

Each item of data is a number labelled by a date and a strike, and so, as with correlation matrices, it can be stored in a rectangular array associated with linear arrays of labels or as a list of data elements, each containing the volatility, date and strike. The first option is simpler to implement but may incur a large space overhead if support for sparse data is required.

Vasicek interest rates This class captures data appropriate to the Vasicek model for stochastic interest rates.

The market observable quantities are implied volatilities (or prices) for instruments such as caps and swaptions. A method is provided for entering the definitions and volatilities of the instruments into the Vasicek yield curve object: auxiliary classes may be defined to capture the data for each instrument type used. The principal service provided by the class is to return calibrated time-dependent mean reversion $\lambda(t)$ and volatility $\sigma(t)$ parameters. The calibration process can be time-consuming and, as for yield curve stripping above, may be deferred until required.

This functionality may be implemented either by subclassing from the Yield Curve or by introducing a separate class, which would then need to reference the corresponding (deterministic) yield curve object to supply discount factors. In either case, a change to the deterministic interest rate information should in principle invalidate the Vasicek $\lambda(t)$ and $\sigma(t)$ parameters, although for performance reasons this might be neglected.

11.2.2 Capturing Security Data

The number and variety of option types is very great, so that we may identify only a very few attributes and methods common to all. We shall require that all option types:

- have an identifier;
- have a currency in which they are denominated;
- can return the identifiers of their underlyings, if any;
- can return their start date, if forward start;
- can indicate whether they are quanto, composite or neither;

In the class hierarchy representing security types, only the above attributes and methods are properties of the (abstract) base class. The placing of attributes for forward start and for (quanto and composite) payoff style at the base class level corresponds to the support for these features provided for most option types (see Sections 1.3.3 and 1.3.4). Beyond that, derivatives are so variable that no attempt is made here to describe a complex class hierarchy: classes representing all concrete security types are assumed to be derived directly from a single base class. The derived classes, of course, have attributes appropriate to their type, in addition to the inherited attributes. In the discussion below of the internal processing carried out by the library, it will be seen that its dependence on the form of the security class hierarchy is very weak. Consequently, a more structured hierarchy could be used with little change to the library's internal design.

Individual securities combine to form structures that we may represent diagrammatically using boxes to represent the securities and lines indicating the relationship "is an underlying of", as in Figure 11.1. The diagram shows a selection of structures, each built up from more elementary securities, each of which is represented by an object of the appropriate class.

No hard distinction is made between options and underlyings: both are considered as particular instances of the general concept of a security. This is necessary since we wish to be able to model, for example, options on baskets and compound options. Since vanilla options and baskets are both structures that we might wish to represent in their own right (and further examples could

Implementation Issues

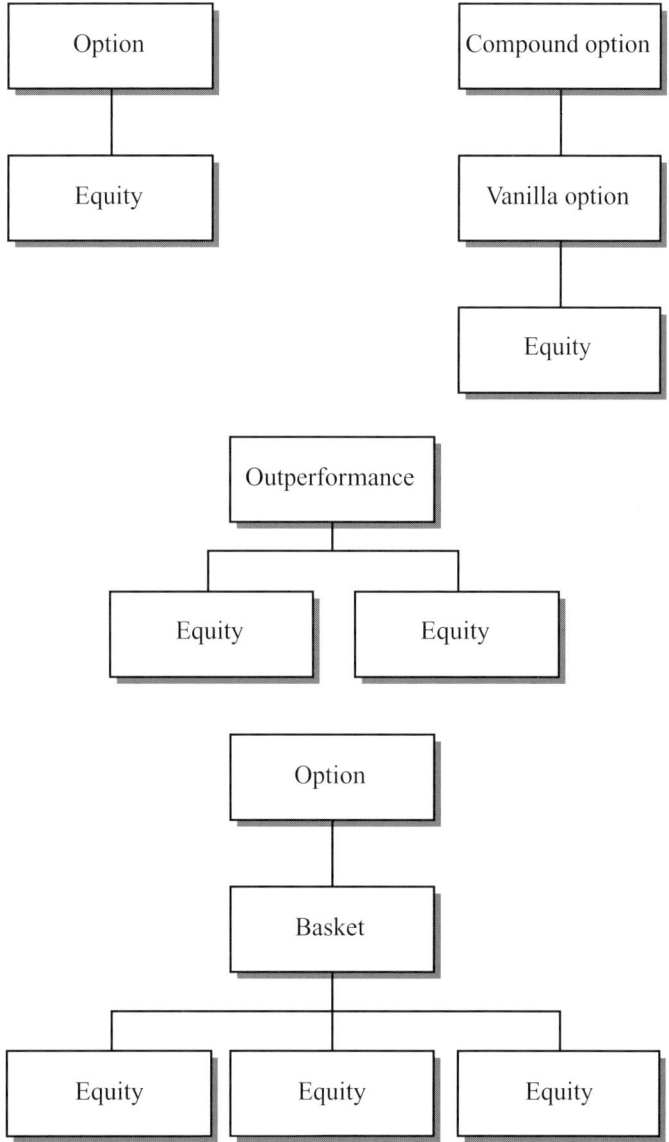

11.1 Structurally different products.

be cited), independent of any higher-level option, coding a split between underlyings and non-underlyings into our class hierarchy is not useful.

Design principle: The definitions of options and their underlyings are represented by different objects. The *identity* of the underlying is part of the option definition, but it is a data modelling error to represent the underlying as *part of* the option. They are separate entities in their own right.

In accordance with this, the option classes do not contain underlying objects: this would give rise to many copies of an underlying definition in memory at one time. The implementation may populate a security object either with identifiers of underlyings, or with references to the corresponding objects. However, the latter solution is only tenable if the lifetimes of the objects are long, and this application-level design decision cannot be anticipated by a library class. The more conservative solution, making no assumptions about how client applications cache data, is the former, which in the interests of brevity will be assumed adopted. Relaxing this assumption simplifies some aspects of the processing in the library, in accordance with the idea that it is the more conservative solution.

11.2.3 *Top-Level Functional Interface*

In addition to the classes required to capture market and static data, the public interface to the library must include a mechanism for making pricing requests and receiving the results. This might be a class method or a function call: we here assume a function call. In either case, the request will consist of:

1. the identity of the option or structure to be priced (the root node in the diagrams of Figure 11.1);
2. a set of market and static data objects;
3. a Control object;

and a Results object will be returned by the library to the application.

Design principle: The pricing request interface shall impose no unnecessary restriction on the amount or type of data that may be passed to the models, as this would prejudge their data requirements. Furthermore, the interface should change only very rarely, as this might impose considerable overheads on client applications upgrading to a higher library version.

These principles follow from requirements 2 and 4 of Section 11.1. To support these, the top-level API of the library must be very general. Since we accept that we cannot foresee the data requirements of models yet to be developed, the interface must admit of change in such a way as not to impact applications that are already successfully using the library.

Container objects such as lists or sets are clearly appropriate for allowing a varying *amount* of information across an interface: a single object is passed across an interface, but that object may itself have been previously populated with an arbitrary number of contained objects.

The same approach may be repeated at a higher level of abstraction to allow us to extend the range of *types* of data crossing the interface. A workable approach is to define a class whose data members are containers for the individual classes of interest (yield curves, volatility surfaces, etc). An object of this "meta-container", or repository, class is then passed across the functional interface. (A design in which the pricing request itself is not a global-scope function but a method of this class is perfectly possible.) Furthermore, to mirror the logical distinction between market and static data, two such classes may be defined, one acting as the market data object repository and the other as the repository for static data. The functional API to the library then accepts an object of each type a parameter.[2]

Although using containers allows the library to accept an indefinite number of objects of types that it already supports, extending this range of types clearly

---

[2] This separation of market data and static data repositories has the useful side-effect of facilitating pricing the same security or portfolio under different market scenarios, each captured in a different market data repository object. This is a common task in client applications, eg in value-at-risk analysis.

requires a new class to capture the corresponding data, plus the enhancement of the repository class to support the new type. None the less, applications already using the library are not affected: only applications needing the new datatype, and the model that it was presumably introduced to support, need change. An example might be the extension of a library that previously supported only deterministic interest rate models to support models requiring Vasicek interest rate information: the repository class would be extended to incorporate Vasicek objects, but applications not needing the new models need not be aware of the change.

Market data and static data repository objects can be compared to databases, in that they contain a large amount of information, only a subset of which is extracted by the library as relevant to the problem in hand. Some applications will indeed populate them from databases, whereas others might use them as a caching mechanism. Clearly, it is possible for an application to establish a data set in these objects and, having done so, to make many calls to the library to price many options – perhaps constituting a portfolio – under the same market conditions.

Finally in this section, the remaining classes in the functional interface are those capturing requests from the client application and presenting to the application the results of those requests: the Control class and Results class respectively. The former is simply a mechanism by which the client application passes to the library a set of requests for calculations, typically of the price and various greeks for a particular structure. The latter is used as the channel through which the library communicates results to the client application. In a distributed application with client and calculation server on different machines, Control and Results objects might be sent across the network.

The use of a Callback class It is worth stating an alternative paradigm for provision of required data by the calling application to the library, which is suitable for some applications but not others. The discussion is applicable both to market and static data objects.

In its simplest implementation, the library raises an error if a required object, such as the yield curve of a currency that appears in the structure, is not present in the repository. It is the task of the implementation classes described below to determine whether such a condition is present, the required data of course depending strongly on the complexity of the model called. This approach requires that the calling application has sufficient "intelligence" to know which data objects are required by each pricing model it supports. While this is not too severe a requirement – the rules are easy to formulate for the vast majority of models – it may none the less be argued that the correct place for these rules is in the pricing library, and that in any event it will certainly contain these rules as it must verify the availability of required data on a per-model basis. Furthermore, in the case of the more complex models, which are not the majority of models but are important nevertheless, it may not be obvious which items of data are mandatory (Vasicek objects, credit objects, etc).

The alternative that answers this objection is that the pricing library allow the calling application, optionally, to supply an object that supports requests by the library to create named objects not available from the repository. Thus, if the library detects that an object required for the current pricing request cannot be retrieved from the repository, instead of reporting an error and returning control to the application, it requests the object – a yield curve in our example – from the Callback object. In effect, it allows the library to call upon services provided by the application, the reverse of the usual relationship between a library and an application.

The implementation of this approach would require the library to define a base class that specifies the range of requests that the library might make to the application: the default implementation would be that these requests for additional objects always fail. The requests include methods for the retrieval of the complete range of objects that the supported models require.

While the library software defines the base Callback class, the application needing to use the facility would implement a derived class supporting only the range of requests to which the application can respond, which is not necessarily the complete set appearing in the Callback class interface; thus an application that did not use, for example, hybrid models (Section 1.5), would not implement the request to supply a Vasicek object, while another application served by the same library would do so. In a typical example, the implementation would retrieve the named object from a database and since the library provides only the base class, without reference to any database or other data source, the independence of the library from the calling application is maintained.

The communications between application and library are shown in Figure 11.2 for the case where no callback is implemented and in Figure 11.3 for the case of an application that uses a callback mechanism. In these diagrams, using a common convention, time is down the page, lifetimes of objects are represented by thick lines, and messages between objects are represented by horizontal arrows.

There is no doubt that this simplifies the introduction of models that require new data items not previously required, in that it does not require the calling

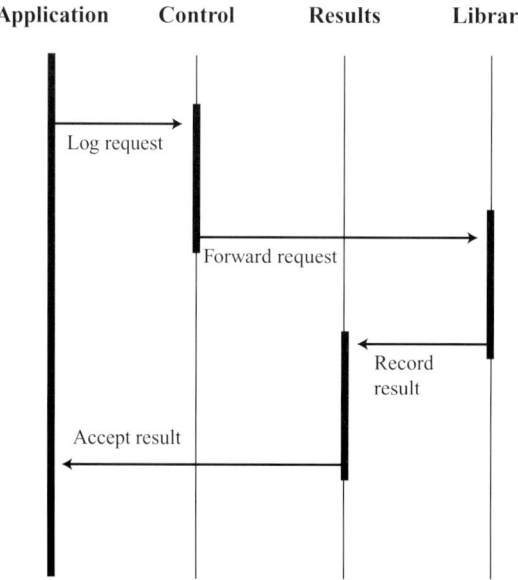

11.2 Simple passing of request and result between application and library.

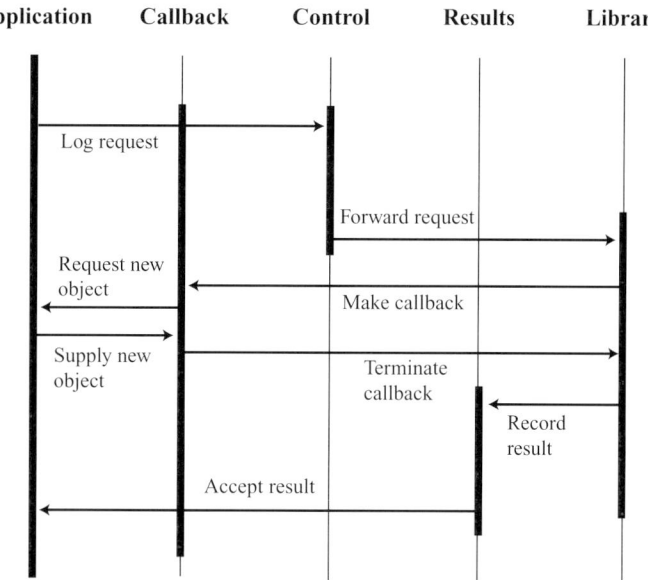

11.3 Callback protocol between application and library.

application to implement any new data retrieval rules. Clearly the benefit is most strongly felt in applications that are designed or required to support a wide and increasing range of derivatives, particularly if required to support a more complex model for an existing option type as the model becomes available in the pricing library. The approach allows the application to implement only a minimal set of rules for retrieving objects prior to pricing, or no rules at all if it is optimal to rely entirely on the callback process for all types of object. Since a callback protocol can exist in parallel with the simpler repository approach, considerable flexibility is available to calling applications at the cost of only a little extra complexity in the library.

Applications that cannot support callback would not supply a callback object and thus the traditional approach of report-error-and-stop would apply: no additional burden is imposed on the entire range of calling applications, only on those willing to undertake it.

In the rest of this section, it is assumed that the calling application provides the required data, whether by a callback method or not. For brevity, few explicit references to callbacks will be made. Where omitted, if it is stated that an error condition is raised if an object is not present, this can be taken to mean, in an implementation that supports callbacks, that the callback has been attempted and has failed.

11.3 Internal Design

This section introduces the classes used internally by the library. Two principal classes are introduced: the first is concerned with capturing the relationships between objects of static and market data; the second, with encapsulating the algorithms that act on this data, ie the models described above.

Design principle: The classes used internally by the library do not represent objects in the external world: they are artefacts of the implementation and, as such, do not have the same status as API classes. They are not part of the library's public interface. The API classes have no dependence on the implementation classes.

The internal processing required of the library can be thought of as layers of software responsible for successive tasks, as follows.

- Build layer: creates a data structure that captures the mappings between static and market data objects.

- Request-Manager layer: makes requests to the pricing layer, adjusting market data objects if necessary.

- Pricing layer: prepares parameters for the individual models.

- Model layer: consists of the implementations of the core analytics.

In object terms, the responsibilities are divided between the library's two main implementation classes. The upper two layers are the responsibility of the first of these, which manages structure and data-handling and has no algorithmic content. The lower two layers are implemented by the Model class, which encapsulates the algorithm. The former class is therefore model-independent and changes less frequently than the latter.

11.3.1 *Data Representation*

The data passed across the interface described above is quite unstructured: it is a pool of data of which some are relevant to a particular pricing request and some not. The market data and static data containers may be regarded as repositories of data from which the library extracts such data as it needs. In particular, the calling application is expected to provide all the objects required by a particular model, while knowing nothing of the processing that model carries out. Applications may adopt conservative or aggressive rules when populating the repository objects, depending on the extent of their reliance on the callback mechanism of Section 11.2.3.

Internally, the library requires a data structure that is sufficiently flexible to capture all the static and market data appropriate to a derivative without imposing, through an inflexible data structure, an arbitrary restriction on the objects that will be available to provide services to the core model routine. (These services might include calculating a required parameter such as a discount factor, or actually passing across the model interface, presumably to be used repeatedly within the model code.)

A suitable choice, prompted by Figure 11.1, is a tree data structure (see eg Knuth's classic text [45]), which allows us to represent structures of arbitrary depth and width. It is clearly related to the Composite pattern described by Gamma *et al* [24], the defining property of which is that it enables clients to treat a component without being aware of whether it is an atomic element or a composite. This property is used in Section 11.3.5 to enable model objects to request information from underlying nodes in the tree (ie underlyings) without

Implementation Issues

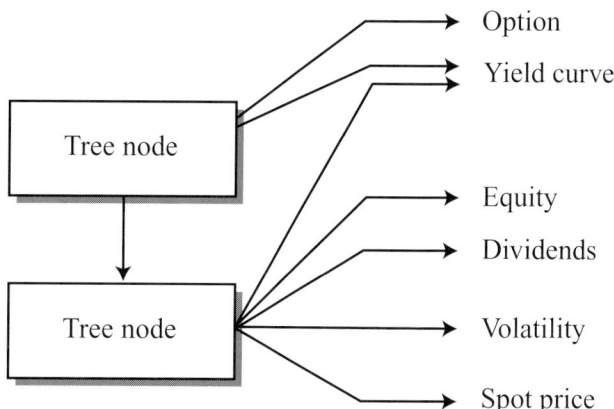

11.4 Tree data structure populated with pointers to security and market objects.

being aware of whether they are primary underlyings (equities) or meta-underlyings (baskets or averages).

The Tree class acts as a placeholder for pointers to security objects and related market data objects. It also contains a list of pointers to underlying tree nodes: see Figure 11.4. It is the association of a node with a Security object that gives the node its identity and governs the remainder of its associations. The relationships thus established between objects mirror real-world mappings. For example, a node associated with an equity object will also be associated with the dividend and volatility objects that are properties of that stock. Conversely, a node associated with an option will be associated with no dividend object since dividends are not a property of an option but of its underlying. Note that most objects are referred to only once in the tree but that yield curves are an important exception: the same yield curve object may be referred to by many tree nodes.

The tree acts to impose a logical structure on otherwise unstructured data. Although this is less flexible than allowing free and unfettered access by the core model routine to the entire market and static data repositories, it has advantages. First, it minimises the risk of accidental access of incorrect objects and renders irrelevant objects completely inaccessible. Second, it records the result of searches for particular objects, which would otherwise have to be repeated each time some service was required of them, such as an interest rate from a Yield Curve object. Third, it simplifies the perturbative greek calculator (see Section 11.3.4), by documenting the complete structure and its market data dependencies.

11.3.2 Building the Tree

The process of building the tree structure of Figure 11.4 may be done in two recursive stages; first the structure is built and populated with pointers to Security objects, and then it is traversed and populated with corresponding market data objects.

Starting with the name of the security to be priced, each node is populated with the corresponding Security object pointer. For each underlying named in that object, a new node is created and this procedure called recursively. The recursion terminates with lowest-level underlyings, normally equities or bonds.

Second, the complete tree is traversed in order to populate it with pointers to market data objects in the market data repository. At each node, the name of the security object is used as a key to search for corresponding dividend, volatility, repo and market price objects. These classes have a single-valued relationship with security objects: for example, a given node will be associated either with one volatility object, if the node is an equity, or none if the node is an option. The currency of the security object establishes a relationship with a Yield Curve object: here the relationship is one-to-many inasmuch as one yield curve may be referred to by many tree nodes (indeed, by all nodes in a single-currency structure).

Multi-underlying options generally require a correlation matrix as a model input. Furthermore, quanto and composite variants on the basic option types require correlations (between underlyings and exchange rates), exchange rate volatilities and spot exchange rates. Correlation and exchange rate data are associated with pairs of identifiers and are therefore treated differently to other market objects. At each tree node a submatrix of the supplied correlation matrix will be created, as will vectors of exchange rates and exchange rate volatilities between the security currency and each underlying currency. These will be degenerate, single-element matrices and vectors for single-underlying, single-currency options.

Validation stages follow both the static data population and market data population traversals of the tree. The static data validation is governed by Security objects, which report on their own internal state and on conditions they expect to be satisfied by their underlyings. Market data validation is governed by model objects as described in the next section.

The tree is populated with pointers to objects rather than copies of objects in the interests of efficiency. This approach also guarantees that a perturbation on a market data object (see Section 11.3.4) acts wherever in the structure that object is referenced.

The timing of the creation of the tree interacts with the lifetimes of the market and static data objects in the calling application. If these are transient, ie they may be destroyed and regenerated between calls to the library, then the tree must be built anew at each call to the library. If they are persistent, ie they remain in memory between calls to the library, even though they might change their internal state, then complete trees or partial trees could also persist, which the calling application would then have to maintain. The library API could be chosen to take account of the efficiency savings available in this case, or else these might have to be sacrificed in the interests of loose coupling between the library and client applications.

11.3.3 *Pricing and Model Layers*

The split between classes representing external-world concepts, such as options and yield curves, and classes that implement the solution is repeated within the library as a split between the Tree class that exists to impose structure on the data but knows nothing about pricing algorithms and the class that encapsulates the pricing algorithm but has no responsibility for assembling the data.

Implementation Issues

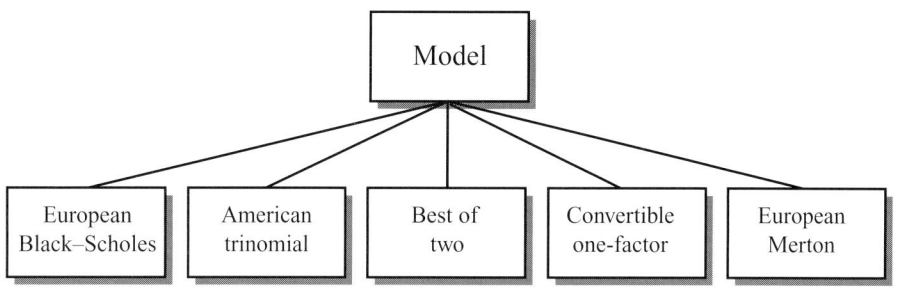

11.5 The Model class hierarchy.

We define a class hierarchy in which the (abstract) base Model class has derived classes implementing each algorithm, as in Figure 11.5. Each tree node will be assigned a derived model object that prices the security object at that node. The creation of this object occurs at the time the tree is built, as soon as the security object is available and before references to market data objects are added to the tree. Since the class of the model object is specific to the algorithm used to price the security at the particular tree node, it may be used by the Tree object to govern which market objects are retrieved and which regarded as mandatory.

If the security classes are to remain strictly independent of the implementation classes, then the creation of the model object must be achieved by some method other than a simple virtual method of the Security class. Fortunately, this may be achieved quite simply by using a model registry similar to the Singleton registry described by Gamma *et al* [24]. The registry is populated with prototypes of all model classes: they are then accessed through a key provided by the security object, returning a clone[3] of the retrieved object. The key, which may be regarded as identifying the option type not the model choice, may be overridden by the Control object to allow the calling application to specify alternative models for pricing the same derivative security. In effect, therefore, the model registry provides the *default* mapping between the type of an option and its pricing algorithm: the calling application can override this default, which will often be the simplest model available for that option type but which is, in any event, a library-level, not an application-level, decision.

When the tree data structure is built, and a model object is available at each node, the mechanism for triggering a calculation of the core algorithm is then a call to a `Price` method of the Tree class. This is implemented by delegation to the model object to which the Tree object refers, passing itself as a parameter. A simplified C++ implementation might be as in Figure 11.6. This is shown as a sequence diagram in Figure 11.7.

Since the `Price` method of the model object accepts a tree node as its argument, the amount of data available in principle to the algorithm is the full set of market and static data objects at its tree node and all underlying nodes. The result is that no artificial restriction is placed on the data available to models which may be developed during the lifetime of the library.

The derived Model class must provide the functionality specific to the model it supports, but there is also generic functionality that must be provided if the

---

[3] As discussed by Stroustrup [69], cloning is an idiom that allows a client of a base class to obtain a copy of a derived class object.

```
class Tree ;

class Model
{
public:
   void RunAlgorithm( const Tree* const ) = 0 ;
   // etc.
} ;

class Tree
{
public:
   void Price() ;
private:
   Model* _model ;
   // etc.
} ;

void Tree::Price()
{
_model-> RunAlgorithm( this ) ;

}
```

11.6 Delegation of a pricing request to a Model object: indicative C++ implementation.

library is to support features such as quanto, composite and forward start across a wide range of option types. This is naturally provided by the base Model class. Gamma *et al* [24] show how to ensure that the base class functionality is always called in addition to the functionality provided by the derived class, naming the approach the Template Method.

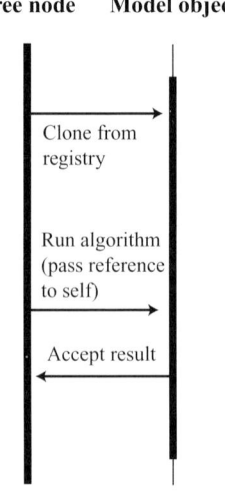

11.7 Delegation of a pricing request to a Model object: sequence diagram.

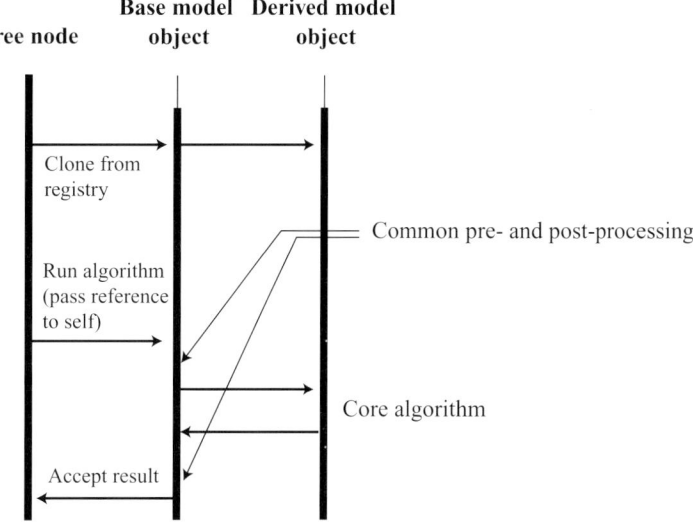

11.8 Common functionality in the base Model class.

The relationship between the base and derived model classes is illustrated in Figure 11.8, in which, for illustration, the model object is split into base and derived parts, even though only one object is present.

Thus the base class provides processing that will support features provided by the library for all or most models: these are the secondary features described below of quanto, composite, forward start and meta-underlyings. The derived classes provide processing specific to the particular model. However, it is found that much of this is identical in many model implementations, leading to code duplication. For example, many closed-form models for European-style options require the discount factor at the maturity date, which is provided by the yield curve object at the tree node of the model object; likewise many of the same models require a volatility σ interpolated at strike and maturity. Code duplication is reduced if the base class provides to the derived classes a set of commonly used functions.

The call to the core analytical routine is made by the derived model object, whose responsibility it is to provide the required parameters. These parameters are given in the formulas stated in the above sections for the individual models. It will be seen that in many cases the parameters are very much simpler than in others: in the particular case of volatility, some models require the single number σ, more complex models require a volatility term structure $\sigma(t)$, and the PDE models require a local volatility surface $\sigma(S, t)$. Interfaces to core model routines are therefore chosen to pass exactly the data required for the calculation. This documents for client code the information that must be provided, encourages reuse and minimises dependencies between code units, all of which is elementary good practice.

11.3.4 *Calculation of Greeks*

Client applications requiring that an analytics library calculate prices for derivative products generally also require greeks. We regard a greek as well defined if it represents the effects of changes to items of market-observable data. The common definitions of some greeks do not satisfy this criterion in a market data model of reasonable complexity, in particular one in which volatility is modelled as more than a single number σ and interest rates as more than just constant time-independent r.

Delta and gamma: These are well defined as they reflect the effect of changes to observable asset prices. In the case of multi-underlying options, however,

$$\delta = \frac{\partial P}{\partial S} \text{ becomes } \delta_i = \frac{\partial P}{\partial S_i}$$

$$\gamma = \frac{\partial^2 P}{\partial S^2} \text{ becomes } \gamma_i = \frac{\partial^2 P}{\partial S_i^2} \text{ and cross-gammas } \gamma_{i,j} = \frac{\partial^2 P}{\partial S_i \partial S_j}.$$

Vega: Defined as $\partial P/\partial \sigma$, it does not map directly onto market observables, since the market observable is an entire set of implied volatilities. We may define vega to be the effect of a parallel perturbation to this entire surface, or an increment applied to the result of interpolation on the original surface. Other variants are possible, in which a perturbation is applied to all surface elements having a particular strike or a particular maturity, or to a single surface element. If this complexity is added, classes to manage the vega requests and the results from those requests will be needed.

Rho: As with vega, the definition $\partial P/\partial r$ does not map onto any market observable, given that the observables form a set of rates R_i for money-market instruments. In view of this, a set of $\rho_i = \partial P/\partial R_i$ may be defined, leading to a definition for a single number representing interest rate risk as the effect of a simultaneous incremental change in all the R_i in a particular yield curve, which probably most closely captures the spirit of the simple definition while remaining well defined. We can also define a ρ that is the effect of a parallel perturbation to the zero curve produced by the stripping process applied to market rates. This definition reduces to $\partial P/\partial r$ in simple cases, but fails the test for being a well-defined greek in that no directly observable item of data is involved.

Although the foregoing are the main greeks, others may be defined and many are important for models of more than minimal complexity. For multi-underlying and multi-currency options, these include sensitivities to correlation matrix elements and exchange rates.

Requests for greek calculations are passed to the library via the control object; indeed, most of the functionality of that class will be concerned with managing these requests. It proves natural to group together requests (in objects used by the control object) that pertain to individual underlyings (deltas and vegas) and to individual currencies (rho and its variants). Other greeks are labelled by pairs of identifiers, such as the gammas and exchange rate and correlation greeks, and may be grouped similarly. While it is not a greek in any sense, implied volatility is a commonly requested calculation. Labelled by the identifier of an individual underlying, it can be treated in a similar way to greek requests in the control object and the request handled by similar code within the library.

In a library supporting many pricing models, many of which are not closed-form solutions, it is completely intractable to have each model implementation provide software for returning the complete range of greeks that may be supported by the library: it will be clear from the above that the number of greeks that may be required by client applications is substantial. The tractable approach requires that approximations to the derivatives be calculated by perturbation of the market data objects in a layer of code that is blind to the nature of the option being priced, and only concerned with identifying which objects are accessible from the tree – this being the complete set that can influence the calculated price. This layer of code will respond to requests in the Control object, perturb corresponding objects and reprice the option.

The algorithm used for estimating greeks from a set of prices evaluated at particular points in parameter space may be chosen independently of the data structures used to manage requests and to record the results. The library may be configured with an object encapsulating the algorithm chosen, or a range of solutions from which client applications or model objects may choose, perhaps using context information such as (in the case of barrier options) the current spot price. This approach is labelled Strategy by Gamma *et al* [24]. Perhaps surprisingly, a simple one- or two-sided perturbation in many cases suffices for practical purposes. In any event, perturbations to the market data objects will be accomplished through calls to methods of the market data classes, so that the set of greek calculations supported by the library will be constrained by the perturbation methods provided by those classes.

Furthermore, it is often the case that clients are less interested in the question "What is the instantaneous rate of change of the price of my position with respect to (say) interest rates?" than in the question "What is my exposure to some modest, plausible (and finite) change in interest rates?", in which case it can be true that for the same effort expended in calculating an approximation to a derivative, or less, the library could return just the finite difference that the client application requests.

It is this emphasis on the necessity of perturbative greek calculations in a layer of software blind to the details of the models that motivates the above definition of the well-defined greek, as one that can be expressed in terms of perturbations to market-observable data. Only such definitions can be uniformly applied across all securities, or applied at all in generic code not specific to a derived model class.

The class by which this functionality can be managed is the Tree class itself, which is already responsible for those aspects of the problem related to the structure of the data, and not at all with those relating to individual pricing algorithms. The complete processing carried out by this class can now be outlined as follows.

1. Establish links between objects in the problem, the links documenting the logical associations among objects, such as the association between an option and its underlyings and that between a stock and its dividends.

2. Call the Model object to calculate the base (unperturbed) price if it is explicitly requested or if any greek is requested.

3. Iterate over the remaining requests in the control object; perturbing,

repricing and un-perturbing market data objects as necessary. (Some care should be taken here to ensure that market data objects are indeed unperturbed in the event of an error occurring, perhaps by cloning the market object rather than perturbing the original, or else (C++) by catching the exception, un-perturbing and rethrowing).[4]

Evaluation of greeks by perturbation is not optimal if the model can do better, either for reasons of efficiency or of stability. Efficiency considerations indicate that tree models should return delta, gamma and possibly theta directly rather than relying on perturbation, since the marginal cost of doing the extra processing is low. Another example is provided by delta and gamma for barrier options (and ladder options, range options, etc), which are undefined at the barrier with the result that a naive perturber gives poor results over a finite region around it. Should a model object calculate a greek, it is the responsibility of that object to match the general definitions, as it is clear that the definitions cannot be allowed to vary between option types.

In general, therefore, it is essential that there be a protocol in place between the Tree object and its associated Model object whereby the latter can indicate to the former which quantities it has calculated. This could be as simple as having the model object populate a structure with results and Boolean flags indicating whether they are valid or not, or more complex such as allowing the Model object to populate the Results object directly. Some such protocol allows the model to assume responsibility for calculation of greeks if necessary or desirable.

A final example is afforded by the vega calculation for tree-based models in which barriers or similar features, including call features in convertible bonds, appear. In this case we find that instabilities result from the fact that changing the volatility causes large changes in the positions and values of the tree nodes, with consequent instability in vega. A possible solution is to make an exception to the usual situation, in which the model neither knows nor cares which quantities have been requested or which perturbations made, and arrange for the tree to fix its node positions between price and vega calculations.

As discussed above, the primary argument in favour of the use of perturbation to calculate greeks is one of tractability: there are many greeks and many models. A further consideration in favour of general perturbation software for most greek calculation is that solutions written for individual models must also be maintained as the library changes around the individual models, and the assumptions under which the original solution was implemented may cease to be valid. This maintenance cost cannot be ignored, and is a potential source of error.

11.3.5 Secondary Features

As described in Sections 1.3.3 and Section 1.3.4, we may reuse many of the model implementations to price forward start, quanto or composite variants of

---

[4] Perturbing the original is certainly more efficient than creating a copy for a one-off pricing, but the possibility of cloning and then perturbing the clone might be optimal for some applications, as we can imagine the market data classes having access to a ready-perturbed clone. This would be particularly interesting if an expensive calibration had to be performed at each perturbation and un-perturbation.

an option and similar options on baskets or averages. The fact that the mathematical treatment of these issues can be treated generally, as opposed to separately for each individual model, strongly suggests that the software implementation of these features can be generic, model-independent code. In the framework described here, this means that it will be in the base Model class, and not in the derived classes that are specific to particular models, or in the Tree class, which is concerned with the data in the problem and its structure but not with the processing carried out on that data. This section therefore outlines the processing carried out by the base Model class.

In summary, the model implementations applicable to standard (single-currency, non-forward starting) variants of the options may be reused for quanto, composite and forward start variants by suitably modifying the inputs to the calculation routines. Similarly, options on baskets and averages may be priced by modifying the information provided by an underlying to its overlying option.

Meta-underlyings Meta-underlyings are synthetic underlyings that may be approximated as lognormal processes and as such may provide to their overlying options first and second moments of lognormal distributions, ie forwards and volatilities. We use the term to distinguish them from the "primary" underlyings, which are equities. They are introduced to extend the range of structures that can be priced with the implementations of the models given in earlier chapters. If users of the library are aware of the approximations being made, then this process is valuable.

We consider two examples of meta-underlyings in Section 9.2, baskets and average processes, the latter herein referred to as "Asian underlyings". This section considers together the two issues that were explicitly separated above, namely the issue of representing the data required to define the structure and the issue of the processing carried out to price it. In the section above, this processing (ie the model implementation) was encapsulated within a model object of a derived class and given in the mathematical treatments of previous chapters: in this section the processing is generic across all models and is treated here.

The data capture for the basket underlying is already covered by the sections above, inasmuch as a basket is a perfectly well-defined security having underlyings which are the basket constituents and whose pricing model is a trivial one, ie just the weighted sum of the prices of the constituents. For an Asian underlying, the issue is more difficult since some new data is required: the fixing dates of the average process and, for past fixings, the values observed at those dates. This information is neither a property of the underlying object nor, in the above data model, is it a property of the option object. An approach is to introduce a class, derived from the base Security class, which explicitly represents an Asian process and whose underlying is the equity in question. This represents something of a departure from the regular philosophy of representing with classes derived from Security only securities that have an existence in their own right. The effect is shown in Figure 11.9, which shows the structure in the same form as in Figure 11.1.

The model object (of a class derived from the base Model class) at the Tree node of the option expects to be able to query its underlyings for their forwards

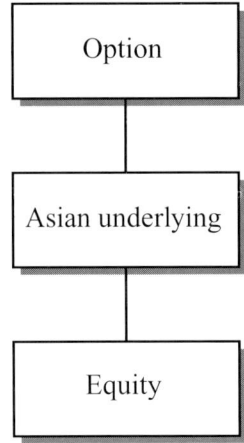

11.9 Tree structure for an option on an average.

and volatilities. If it is not to assume that the underlyings are primary underlyings, meta-underlyings must be able to provide these parameters using the same interface as primary underlyings. The calculation of a forward by an underlying is relatively simple and amounts to calling the model with some horizon date beyond today. This is well defined for an equity – it is just the forward calculation $F_T = S_0 \exp(\mathcal{M}_T)$ – and thus for baskets that just combine the results for their underlyings. For Asian underlyings, we state that the forward is undefined until the averaging process is complete,[5] at which times beyond the last averaging date it is given by the average of the forwards of its underlying at the fixing dates, the first moment of the average given in Section 9.2.1. Thus models (and they are a wide class covering all the closed-form models including Europeans, Asians, lookbacks and barriers) that require the forward of their underlyings only at discrete dates are catered for by requiring that they acquire these forwards only by calling the Price method of their underlyings with a suitable horizon date parameter. This avoids any explicit assumption that the underlyings are primary rather than meta-underlyings, the Tree::Price method being an interface that does not depend on the type of security object attached.

Volatility is similar. It may also be calculated at the same dates as the forward is calculated; models that accept piecewise constant drift generally also accept piecewise constant volatility. Again, the same interface must be available to overlying model objects, irrespective of the nature of the underlying.

Models that require the drift or volatility of an underlying as a function of time, notably tree and PDE models, accept as parameters objects giving the drift, volatility and dividends of their underlyings. Internally they query these objects at each time step. In order for these to be blind to the distinction between primary underlyings and meta-underlyings, classes must be available that provide this information in a way that is neutral as to this question.

We introduce the concept of "secondary market data objects" to achieve this. These are objects that exist at the tree node of an underlying (either a

---

[5] We may, of course, define the average as a kind of rolling average, in which case it certainly exists at all times after the first fixing; but we do not treat that extension here.

primary underlying or a meta-underlying) in order to furnish to the overlying model object the information it needs without its having to know the nature of the underlying. We find that tree models, for instance, require a Dividends object, a Drift object and a Volatility object. Of these three, we only need to introduce one new class, Drift, into the model interface. The other two are related by inheritance to the primary market data classes, with the result that model development may proceed without any explicit acknowledgement of these implementation classes. Thus there are three secondary market data classes: Drift, Secondary Volatility and Secondary Dividends.

An example is afforded by basket secondary dividends: these are just the superposition of the dividends, suitably weighted, of the basket constituents. Thus a basket secondary dividend object, which provides the aggregated dividends to an overlying option, holds references to the underlying dividend objects, as shown in Figure 11.10 (other objects being omitted for simplicity). Solid lines in the diagram indicate references from one object to others; the dashed lines indicate that the option pricing model object may make use of services provided by the basket dividends object and by the basket tree node.

Note that, although equities are primary underlyings, they still create secondary market data objects, even if they are particularly simple ones that refer directly to primary market data objects at the same tree node. The Secondary Dividends object refers internally to a real Dividends object, forwards to that object all requests, and returns the results unchanged. This leads to the simplified version of Figure 11.10 for an equity option, given in Figure 11.11, in which an equity secondary dividends object delegates all its operations to a primary dividends object. The Secondary Volatility object likewise refers internally to a Volatility object, and likewise forwards to that object all requests. The Drift object refers internally to a Yield Curve object and a Repo object and uses them to return rates. Thus the secondary market data objects of an equity do no processing, but delegate it to the corresponding primary objects.

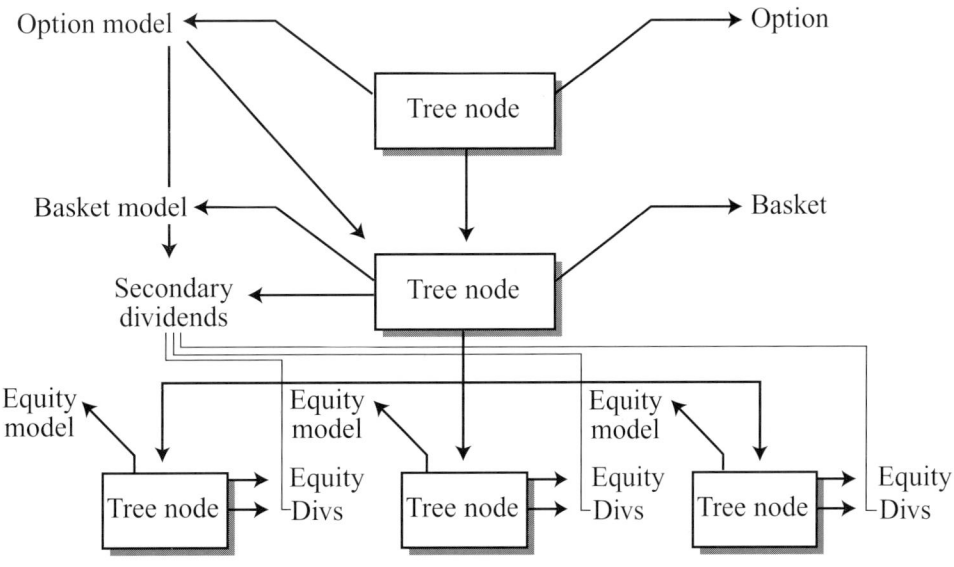

11.10 Basket dividends by aggregation of underlying dividends.

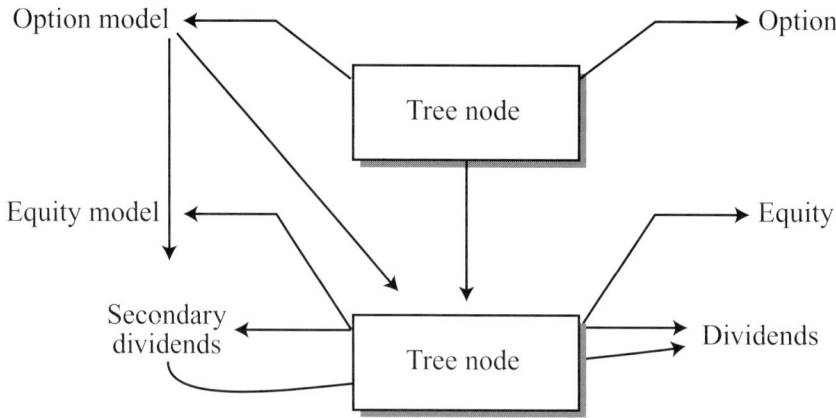

11.11 Equity dividends by delegation to primary dividends.

The creation of secondary market data objects is the responsibility of the model object and is introduced as a virtual method in the base Model class. The default behaviour is, of course, for the class to create no secondary market data objects, nor will this be overridden in most derived classes, including all those associated with options and other derivatives. The derived model classes for equities, baskets and Asian underlyings, however, will create secondary market objects of the above three types. Creation is carried out as part of building the Tree data structure; since the secondary objects may refer to primary objects, it must be carried out after these are available. The secondary objects do no processing at construction, so their construction imposes no real overhead.

In order to ensure that a model object associated with a derivative security responds correctly to the presence of meta-underlyings, we introduce the rule that these model objects will reference only the secondary market data objects of their underlyings, which might be grouped together in a "model parameters" structure, and not the primary objects. In this way the derivative's model object is blind to whether its underlyings are primary underlyings (equities) or meta-underlyings. Furthermore, the rule ensures that an attempt to price an option whose underlying identifier is incorrect, and identifies not a *bona fide* underlying but some other option, fails automatically even if no earlier processing detected the error.

In the case of basket and Asian underlyings, the secondary market objects they create embody the algorithms required to calculate the dividends, volatility and drift of the process. Basket dividends are just the ordered, weighted sum of the constituents' dividends; its volatility is given in Section 9.2 and its effective drift follows from its spot and forward prices. Asian underlyings are taken to have no dividends (and so a special secondary dividend class that never returns any dividends), a volatility also given in Section 9.2 and a drift undefined before the last fixing date.

Note that secondary market data objects are regarded as a service made available by a model object to overlying model objects. The secondary market objects at a given tree node are not used in the course of pricing the security object at that node. The reason for this is that secondary market objects might well make use of the `Price` method at that tree node in order to calculate results, as will be true for the drift object of baskets and Asian underlyings. This would enter an infinite loop if the `Price` method itself referred back to the secondary market objects.

Finally, we note that, although secondary market data objects are available for use by overlying model objects, the primary objects are still available unless explicitly made unavailable through interface design. Thus models can, should they need to, violate the above rule, bypass secondary objects and refer directly to primary objects. This they will do if the facilities provided by secondary market data objects are inappropriate for the particular model. Designing interfaces to prevent this makes the strong assumption that the same processing is appropriate for all conceivable models. Naturally, if a model bypasses secondary market data objects and yet needs to support meta-underlyings, then it must itself provide appropriate functionality. Without additional protocols, which might carry a maintenance overhead, option model objects cannot influence the classes of the secondary market data objects created by their underlying model objects. They can, however, parametrise the behaviour, and this is the mechanism by which quanto and composite corrections are made, as discussed below.

Quanto and composite The secondary market data classes introduced above are referred to by code in the base and derived model classes in order to insulate the model object from the nature of the underlying: it is the responsibility of the underlying model object to provide these services to the model object at the overlying tree node. These classes can be usefully extended to apply corrections to the raw market data to yield composite and quanto pricing.

Composite pricing requires that the volatility used by the model is modified according to $(\tilde{\sigma}^S)^2 = (\sigma^S)^2 - 2\rho\sigma^S\sigma^X + (\sigma^X)^2$ in the notation of Section 1.3.4. This processing can be allocated to the base Secondary Volatility class. The model object of the overlying option must register its interest in applying this correction with the secondary volatility object and provide an exchange rate volatility object and a correlation (these data being extracted from those populating the tree).

Composite pricing also requires a correction to the drift of the underlying, so that the effective drift is as though the asset were denominated in the composite currency, not its native currency. This is achieved similarly by the overlying model object registering with the underlying drift objects its interest in having a composite correction applied.

Finally, composite pricing requires that all discounting be done in the payoff currency. This is, in fact, automatically satisfied if:

1. the yield curve used for calculating discount factors, either within or external to the core calculation routines, is always that of the option currency, ie the currency named in the base Security class;

2. the option currency is, by definition, that currency in which the price is denominated;

3. composite options return prices in the payoff (composite) currency.

Quanto pricing requires the correction of the drift of the underlying asset by the amount $\rho\sigma^S\sigma^X$. This is again achieved simply by delegating this functionality to the Drift class. As before, the overlying model object registers its interest with the underlying secondary market object and provides the appropriate information with which to make the correction, in this case a correlation and a volatility term structure object (itself just a reference to a secondary volatility object and an interpolation level identified by the model object – frequently the option's strike).

As with composite pricing, quanto pricing requires that discounting is done in the payoff currency and, as in that case, this requirement is automatically satisfied.

We finally note that the requirement to support quanto and composite options imposes a requirement on core calculation routine interfaces if this is to be done in a straightforward way in generic code (ie in the base Model class). This requirement is that the core interface explicitly separate the drift of the underlying from the discounting applied to payoffs arising from exercise of the option: a single-currency implementation might use the same interest rate in both the drift of the asset and in the discounting.

Forward start Forward start options are priced using forward interest rates and forward volatilities, over the interval between the start date and maturity. This is straightforwardly achieved without reference to secondary market data objects by defining a start date that is equal to evaluation date in the standard case and to the forward start date where this exists. Thereafter, it is essential to adhere to the rule that this date is used as the start date in all calls to methods returning interest rates and volatilities. Calculation routines that require evaluation date as an input are therefore in general passed this date in lieu of the true evaluation date and return a price discounted to this date. Post-processing in the base Model class is then responsible for applying the extra discounting.

11.3.6 Layering Abstractions

In a library or application of any complexity, partition of the whole into parts identified and delimited by their function is essential to mitigate the tendency towards evolution of the implementation into a monolith. A monolith is a tightly coupled collection of code in which all parts depend on all other parts for their correct functioning. Maintaining such a monolith as its complexity increases become wholly intractable, resulting in high costs and errors. In every software project, steps are taken to avoid this.

Classes themselves are a key weapon in the struggle against the monolith. Their key feature is that they group together data and functionality into an entity with well-defined behaviour and responsible for maintaining its own state. The methods of a class form a set of interactions between an object and its environment; since this set is finite, it strictly limits the coupling between the data of an object and is environment. Any change to this set of interactions can be analysed in the context that the the complete set of interactions of the object with the environment is known, there being no possibility that some other part of the application interacts in some way not declared by the class interface.

A large collection of classes, however, can become monolithic, as discussed by Gamma *et al* [24]. A higher-level weapon in the war is the design patterns that deliberately seek to minimise the coupling between classes, principally by allowing one class to refer only to a base class and hiding complexity from the client class by an inheritance class hierarchy in the service-providing class. A base class can be regarded as an *aspect* of a derived class [14], in that it may capture a subset of behaviours needed by a client that need not be aware of the full nature of a service-providing object. For example, we might define the Yield Curve and Vasicek classes to have a common (abstract) base Discount Curve that furnishes (deterministic) discount factors as seen at evaluation date: many clients who require only discount factors can in this way be protected from the fact that interest rates can be modelled either deterministically or as stochastic processes. Other inheritance relationships between deterministic and stochastic interest rate classes are also possible.

Finally, we note here that whole collections of classes can be grouped into subsystems or layers with defined interactions between them. It is often a matter of discipline in the development team to enforce these rules: C++, for example, makes class names global unless steps are deliberately taken to prevent this, such as by using nested classes or namespaces.

The remainder of this section briefly discusses ways in which we can manage (but not reduce) the complexity of the pricing library, which will often be one of the more complex areas of any derivatives application.

Abstraction 1: Public interface. The interface classes, namely those that capture market data and static data, are free of dependence on the internal classes, namely those that encapsulate the solution. Many elements of the solution can therefore vary without impact – even recompilation – on client applications.

Abstraction 2: Project layering. Throughout this discussion it has been assumed that a library is implemented in such a way as to be portable between applications; in other words, it has as little dependence as possible on the applications that will use it. This is itself a layering, and perhaps the most valuable of all.

Within the pricing library project, the above point leads to a two-layer split between API classes and implementation classes, where there is a strictly one-way dependence. The solution depends very much on the details of the data in the external world, but the reverse is not true. Finally, we use utilities that are independent of the problem domain, including externally supplied class libraries and mathematical routines. These form a third, lowest-level layer providing resources for higher-level routines.

Abstraction 3: Uniform pricing interfaces. In order for the request manager layer to function without intractable complexity, it must rely on a single method call for pricing all option types. This is the `Tree::Price` method. This allows the request manager, and the Tree class overall, to concentrate on the structure of a security, while remaining blind to its actual type. Thus the request manager uses only those relationships between objects that are captured by the Tree class, and no other information.

The other uniform pricing interface is, of course, that presented by the library to client applications, the purpose of which is to limit their complexity by trading it for complexity within the library. This is a good trade because the library is implemented once and made available to many clients. Furthermore, it is generally implemented by a specialist team with detailed knowledge of pricing algorithms, who are then in a position to manage the complexity which does, inevitably, result.

Appendix
Useful Formulas

A.1 Useful Integrals I

The following integral is used in computing hitting probabilities. Let $b > 0$ and let a be any real number. We then have

$$\int_0^T \frac{b}{\sqrt{2\pi t^3}} \exp\left(-\frac{(at+b)^2}{2t}\right) dt$$

$$= \int_0^T \frac{at+b}{2\sqrt{2\pi t^3}} \exp\left(-\frac{(at+b)^2}{2t}\right) dt - \int_0^T \frac{at-b}{2\sqrt{2\pi t^3}} \exp\left(-\frac{(at+b)^2}{2t}\right) dt.$$

Substituting $(at-b)/\sqrt{t}$ in the first and $(at+b)/\sqrt{t}$ in the second integral gives

$$\int_0^T \frac{b}{\sqrt{2\pi t^3}} \exp\left(-\frac{(at+b)^2}{2t}\right) dt = N\left(\frac{-aT-b}{\sqrt{T}}\right) + e^{-2ab} N\left(\frac{aT-b}{\sqrt{T}}\right). \quad (A.1)$$

In the case $b < 0$, we can deduce that

$$\int_0^T \frac{b}{\sqrt{2\pi t^3}} \exp\left(-\frac{(at+b)^2}{2t}\right) dt = -N\left(\frac{aT-b}{\sqrt{T}}\right) - e^{-2ab} N\left(\frac{-aT+b}{\sqrt{T}}\right).$$

A.2 Useful Integrals II

In this section we give some useful formulas for integrals over one- and two-dimensional normal distributions and densities.

We first list the identities, and then give some hints on their derivation.

$$\int_{-\infty}^{\gamma} \varphi(x) N(\alpha x + \beta) \, dx = N_2\left(\gamma, \frac{\beta}{\sqrt{1+\alpha^2}}, \frac{-\alpha}{\sqrt{1+\alpha^2}}\right),$$

$$\int_{-\infty}^{\infty} \varphi(x) N(\alpha x + \beta) \, dx = N\left(\frac{\beta}{\sqrt{1+\alpha^2}}\right),$$

$$\int_{-\infty}^{\gamma} e^{\delta x} \varphi(x) N(\alpha x + \beta) \, dx = e^{\delta^2/2} N_2\left(\gamma - \delta, \frac{\alpha\delta + \beta}{\sqrt{1+\alpha^2}}, \frac{-\alpha}{\sqrt{1+\alpha^2}}\right),$$

$$\int_{-\infty}^{\infty} e^{\delta x} \varphi(x) N(\alpha x + \beta) \, dx = e^{\delta^2/2} N\left(\frac{\alpha\delta + \beta}{\sqrt{1+\alpha^2}}\right),$$

$$\int_{-\infty}^{\gamma} \varphi_2(x, y, \rho) \, dx = \varphi(y) N\left(\frac{\gamma - \rho y}{\sqrt{1-\rho^2}}\right),$$

$$\int_{\gamma}^{\infty} \varphi_2(x, y, \rho) \, dx = \varphi(y) N\left(-\frac{\gamma - \rho y}{\sqrt{1-\rho^2}}\right),$$

$$\int_{-\infty}^{\gamma} e^{\delta x} \varphi_2(x, y, \rho) \, dx = \frac{1}{\sqrt{2\pi}} \exp\left(-\frac{y^2 - \delta^2(1-\rho^2) - 2\rho\delta y}{2}\right)$$
$$\cdot N\left(\frac{\gamma - \rho y - \delta(1-\rho^2)}{\sqrt{1-\rho^2}}\right),$$

$$\int_{\gamma}^{\infty} e^{\delta x} \varphi_2(x, y, \rho) \, dx = \frac{1}{\sqrt{2\pi}} \exp\left(-\frac{y^2 - \delta^2(1-\rho^2) - 2\rho\delta y}{2}\right)$$
$$\cdot N\left(-\frac{\gamma - \rho y - \delta(1-\rho^2)}{\sqrt{1-\rho^2}}\right).$$

For the proof, let X, Y and Z denote three pairwise independent standard normally distributed random variables. Using a Girsanov-type argument, we get

$$\mathbb{E}\left[e^{\delta X} 1_{\{X \leq \gamma\}} 1_{\{Y \leq \alpha X + \beta\}}\right] = e^{\delta^2/2} \mathbb{E}\left[e^{\delta X - \delta^2/2} 1_{\{X \leq \gamma\}} 1_{\{Y \leq \alpha X + \beta\}}\right]$$
$$= e^{\delta^2/2} \mathbb{P}\left[X \leq \gamma - \delta, Y \leq \alpha X + \alpha\delta + \beta\right]$$
$$= e^{\delta^2/2} \mathbb{P}\left[X \leq \gamma - \delta, \frac{Y - \alpha X}{\sqrt{1+\alpha^2}} \leq \frac{\alpha\delta + \beta}{\sqrt{1+\alpha^2}}\right],$$

and thus

$$\int_{-\infty}^{\gamma} e^{\delta x} \varphi(x) N(\alpha x + \beta) \, dx = e^{\delta^2/2} N_2\left(\gamma - \delta, \frac{\alpha\delta + \beta}{\sqrt{1+\alpha^2}}, -\frac{\alpha}{\sqrt{1+\alpha^2}}\right).$$

A.3 Useful Integrals III

In three dimensions, we have

$$\mathbb{E}\left[e^{\delta X} 1_{\{X \leq \gamma\}} 1_{\{Y \leq \alpha_1 X + \beta_1\}} 1_{\{\rho Y + \sqrt{1-\rho^2} Z \leq \alpha_2 X + \beta_2\}}\right] =$$

$$e^{\delta^2/2} \mathbb{P}\left[X \leq \gamma - \delta, \frac{Y - \alpha_1 X}{\sqrt{1+\alpha_1^2}} \leq \frac{\alpha_1 \delta + \beta_1}{\sqrt{1+\alpha_1^2}}, \frac{\rho Y - \sqrt{1-\rho^2} Z - \alpha_2 X}{\sqrt{1+\alpha_2^2}} \leq \frac{\alpha_2 \delta + \beta_2}{\sqrt{1+\alpha_2^2}}\right].$$

So, we get the identity

$$\int_{-\infty}^{\gamma} e^{\delta x} \varphi(x) N_2(\alpha_1 x + \beta_1, \alpha_2 x + \beta_2, \rho)\, dx$$

$$= e^{\delta^2/2} N_3\Bigg(\gamma - \delta, \frac{\alpha_1 \delta + \beta_1}{\sqrt{1+\alpha_1^2}}, \frac{\alpha_2 \delta + \beta_2}{\sqrt{1+\alpha_2^2}};$$

$$-\frac{\alpha_1}{\sqrt{1+\alpha_1^2}}, -\frac{\alpha_2}{\sqrt{1+\alpha_2^2}}, -\frac{\rho + \alpha_1 \alpha_2}{\sqrt{1+\alpha_2^2}\sqrt{1+\alpha_2^2}}\Bigg) \quad \text{(A.2)}$$

A.4 Useful Integrals IV

In this section we summarise some formulas used throughout the derivation of the pricing formulas.

Let φ and N denote respectively the density and distribution function of a normally distributed random variable with mean 0 and variance 1. Let N_2 denote the distribution function of the bivariate normal distribution, ie

$$N_2(a, b, \rho) = \mathbb{P}[X < a, Y < b],$$

where X and Y are standardised, normalised normally distributed random variables with correlation ρ. Throughout, we assume $\alpha \neq 0$.

We have

$$\int_{-\infty}^{\gamma} e^{\delta y} \varphi(\alpha y + \beta) \, dy = \frac{1}{\alpha} \exp\left(\frac{\delta^2}{2\alpha^2} - \frac{\delta \beta}{\alpha}\right) N\left(\alpha \gamma + \beta - \frac{\delta}{\alpha}\right),$$

$$\int_{\gamma}^{\infty} e^{\delta y} \varphi(\alpha y + \beta) \, dy = \frac{1}{\alpha} \exp\left(\frac{\delta^2}{2\alpha^2} - \frac{\delta \beta}{\alpha}\right) N\left(-\alpha \gamma - \beta + \frac{\delta}{\alpha}\right),$$

$$\int_{-\infty}^{\gamma} y \varphi(y) N(\alpha y + \beta) \, dy = \frac{\alpha}{\sqrt{1 + \alpha^2}} \varphi\left(\frac{\beta}{\sqrt{1 + \alpha^2}}\right) N\left(\sqrt{1 + \alpha^2}\, \gamma + \frac{\alpha \beta}{\sqrt{1 + \alpha^2}}\right)$$
$$- \varphi(\gamma) N(\alpha \gamma + \beta),$$

$$\int_{-\infty}^{\gamma} e^{\delta y} y \varphi(y) N(\alpha y + \beta) \, dy = e^{\delta^2/2} \left[\frac{\alpha}{\sqrt{1 + \alpha^2}} \varphi\left(\frac{\beta + \alpha \delta}{\sqrt{1 + \alpha^2}}\right) N\left(\frac{(1 + \alpha^2)\gamma - \delta + \alpha \beta}{\sqrt{1 + \alpha^2}}\right)\right.$$
$$- \varphi(\gamma - \delta) N(\alpha \gamma + \beta)$$
$$\left.+ \delta N_2\left(\gamma - \delta, \frac{\beta + \alpha \delta}{\sqrt{1 + \alpha^2}}, \frac{-\alpha}{\sqrt{1 + \alpha^2}}\right)\right],$$

$$\int_{-\infty}^{\gamma} \varphi(y) \varphi(\alpha y + \beta) \, dy = \frac{1}{\sqrt{1 + \alpha^2}} \varphi\left(\frac{\beta}{\sqrt{1 + \alpha^2}}\right) N\left(\sqrt{1 + \alpha^2}\, \gamma + \frac{\alpha \beta}{\sqrt{1 + \alpha^2}}\right),$$

$$\int_{-\infty}^{\gamma} e^{\delta y} \varphi(y) \varphi(\alpha y + \beta) \, dy = e^{\delta^2/2} \frac{1}{\sqrt{1 + \alpha^2}} \varphi\left(\frac{\beta + \alpha \delta}{\sqrt{1 + \alpha^2}}\right) N\left(\frac{(1 + \alpha^2)\gamma - \delta + \alpha \beta}{\sqrt{1 + \alpha^2}}\right).$$

Bibliography

1. Andersen, L. B. G. and R. Brotherton-Ratcliffe, 1997, "The Equity Option Volatility Smile: An Implicit Finite-Difference Approach", *Journal of Computational Finance*, 1(2), pp. 5–37.

2. Barone-Adesi, G. and R. E. Whaley, 1987, "Efficient Analytic Approximation of American Option Values", *Journal of Finance*, 42, pp. 301–20.

3. Baxter, M. and A. Rennie, 1996, *Financial Calculus*, Cambridge University Press.

4. Bismut, J. M., 1994 *Large Deviations and the Malliavin Calculus*, Birkhäuser.

5. Black, F. and M. Scholes, 1973, "The Pricing of Options and Corporate Liabilities", *Journal of Political Economy*, 81, pp. 637–59.

6. Black, F., 1976, "The Pricing of Commodity Contracts", *Journal of Financial Economics*, 3, pp. 167–79.

7. Borodin, A. N. and P. Salminen, 1996, *Handbook of Brownian Motion – Facts and Formulae*, Birkhäuser Verlag, Basle.

8. Brennan, M. and E. Schwartz, 1977, "The Valuation of American Put Options", *Journal of Finance*, 32, pp. 449–62.

9. Broadie, M. and P. Glasserman, 1997, "Pricing American-Style Securities Using Simulation", *Journal of Economic Dynamics and Control*, 21(8–9), pp. 1323–52.

10. Bunch, D. S. and H. E. Johnson, 1992, "A Simple and Numerically Efficient Valuation Method for American Puts Using a Modified Geske–Johnson Approach", *Journal of Finance*, 47, pp. 809–16.

11. Cheuk, T. H. F. and T. C. F. Vorst, 1997, "Currency Lookback Options and Observation Frequency: A Binomial Approach", *Journal of International Money and Finance*, 16, pp. 173–87.

12. Chriss, N. A. and K. Tsiveriotis, 1998, "Pricing with a Difference", *Risk*, pp. 80–83.

13. Colwell, D. B., Elliott, R. J. and Kopp, P. E., 1991, "Martingale Representation and Hedging Policies", *Stochastic Processes and their Applications*, 38, pp. 335–45.

14. Coplien, J. O., and D. C. Schmidt (eds), 1995, *Pattern Languages of Program Design*, Addison-Wesley.

15. Corrado, C. J. and T. Su, 1997, "Implied Volatility Skews and Stock Index Skewness and Kurtosis Implied by S&P500 Index Option Prices", *Journal of Derivatives*, 2, pp. 8–19.

16. Delbaen, F., and W. Schachermaer, 1994, "A General Version of the Fundamental Theorem of Asset Pricing", *Mathematische Annalen*, 300, pp. 463–520.

17. Derman E. and I. Kani, 1994, "Riding on a Smile", *Risk*, 7, pp. 32–29.

18. Doob, J. L., 1984, *Stochastic Processes*, Wiley.

19. Dothan, M., 1990, *Prices in Financial Markets*, Oxford University Press, New York.

20. Duffie, D., 1992, *Dynamic Asset Pricing Theory*, Princeton University Press.

21. Dupire, B., 1998, "Hedging European and Exotic Options with Alternative Strategies", *Global Derivatives*.

22. Dupire, B., 1994, "Pricing with a Smile", *Risk*, 7, pp. 18–20.

23. Föllmer, H., 1989, Private communication.

24. Gamma, E., R. Helm, R. Johnson and J. Vlissides, 1995, *Design Patterns*, Addison-Wesley.

25. Geman, H., N. E. Karoui, and J. Rochet, 1995, "Changes of Numéraire, Changes of Probability Measure and Option Pricing", *Journal of Applied Probability*, 32, pp. 443–58.

26. Geman, H. and M. Yor, 1993, "Bessel Processes, Asian Options, and Perpetuities", *Mathematical Finance*, 3, pp. 349–75.

27. Geske, R., 1979, "The Valuation of Compound Options", *Journal of Financial Economics*, 7, pp. 63–81.

28. Geske, R. and H. E. Johnson, 1984, "The American Put Valued Analytically", *Journal of Finance*, 39, pp. 1511–42.

29. Gilks, W. R., S. Richardson and D. J. Spiegelhalter (eds), 1996, *Markov Chain Monte Carlo in Practice*, Chapman and Hall.

30. Goldman, M. B., H. B. Sosin, and M. A. Gatto, 1979, "Path Dependent Options: Buy at the Low, Sell at the High", *Journal of Finance*, 34, pp. 1111–27.

31. Harrison, J. M. and D. M. Kreps, 1979, "Martingales and Arbitrage in Multiperiod Securities Markets", *Journal of Economic Theory*, 20, pp. 381–408.

32. Harrison, J. M., and S. R. Pliska, 1981, "Martingales and Stochastic Integrals in the Theory of Continuous Trading", *Stochastic Processes and their Applications*, 11, pp. 215–60.

33. Heston, S. L., 1993, "A Closed-Form Solution for Options with Stochastic Volatility with Applications to Bond and Currency Options", *The Review of Financial Studies*, 6, pp. 327–43.

34. Heynen, R., and H. Kat, 1994, "Selective Memory", *Risk*, 7, pp. 46–51.

35. Ho, T. S., R. C. Stapleton, and M. G. Subrahmanyam, 1994, "A Simple Technique for the Valuation and Hedging of American Options", *The Journal of Derivatives* (Fall), pp. 52–66.

36. Hull, J., 1997, *Options, Futures and Other Derivative Securities*, Prentice-Hall.

37. Hull, J., and A. White, 1994, "Numerical Procedures for Implementing Term Structure Models. I: Single Factor Models", *The Journal of Derivatives* (Fall), pp. 7–16.

38. Hull, J., and A. White, 1993, "One-Factor Interest-Rate Models and the Valuation of Interest Rate Derivatives", *Journal of Financial and Quantitative Analysis*, 28, pp. 235–54.

39. Ikeda, N. and S. Watanabe, 1981, *Stochastic Differential Equations and Diffusion Processes*, North-Holland.

40. Jarrow, R. A., D. Lando, and S. M. Turnbull, 1997, "A Markov Model for the Term Structure of Credit Risk Spreads" *The Review of Financial Studies*, 10, pp. 481–523.

Bibliography

41. Jarrow, R. A., and A. Rudd, 1992, "Approximate Option Valuation for Arbitrary Stochastic Processes", *Journal of Financial Economics*, 10, pp. 347–69.

42. Johnson, H., 1987, "Options on the Maximum or the Minimum of Several Assets", *Journal of Financial and Quantitative Analysis*, 22, pp. 277–83.

43. Johnson, R. A., and D. W. Wichern, 1988, *Applied Multivariate Statistical Analysis*, Prentice-Hall.

44. Karatzas, I. and S. E. Shreve, 1991, *Brownian Motion and Stochastic Calculus*, 2nd edn, Springer.

45. Knuth, D. E., 1997, *The Art of Computer Programming*, Vol. 1, 3rd edn, Addison-Wesley.

46. Lagnado, R. and S. Osher, 1997, "A Technique for Calibrating Derivative Security Pricing Models: Numerical Solution of an Inverse Problem", *Journal of Computational Finance*, 1(1), pp. 13–25.

47. Levy, E. and F. Mantion, 1997, "Discrete by Nature", *Risk*, 10, pp. 74–75.

48. Levy, E. and F. Mantion, 1997, "Approximate Valuation of Discrete Lookback and Barrier Options", *NetExposure*, 2.

49. Liu, R. Y., 1995, "The Alchemy of Asian Exotics", *AsiaRISK*, November.

50. Malliavin, P., 1976, "Stochastic Calculus of Variations and Hypoelliptic Operators", *Proc. Int. Symp. on Stochastic Differential Equations*, pp. 195–263.

51. Margrabe, W., 1978, "The Value of an Option to Exchange One Asset for Another", *Journal of Finance*, 33, pp. 177–86.

52. Merton, R. C., 1973, "Theory of Rational Option Pricing", *Bell Journal of Economic and Management Science*, 4, pp. 141–83.

53. Meyers, S., 1996, *More Effective C++*, Addison-Wesley.

54. Musiela, M. and M. Rutkowski, 1997, *Martingale Methods in Financial Modelling*, Springer.

55. Myneni, R., 1992, "The Pricing of American Options", *The Annals of Applied Probability*, 2, pp. 1–23.

56. Nualart, D., 1995, *The Malliavin Calculus and Related Topics*, Springer.

57. Ocone, D. L., 1984, "Malliavin's Calculus and Stochastic Integral Representations of Functions of Diffusion Processes", *Stochastics*, 12, pp. 161–85.

58. Ocone, D. L. and Karatzas, I., 1991, "A Generalized Clark Representation Formula, with Application to Optimal Portfolios", *Stochastics*, 34, pp. 187–220.

59. Oksendal, B., 1995, *Stochastic Differential Equations*, Springer.

60. Parthasarathy, K. R., 1967, *Probability Measures on Metric Spaces*, Academic Press, New York.

61. Press, W. H., S. A. Teukolsky, W. T. Vetterling and B. P. Flannery, 1992, *Numerical Recipes in C: The Art of Scientific Computing*, 2nd edn, Cambridge University Press, Cambridge.

62. Revuz, D. and M. Yor, 1991, *Continuous Martingales and Brownian Motion*, Springer.

63. Rogers, L. C. G. and D. Williams, 1987, *Diffusions, Markov Processes and Martingales*, Wiley.

64. Rubinstein, M., 1991, "Options for the Undecided", *Risk*, 4, p. 43.

65. Rubinstein, M. and E. Reiner, 1991, "Breaking Down the Barrier", *Risk*, 4, pp. 28–35.

66. Rubinstein, M., 1994, "Implied Binomial Trees", *Journal of Finance*, 69, pp. 771–818.
67. Shiryayev, A. N., 1984, *Probability*, Springer.
68. Smith, G. D., 1985, *Numerical Solution of Partial Differential Equations: Finite Difference Methods*, Oxford University Press.
69. Stroustrup, B., 1997, *The C++ Programming Language*, 2nd and 3rd edns, Addison-Wesley.
70. Stuart, A., and K. Ord, 1994, *Kendall's Advanced Theory of Statistics*, 6th edn, Vol. 1, Edward Arnold.
71. Stulz, R., 1982, "Options on the Minimum or the Maximum of Two Risky Assets: Analysis and Applications", *Journal of Financial Economics*, 10, pp. 161–85.
72. Tezuka, S., 1995, *Uniform Random Numbers: Theory and Practice*, Kluwer.
73. Tilley, J. A., 1993, "Valuing American Options in a Path Simulation Model", *Transactions of the Society of Actuaries*, 45, pp. 83–104.
74. Turnbull, S. M., and L. M. Wakeman, 1991, "A Quick Algorithm for Pricing European Average Options", *Journal of Financial and Quantitative Analysis*, 26, pp. 377–89.
75. Wilmott, P., J. Dewynne and S. Howison, 1993, *Option Pricing: Mathematical Models and Computation*, Oxford Financial Press.
76. Wilmott, P., J. Dewynne and S. Howison, 1995, *The Mathematics of Financial Derivatives*, Cambridge University Press.

Index

A
Abstractions, layering 274–6
American depository receipt (ADR) 247
American digital options 61
American double digital options 66–70
American options 52–4
  Barone–Adesi–Whaley approximation 54–6
  Broadie and Glasserman's method 182
  Geske–Johnson technique 57–8
  Monte Carlo method 181–4
  partial differential equation method 196–7
  perpetual American options 56–7
  Tilley's method 181–2
Antithetic paths 175
Arbitrage 21
Asian options 89, 92–6, 175
Asian strike options 95–6
Asian underlyings 213–15, 269–70, 272
Asset-or-nothing options 60
Average strike reset options 182–4

B
Barone–Adesi–Whaley approximation 54–6
Barrier options 70–72, 241–3
  complex barrier conditions 193
  knock-in options 157, 190
  knock-out options 156, 190
  moving barriers 192–3
  outside barrier options 135–7
  outside digital options 137–8
  pricing 156–61
  probability fitting 157–60
  probability fitting, near the barrier 157–60
  tree correlation 161
Basket dividends 271
Basket options 128–32
  lognormal approximation 215–16

Bermudan Asian options 161–6
  discrete dividends 165
  greeks 166
Bermudan options 57–8
  Monte Carlo method 181
Binomial tree
  for a Gaussian process 148–9
  for a lognormal process 149
Black–Scholes equation 35, 54, 112, 113, 224
Black–Scholes equity model 19–25
  basic model 25–6
  domestic approach 29–31
  extensions
  discrete dividends 27–8
  forward start options 28
  multiple assets 26–7
  quanto and composite options 28–9
  foreign approach 31
  one-dimensional Black–Scholes model 22–3
  option pricing 29
Black–Scholes partial differential equation (PDE) 23–5, 185
Black's model 142–3, 211, 241
Bond floor 167, 171
Bond options 140–41
Borel measure 2
Boundary conditions 189–92
  Dirichlet boundary conditions 190
  Discrete dividends 198–9
  Neumann boundary conditions 190
Broadie and Glasserman's method 182
Brownian bridges 77
  hitting times 15
  Monte Carlo method 178–9
Brownian motion 8–9, 11, 12, 22–3, 85, 148, 149, 223, 224
  hitting times 14–15

C
Callback class 257–9
Caps 141–3, 212
  caplets 141–2, 212
  and floors 141–3
Central limit theorem 12
Chain rule 9
Cholesky decomposition 174
Chooser options (as-you-like-it options) 109–111
  hedging 237–8
Clark's formula 32–5
Composite equity swap, hedging 240
Composite options 47
  composite pricing 273
Compound barriers 84–7
Compound options 115–16
  hedging 238
Conditional expectations 4
Control variate technique 175
Convertible bonds 166–9
  conversion feature 167
  hedging 245–7
  issuer call feature 167–8
  put feature 168
  redemption 168
  valuing on a tree 169–70
  incorporating credit spread 170–72
Correlated extremum, and terminal value 18–19
Correlation 252–3
Coupon bonds, options 140–41
Crank–Nicholson scheme 187–8
Credit risk 247
Credit spread 170–72

D
Data representation 260–61
Delta 206, 266
Delta hedging 221
Derivative equity products 49–51
Derivative markets, evolution 221–2
Deterministic volatility models 200–10
Digital options 58–70
  hedging 237

partial American digital options 63–6
partial barriers 76–84
Dirichlet boundary conditions 190
Discrete dividends 27–8, 197–8
Distribution function 2–3
Dividends 252
Donsker's invariance principle 12
Double barriers 75–6
Double time-varying linear barrier 193
Dupire surface equation 207

E
Edgeworth expansion 210, 216–20
Elementary events 1–2
Equity, modelling with stochastic interest rates 40–41
Equity swaps 51–2
  hedging 240–41
Equity/bond outperformance options 143–6
European digital options 58–60
  on best or worst of several assets 124–28
  on best or worst of two assets 122–4
European options 185
  (Merton formula) 143
  pricing and hedging under transaction costs 233–5
Exchange options 117–18
Exchange rate volatility 253
Exchange rates 253
Expansion, and option pricing 217–19
Expectation, and higher moments 3
Extended Vasicek model (Hull and White model) 35, 36–40
  calibration 38–9, 211–13
Extremum, and terminal value 15–16

F
Fade–in barrier options 106–9, 196
Fade–in options 104–6, 195–6
Faure sequences 178
Filtration 6
Fixed notional 52
Floating notional 52
Floors 141–3
Forward contracts 49–50
Forward measure 46–7, 180–81
Forward rates 37–8
Forward start options 274
Fully hedged portfolio 224

G
Gambling system 6
Gamma 206, 221, 226, 241–2, 266
Gauss–Hermite numerical quadrature technique 122
Gaussian process 148–9
Generic Monte Carlo pricing 179–80
Geske–Johnson technique 57–8
Gibbs sampler 179
Girsanov's theorem 11, 13, 14, 19, 73, 85
Global vega 206
Gray code 177–8
Greeks 115, 147
  Bermudan Asian options 166
  calculated using trees 155–6
  calculation 265–8
  partial differential equation 191–2

H
Halton sequences 176
Hamilton–Jacobi–Bellman equation 233, 234
Heath–Jarrow–Morton yield curve models 224
Hedging 59, 222–6
  continuous-time hedging 227–9
  delta hedging 221
  discrete-time hedging 229–32
  mathematics 226–33
  and risk management 221–6
  of specifc products 236–47
  static hedging 236–7
Hedging strategies 33, 205–6
Hindsight options 132–4
Hitting times
  of Brownian bridges 15
  of Brownian motion 14–15
Hull and White see extended Vasicek model
Hybrid Monte Carlo 180–81

I
IBVPs 199
Implied diffusion theory 201–5
Implied tree models 147, 209
Implied volatility surface 253
Improving options 112–13
Independence 3–4
Itô's lemma 9–10, 23

J
Jamshidian's decomposition 146
Jensen's inequality 3, 93, 129

K
Knock-in options 157, 190
Knock-out options 156, 190
Koksma and Hlawka's theorem 177

L
Ladder options 61
Lattice generation 191
Law of the integrated logarithm 9
Locally hedged portfolio 224
Lognormal process 10–11, 149
Long barriers 89–92
Lookback options 97–104
  hedging 243–4
Low-discrepancy sequences 175–8

M
Malliavin calculus 33–5
Market data, capturing 251–4
Market price 252
Market values, fitting 219–20
Markov chain Monte Carlo methods (MCMC) 179
Markov process 157, 172
Markov property 13
Martingales 6–7
Meta-underlyings 269–73
Metropolis–Hastings algorithm 179
Model library 249–50
  interface design 251
  internal design 259–76
Monte Carlo method 173–84
  for American options 181–4
  antithetic paths 175
  Brownian bridges 178–9
  control variate technique 175
  Faure sequences 178
  generic Monte Carlo pricing 179–80
  Halton sequences 176
  hybrid Monte Carlo 180–81
  low-discrepancy sequences 175–7
  pseudo-random numbers 174, 175
  Sobol numbers 177–8
Moving barriers 192–3
  double time-varying linear barrier 193
  single exponential barrier 192
  single time-varying linear barrier 192–3
Moving boundaries 199–200
Multi-currency hybrid model 43–8

Index

N

Neumann boundary conditions 190
Newton–Raphson method 115, 121
No free lunch with vanishing risk (NFLVR) 21, 32
Normal distribution 4–5

O

One-dimensional Black–Scholes model 22–3
Option pricing 29, 32
  and expansion 217–19
  using trees 154–6
Outperformance options 120–22
Outside barrier options 135–7
Outside digital options 137–8

P

Partial American digital options 63–6
Partial barriers 76–84
Partial differential equation (PDE) 185
  barrier example 208–9
  calibration strategy 207–8
  discretisation 186–9
  finite difference methods 188–9
  and greeks 191–2
  model calibration 200–201
Partial extremum, and terminal value 15–18
Perpetual American options 56–7
Piecewise constant parameters 41–2, 47–8
Plain vanilla European options 50–51
Power and powered options 113, 113–15
  hedging 239–40
Price method 263
Pricing, and model layers 262–5
Probability fitting, near the barrier 157–60

Probability spaces 1–2
Probability theory 1–19
Prolongation options 111–12
Pseudo-random numbers 174, 175

Q

Quadratic variation 7
Quanto options 47
Quanto pricing 274
Quantoed equity swap, hedging 239–40

R

Radon–Nikodym theorem 4, 50, 73, 85
Random variables 2–3
Random walks 6
Range options 87–9, 194–5
Reflection principle 12–14
Relative digital options 118–19
Relative outperformance options 119–20
Repo curve 252
Rho 266
Risk management 221–2

S

Security data 254–6
Single barriers 72–5
Single exponential barrier 192
Single time-varying linear barrier 192–3
Singleton registry 263
Sloping barriers 160
Sobol numbers 177–8
Standard Asian options 94–5
Stochastic differential equation (SDE) 10–11, 43–6, 185
Stochastic integration 7–8
Stochastic processes 5–6
Stochastic volatility models 210
Straddles 244
Swaptions 141
  hedging 241–3

T

Taylor expansions 186, 229
Template method 264
Terminal value
  and correlated extremum 18–19
  and extremum 15–16
  and partial extremum 15–18
Tilley's method 181–2
Top–level functional interface 256–7
Trading (portfolio) strategy 20
Transaction costs model 233–5
Tree correlation, for barriers 161
Tree data structure 260
  building 261–2
Tree greeks 155–6
Trees 147–54
  binomial trees 148–9
  finite difference methods 188–9
  and option pricing 154–6
  running through 154–5
  thick trees 157, 171
  trinomial tree 149–50
  two-factor tree 151–4
  valuing convertible bonds 169–70
  incorporating credit spread 170–72

U

Underlyings 260–61

V

Vasicek interest rates 253–4
Vega 221, 266
Volatility 232
Volatility smile 59, 185, 205, 212, 216–20

Y

Yield curve 252

Z

Zero bonds 37, 38
  options 140–41